Springer Japan KK

A.T. Fomenko, T.L. Kunii

Topological Modeling for Visualization

With 337 Figures, Including 7 in Color

 Springer

Anatolij T. Fomenko
Professor
Department of Differential Geometry and Applications
Faculty of Mathematics and Mechanics
Moscow State University
Moscow 119899, Russia

Tosiyasu L. Kunii
Director
Laboratory of Digital Art and Technology
1-25-21-602 Hongo, Bunkyo-ku,
Tokyo, 113 Japan
and
Chairman of Digital Planet Institute
Senior Partner of MONOLITH Co. Ltd.
1-7-3 Azabu-Juban, Minato-ku,
Tokyo, 106 Japan

ISBN 978-4-431-66958-6   ISBN 978-4-431-66956-2 (eBook)
DOI 10.1007/978-4-431-66956-2

Printed on acid-free paper

© Springer Japan 1997
Originally published by Springer-Verlag Tokyo Berlin Heidelberg New York in 1997
Softcover reprint of the hardcover 1st edition 1997

Typesetting: Camera-ready by authors

Preface

The flood of information through various computer networks such as the Internet characterizes the world situation in which we live. Information worlds, often called virtual spaces and cyberspaces, have been formed on computer networks. The complexity of information worlds has been increasing almost exponentially through the exponential growth of computer networks. Such nonlinearity in growth and in scope characterizes information worlds. In other words, the characterization of nonlinearity is the key to understanding, utilizing and living with the flood of information. The characterization approach is by characteristic points such as peaks, pits, and passes, according to the Morse theory. Another approach is by singularity signs such as folds and cusps. Atoms and molecules are the other fundamental characterization approach. Topology and geometry, including differential topology, serve as the framework for the characterization. *Topological Modeling for Visualization* is a textbook for those interested in this characterization, to understand what it is and how to do it. Understanding is the key to utilizing information worlds and to living with the changes in the real world.

Writing this textbook required careful preparation by the authors. There are complex mathematical concepts that require designing a writing style that facilitates understanding and appeals to the reader. To evolve a style, we set as a main goal of this book the establishment of a link between the theoretical aspects of modern geometry and topology, on the one hand, and experimental computer geometry, on the other. There are many excellent books on modern geometry and topology (generally speaking, "theory"), and many excellent books on modern computer and experimental geometry. As far as we know, however, there is no book that bridges the gap between these two branches of modern science, that is, between theory and practice. We have tried to fill this gap. Our intention was to write a book that will be useful to both communities of scientists. Of course, we realize that this separation between theoretical science and experimental science is not clear-cut, and we use this language and these images only for easier description of our main idea. We collect in the book some basic elements of theoretical geometry and topology that are used today in different branches of experimental computer geometry. We do not give detailed proofs because of lack of space, but we give references that can help the reader find the proofs. The advantage of such a style is this:

We collect in one book a short description of the most powerful theoretical tools, and experts in experimental science can use this material in their work. Certainly, as we know from our own experience, modern topological methods can improve the results of experimental computer geometry. On the other hand, experts in theoretical geometry and topology can find in our book possible applications of those fields to very interesting computer experiments in the world of geometrical computer methods, medicine, the automobile industry, architecture, and so on. Many pure mathematicians will also find here material for development of new theoretical ideas. Each chapter consists of two layers: first theoretical ideas, then applications to the different branches of modern experimental computer geometry. We have tried to make chapters as independent as possible to help the reader use each chapter as an individual research tool without a complete study of other sections of the book. Consequently, we sometimes include in some chapters a summary of material from another section to recall important ideas.

Anatoly T. Fomenko
Tosiyasu L. Kunii

Acknowledgments. The project to publish a textbook on topological modeling for visualizing complex objects and phenomena has been a long-standing desire of the authors. We now have the initial version. Many people have contributed to the project. Springer-Verlag has done an excellent job of improving the text and the style. Multiple sources of support have been helpful, including the MOVE Project sponsored by Fujitsu, Microsoft Japan, and Ricoh; the University of Aizu; the University of Tokyo; Moscow State University; and the Aizu Area Foundation for Promotion of Science and Education.

Table of Contents

Part I

Foundation

1. Curves

1.1. Curves and Coordinate Systems

Smooth curves and length in a Euclidean coordinate system. We consider Euclidean space $\mathbb{R}^n$ and define in it the standard Euclidean scalar product

$$\langle \xi. \eta \rangle = \xi^1 \eta^1 + \cdots + \xi^n \eta^n,$$

for any vectors ξ and η. With each vector ξ we associate a real number, called its *length* and defined as

$$|\xi| = \langle \xi, \xi \rangle^{\frac{1}{2}}.$$

This is the length (in the usual sense) of the vector going from the origin 0 to the point ξ in $\mathbb{R}^n$. The *distance between two points* ξ and η is the length of the vector $\xi - \eta$. The *angle* φ between two vectors ξ and η can be expressed in terms of the scalar product as

$$\cos \varphi = \frac{\langle \xi, \eta \rangle}{|\xi|\,|\eta|}.$$

All the fundamental properties of Euclidean geometry are based on this notion of Euclidean scalar product. In this sense, the scalar product is the foundation of Euclidean geometry. When defining other geometrical concepts we shall often proceed from the scalar product of vectors.

Definition 1.1.1. A *smooth curve* $\gamma(t)$ in Euclidean space is a vector-valued function of the form

$$\gamma(t) = \left(x^1(t), \ldots, x^n(t) \right),$$

where the components represent Cartesian coordinates in space, and $x^1(t)$, $\ldots, x^n(t)$ are *smooth* functions of the time parameter t, which runs through either the entire real axis or a segment $[a, b]$. More precisely, what we have just introduced is the *parametric form* (or *parametric representation*) of a smooth curve γ.

Remark. A *smooth* function is one that has continuous derivatives of any order.

We stress that the concept of a smooth parametrized curve includes the notion of a time parameter along the curve. The same set of the points (trajectory) can be parametrized in different ways, leading in general to different smooth parametrized curves. In other words, any change in time along the trajectory changes the parametrized curve.

When the time parametrization is not important for an application, we can ignore the time t and consider the curve simply as a set of points in the ambient space (without parametrization).

Definition 1.1.2. The vector

$$\dot{\gamma}(t) = \left(\frac{dx^1}{dt}(t), \ldots, \frac{dx^n}{dt}(t) \right)$$

is called the *velocity vector* of the smooth curve $\gamma(t)$, or the *tangent vector* to $\gamma(t)$. We call a smooth curve $\gamma(t)$ *regular* if its tangent vector is nonzero at any point of the curve (Figure 1.1).

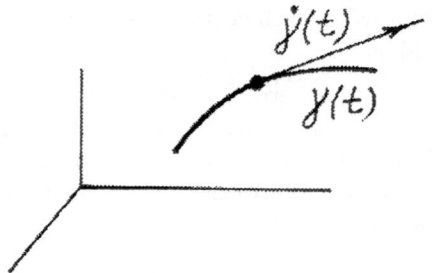

FIGURE 1.1.

If the tangent vector of a smooth curve γ is nonzero at some point $\gamma(t_0)$, it follows from the implicit function theorem that locally, near this point, the trajectory of γ is a smooth *one-dimensional manifold*, without singularities and cusps. By contrast, at points where the velocity vector of a smooth curve vanishes, the curve may have a *cusp* and abruptly change its direction.

The existence of a cusp on a smooth curve at parameter values where the velocity vector vanishes by no means contradicts the condition that the curve is smooth, according to Definition 1.1.1. Figure 1.2 gives an example of a smooth curve with a cusp, with *cusp angle* $\pi/2$. It is also very simple to construct a smooth (but nonregular) curve with a cusp angle of π at the singular point (Figure 1.3). We suggest to the reader an exercise: Write a parametric representation for the curve in Figure 1.3 and decide whether this curve can be defined by means of analytic functions $x(t)$ and $y(t)$ in the Euclidean plane.

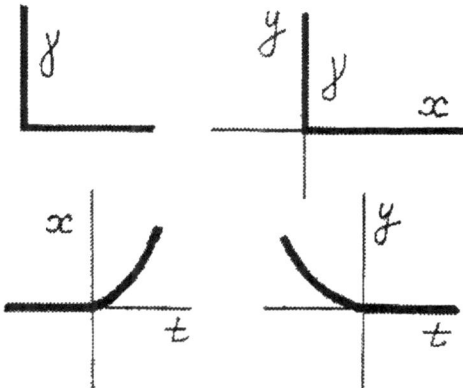

FIGURE 1.2.

FIGURE 1.3.

Remark. In our book we will consider mainly *smooth regular curves,* and because of this we will usually omit the term *regular.* Thus, when we speak about a smooth curve, we assume it regular.

Definition 1.1.3. The *length* of a curve $\gamma(t)$ from the point $\gamma(a)$ to the point $\gamma(b)$ (that is, from the parameter value $t = a$ to $t = b$) is the integral of the length of the velocity vector of γ:

$$l(\gamma) = \int_a^b \langle \dot{\gamma}(t), \dot{\gamma}(t) \rangle^{\frac{1}{2}} \, dt = \int_a^b |\dot{\gamma}| \, dt.$$

In coordinates,

$$l(\gamma) = \int_a^b \left(\left(\frac{dx^1}{dt}(t) \right)^2 + \cdots + \left(\frac{dx^n}{dt}(t) \right)^2 \right)^{\frac{1}{2}} dt.$$

Proposition 1.1.4. *The length of a smooth curve does not depend on the choice of smooth parametrization of the curve.*

To rephrase this more precisely, consider a smooth curve $\gamma(t)$ and two points $\gamma(a)$ and $\gamma(b)$ corresponding to the parameter values $t = a$ and $t = b$.

Let $t = t(\tau)$ be an arbitrary smooth transformation of the time parameter t into a new one τ satisfying $dt/d\tau > 0$ for every τ (Figure 1.4). Then the length of the curve remains unchanged, that is, *the length of* $\gamma(t)$ *from* $t = a$ *to* $t = b$ *is equal to the length of* $\gamma(t(\tau))$ *from* $\tau = \alpha$ *to* $\tau = \beta$, *where* $a = t(\alpha)$ *and* $b = t(\beta)$ (Figure 1.4).

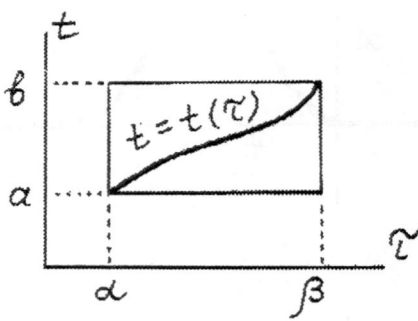

FIGURE 1.4.

The proof follows from a straightforward calculation (check it!). Exercise: what happens if we consider time transformations $t = t(\tau)$ where the derivative $dt/d\tau$ can change its sign when t runs along the time interval?

Now assume that we have two smooth curves $\gamma_1(t)$ and $\gamma_2(t)$ intersecting at a point P, as in Figure 1.5.

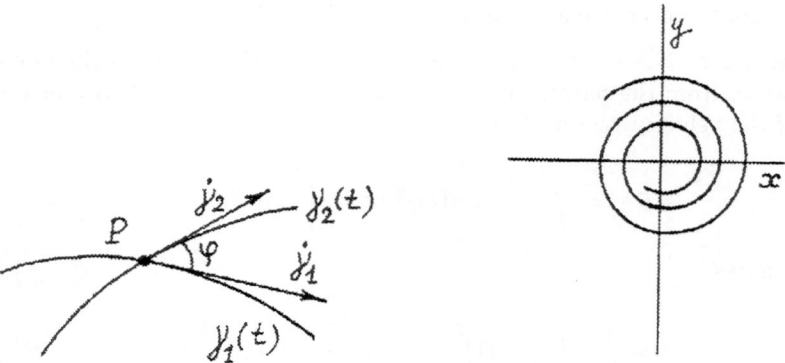

FIGURE 1.5.   FIGURE 1.6.

Definition 1.1.5. The *angle* φ between the two smooth curves γ_1 and γ_2 at P is defined by

$$\cos \varphi = \frac{\langle \dot{\gamma}_1, \dot{\gamma}_2 \rangle}{|\dot{\gamma}_1| \, |\dot{\gamma}_2|},$$

provided that $\dot{\gamma}_1$ and $\dot{\gamma}_2$ are both nonzero at P.

This formula defines not one, but two angles differing by a sign. If, however. we are working in two dimensions and we assume that the order in which the curves are given matters, we can define a notion of *oriented angle*, by decreeing that the angle is positive if and only if $|\gamma_1|$ needs to be rotated counterclockwise by $\varphi \in (0, \pi)$ in order to point in the direction of γ_2.

Curvilinear coordinate systems in Euclidean space. Cartesian coordinates are not always the most convenient ones to use in solving analytically many problems in physics and other fields of science. We often deal with smooth curves (say, trajectories of a particle in a force field) whose equations in Cartesian coordinates are rather cumbersome. For example, the Cartesian equation

$$(x^2 + y^2)^{\frac{1}{2}} - \exp(\lambda(\tan^{-1}(y/x))) = 0$$

in the plane determines the *spiral* shown in Figure 1.6. This equation could be written in the simpler form $r = \exp(\lambda \varphi)$ in the *polar coordinate system* (r, φ). The polar equation clearly demonstrates the character of the trajectory. Thus, the introduction of such curvilinear coordinates is not a caprice of mathematicians. We demonstrate this further with an important example.

The motion of a particle in the plane can be described in polar coordinates by two functions, $r = r(t)$ and $\varphi = \varphi(t)$. *Kepler's second law*, published by Johannes Kepler in 1609 as a result of his studies on the motion of planets around the Sun, says that when a material particle moves in a central force field the quantity $r^2 \dot{\varphi}$ is conserved. This law is much easier to state in polar than in Cartesian coordinates.

The solution of particular problems in fields such as mechanics, chemistry. and computer geometry has called for the invention of other curvilinear coordinate systems: cylindrical, spherical, and so on.

We now give a formal definition of a curvilinear coordinate system. Consider an arbitrary domain in Euclidean space $\mathbb{R}^n$. (A *domain* or *open set* in $\mathbb{R}^n$ is a subset $C \subset \mathbb{R}^n$ such that every point P of C is the center of some ball that is entirely contained in C: see Figure 1.7.)

Suppose we associate with each point P of C an ordered set of n real numbers, or *coordinates*. The simplest way to do this is to use Cartesian coordinates, which come from the definition of $\mathbb{R}^n$ itself. However, any scheme for associating n-tuples of real numbers to points can be considered instead. Clearly. any such scheme gives rise to a set of n functions $x^1(P), \ldots, x^n(P)$ defined in C. We can consider $x^1, \ldots, x^n$ as Cartesian coordinates in another copy $\mathbb{R}_1^n$ of Euclidean space (Figure 1.7). We will usually require these

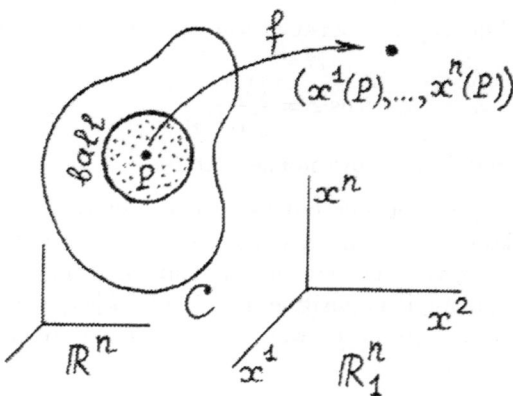

FIGURE 1.7.

functions to be *continuous* and even *smooth*. Continuity means that a small change in the position of P should lead to a small change in its coordinate values. Smoothness means that this dependence should be differentiable to any order (this will soon be explained more precisely).

Thus, we are formally considering two copies of Euclidean space:

$\mathbb{R}^n$, with Cartesian coordinates $y^1, \ldots, y^n$, and
$\mathbb{R}^n_1$, with Cartesian coordinates $x^1, \ldots, x^n$.

The domain C lies in $\mathbb{R}^n$.

Definition 1.1.6. A *continuous coordinate system* in $C \subset \mathbb{R}^n$ is an n-tuple of continuous functions

$$x^1 = x^1(y^1, \ldots, y^n), \quad \ldots, \quad x^n = x^n(y^1, \ldots, y^n)$$

that maps C bijectively and continuously in both directions onto some domain A of Euclidean space $\mathbb{R}^n_1$. In other words, the map f defined by

$$f(P) = (x^1(P), \ldots, x^n(P))$$

is a *homeomorphism* of C onto A (recall that this means a bijective map that is continuous and has a continuous inverse; two spaces are *homeomorphic* if there is a homeomorphism between them). The values $x^1(P), \ldots, x^n(P)$ are called the *coordinates* of P relative to the *coordinate map* $f : C \to A$ (Figure 1.8); the component functions $x^1, \ldots, x^n$ of f are the *coordinate functions*.

Sometimes we shall write a point P with coordinates $x^1(P), \ldots, x^n(P)$ in the short form $P(x^1, \ldots, x^n)$, assuming that the coordinate map f has already been fixed.

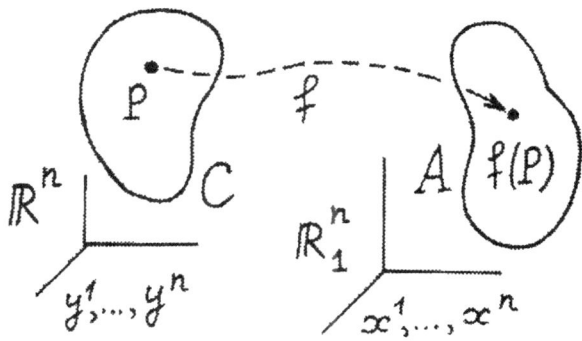

FIGURE 1.8.

Among all *continuous* coordinate maps, we are especially interested in *smooth* ones, namely those for which the functions $x^i(y^1, \ldots, y^n)$ *and* the inverse functions $y^i(x^1, \ldots, x^n)$ are smooth (have continuous partial derivatives of all orders). Let $f : C \to A$ be a smooth map defined by the coordinate functions

$$x^1(y^1, \ldots, y^n), \quad \ldots, \quad x^n(y^1, \ldots, y^n).$$

Definition 1.1.7. The *Jacobian matrix* $df = \partial x / \partial y$ of f is the functional matrix

$$\begin{pmatrix} \dfrac{\partial x^1}{\partial y^1} & \cdots & \dfrac{\partial x^1}{\partial y^n} \\ \vdots & & \vdots \\ \dfrac{\partial x^n}{\partial y^1} & \cdots & \dfrac{\partial x^n}{\partial y^n} \end{pmatrix}$$

whose entries are the partial derivatives of the functions $x^i(y^1, \ldots, y^n)$. The determinant of this matrix is denoted by $J(f)$ and called the *Jacobian* of f.

The Jacobian matrix is associated with a linear approximation to the map f that is a natural extension of the differential of a smooth function of one variable. The notation df is applied to this linear map, as well as to Jacobian matrix.

The entries of the Jacobian matrix are functions of P, so we may think of df as a smooth matrix-valued function on the domain C. The Jacobian $J(f)$ is a smooth (real-valued) function on C.

Definition 1.1.8. A *regular smooth coordinate system* in a domain C of Euclidean space $\mathbb{R}^n$ is an n-tuple of smooth functions

$$x^1(y^1, \ldots, y^n), \quad \ldots, \quad x^n(y^1, \ldots, y^n)$$

such that the map $f = (x^1, \ldots, x^n)$ is a smooth bijection from C onto a domain A in Euclidean space $\mathbb{R}^n_1$, and such that the Jacobian $J(f)$ is nonzero at all points of C.

The condition that the Jacobian does not vanish anywhere in C means that the *inverse map* f^{-1} is not only continuous, but also *smooth*. This follows immediately from the well-known implicit function theorem. In other words, f is a *diffeomorphism* between C and A.

A regular smooth coordinate system in a domain C is also called a *curvilinear coordinate system* in C.

Every curvilinear coordinate system in C specifies families of *coordinate lines* defined as follows: the i-th coordinate line γ_i going through a point P_0 is the curve formed by points $P = P(t)$ such that

$$x^i(P) = t, \qquad x^j(P) = x^j(P_0) \quad \text{for } j \neq i.$$

Here t is a parameter; as it varies, the corresponding point $P(t)$ describes a smooth trajectory in C (Figure 1.9), naturally parametrized by t. Thus, n smooth trajectories go through each point P_0 in C. For another point P_0, we obtain another family of n coordinate lines, varying smoothly with P_0 (Figure 1.10).

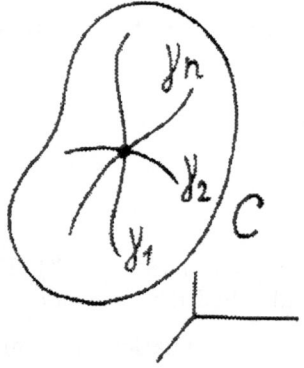

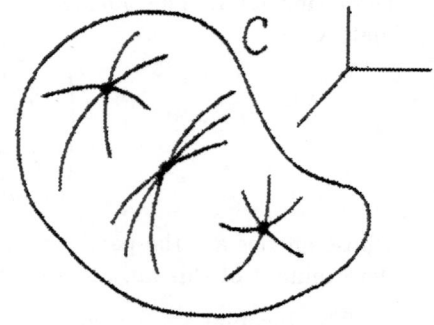

FIGURE 1.9.   FIGURE 1.10.

Example 1.1.9. In the Cartesian coordinate system, the coordinate lines are straight lines through P_0, parallel to the coordinate axes.

In computer geometry and in the graphic representation of a curvilinear coordinate system it is often useful to *draw* coordinate lines (for example on the screen of a computer). In particular, the transformation to a curvilinear coordinate system is especially clear if the coordinate network is depicted (Figure 1.11).

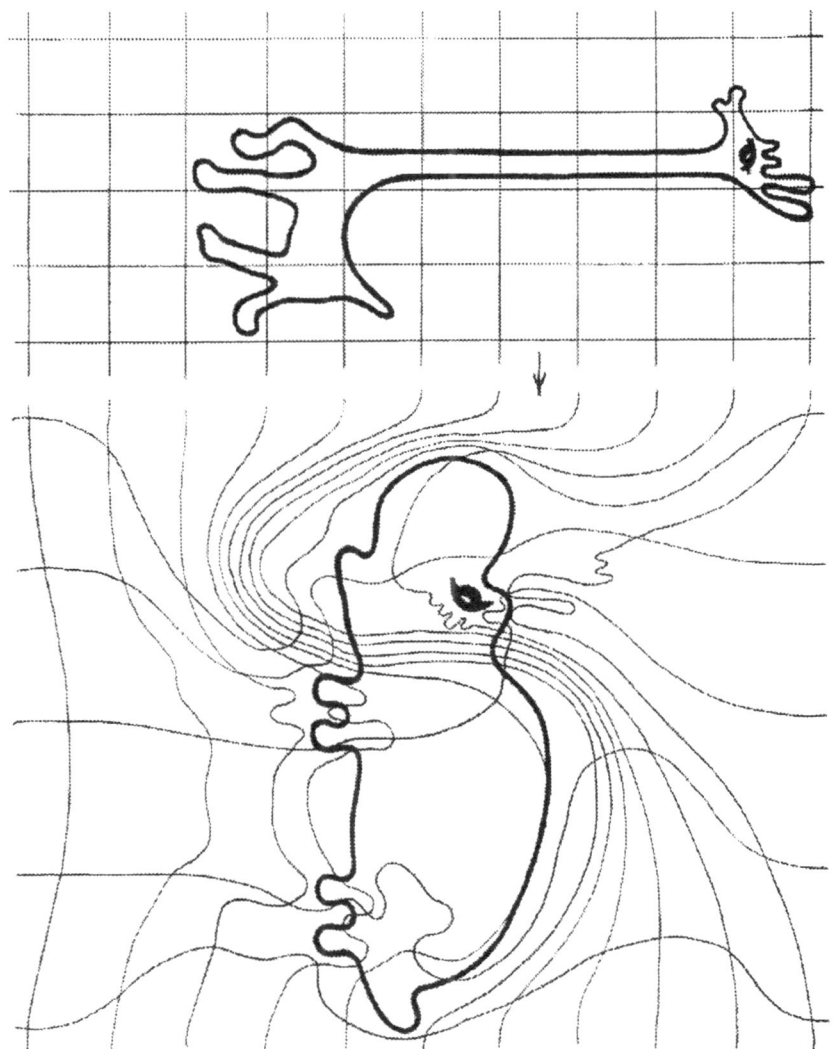

FIGURE 1.11.

We now turn to some basic examples of curvilinear coordinate systems in the Euclidean plane and in Euclidean three-space.

Polar coordinates on the plane. The *polar coordinate system* (r, φ) is related to the Cartesian coordinates x and y on the plane by the formulas

$$x = r \cos \varphi, \quad y = r \sin \varphi,$$

(see Figure 1.12). According to the preceding definition, this is not a reg-

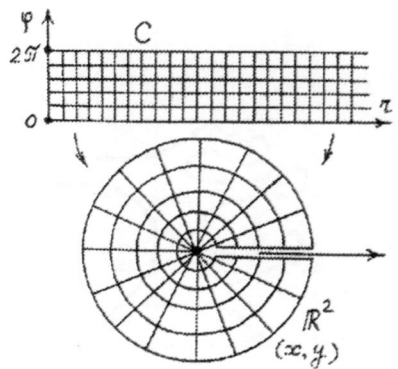

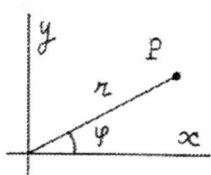

FIGURE 1.12.   FIGURE 1.13.

ular coordinate system on the whole Euclidean plane $\mathbb{R}^2$. Indeed, if we set $f(r,\varphi) = \big(x(r,\varphi),\ y(r,\varphi)\big)$, a direct calculation shows that the Jacobian matrix and the Jacobian of f are

$$df = \begin{pmatrix} \cos\varphi & \sin\varphi \\ -r\sin\varphi & r\cos\varphi \end{pmatrix}, \quad J(f) = r.$$

Thus, the Jacobian is zero at the origin, contradicting the condition in Definition 1.1.8. And this is not the only disadvantage of the polar system: the map f is not a bijection of $\mathbb{R}^2$, since the points (r,φ) and $(r,\ \varphi + 2\pi)$ are transformed into the same point.

We can restrict the domain of the polar coordinate system so that it becomes regular, as follows. In the (r,φ)-plane, let C be the infinite strip $\{(r,\varphi) : r > 0,\ 0 < \varphi < 2\pi\}$, as shown in Figure 1.13. Then the image A of C in the (x,y)-plane is the entire plane minus the ray $x > 0$, $y = 0$. Figure 1.13 shows the coordinate lines of the polar system and the transformation of a few coordinate lines of the Cartesian coordinate system.

Cylindrical coordinates in three-dimensional space. The *cylindrical coordinate system* $(r,\varphi.t)$ is related to the Cartesian coordinates x,y,z in three-space by the relations

$$x = r\cos\varphi, \quad y = r\sin\varphi, \quad z = t.$$

It is a regular coordinate system on the domain C given by $\{(r,\varphi,\theta) : r > 0,\ 0 < \varphi < 2\pi,\ -\infty < t < +\infty\}$, as shown in Figure 1.14. The domain A is $\mathbb{R}^3$ minus the half-plane $x = 0$, $y > 0$. The Jacobian matrix is

$$df = \begin{pmatrix} \cos\varphi & \sin\varphi & 0 \\ -r\sin\varphi & r\cos\varphi & 0 \\ 0 & 0 & 1 \end{pmatrix},$$

and the Jacobian $J(f)$ is equal to r.

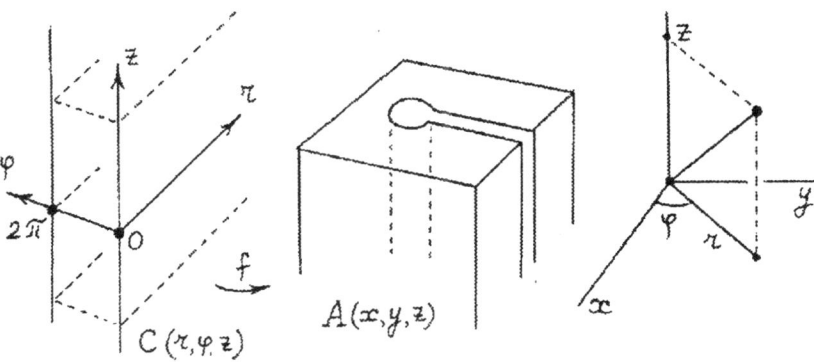

FIGURE 1.14.

Spherical coordinates in three-dimensional space. Spherical coordinates are usually denoted by (r, θ, φ), and the transformation formulas are

$$x = r \sin \theta \cos \varphi, \quad y = r \sin \theta \sin \varphi, \quad z = r \cos \theta.$$

To make the coordinate system regular, we can take $0 < \theta < \pi$, $0 < \varphi < 2\pi$, $r > 0$. The domains C and A is shown in Figure 1.15. The Jacobian $J(f)$ is $r^2 \sin \theta$. The Jacobian matrix is

$$df = \begin{pmatrix} \sin \theta \cos \varphi & r \cos \theta \cos \varphi & r \sin \theta \sin \varphi \\ \sin \theta \sin \varphi & r \cos \theta \sin \varphi & r \sin \theta \cos \varphi \\ \cos \theta & -r \sin \theta & 0 \end{pmatrix}.$$

Figure 1.16 shows θ- and φ-coordinate lines for a fixed r. These angular parameters are associated with the usual notions of *latitude* and *longitude* on the globe: the longitude is φ, and the latitude is $90° - \theta$.

Length of a curve and metric form in curvilinear coordinates. Consider a regular curvilinear coordinate system in a Euclidean domain C, and let $\gamma(t)$ be a regular curve in C. What is the length of the curve in this coordinate system? Assume that $x^1, \ldots, x^n$ are the Cartesian coordinates and that $y^1, \ldots, y^n$ are the curvilinear coordinates, that is, $x^i = x^i(y^1, \ldots, y^n)$. Recall that

$$\frac{dx^i}{dt} = \frac{\partial x^i}{\partial y^1} \frac{dy^1}{dt} + \frac{\partial x^i}{\partial y^2} \frac{dy^2}{dt} + \cdots + \frac{\partial x^i}{\partial y^n} \frac{dy^n}{dt} = \sum_{k=1}^{n} \frac{\partial x^i}{\partial y^k} \frac{dy^k}{dt}.$$

Since the length of the curve in Cartesian coordinates is

$$l(\gamma) = \int_a^b \left(\left(\frac{dx^1}{dt}(t) \right)^2 + \cdots + \left(\frac{dx^n}{dt}(t) \right)^2 \right)^{\frac{1}{2}} dt,$$

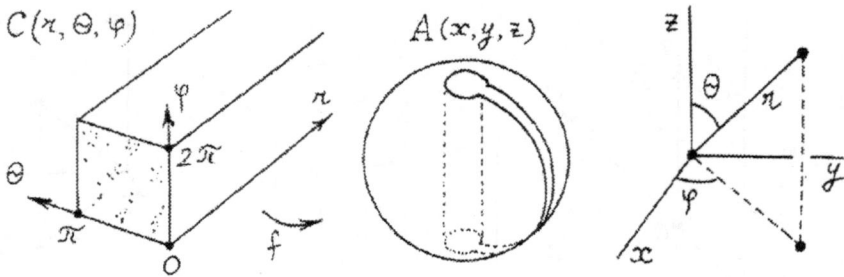

FIGURE 1.15.

FIGURE 1.16.

we obtain, by direct calculation, the following formula for the length of the smooth curve in curvilinear coordinates:

$$l(\gamma) = \int_a^b \left(\sum_{m,p} g_{mp}(y) \frac{dy^m}{dt} \frac{dy^p}{dt} \right)^{\frac{1}{2}} dt, \qquad \text{where} \quad g_{mp}(y) = \sum_i \frac{\partial x^i}{\partial y^m} \frac{\partial x^i}{\partial y^p}.$$

Obviously, the latter sum is *symmetric* in m and p, that is, $g_{mp} = g_{pm}$. Consequently, they give rise to a symmetric matrix $G = (g_{mp})$. For example, for $n = 2$ we have

$$G = \begin{pmatrix} g_{11} & g_{12} \\ g_{21} & g_{22} \end{pmatrix}.$$

This matrix depends on the curvilinear coordinate system, and it is useful to know how it varies under coordinate transformations. In other words, what is the law of transformation for G? Take another coordinate transformation $y \rightarrow z$, that is, consider regular coordinate functions $y^i = y^i(z^1, \ldots, z^n)$, where $z^1, \ldots, z^n$ are new regular curvilinear coordinates in a region of Euclidean space. Then a simple direct calculation shows that the resulting formula is

1.1.10.
$$g_{kl}(z) = \sum_{m,p} \frac{\partial y^m}{\partial z^k} g_{mp}(y) \frac{\partial y^p}{\partial z^l}.$$

In terms of the corresponding matrices, this becomes $G(z) = AG(y)A^T$, where A is the Jacobian matrix for the coordinate transformation $y \to z$.

Important Convention. To simplify the notation, we adopt the *Einstein summation convention*: In an expression such as $\sum_i a^i b_i$, where the sum is taken over an index that appears once as a subscript and once as a superscript, we omit the symbol $\sum$ and write $a^i b_i$. The implicit summation can be over more than one symbol: for example, formula 1.1.10 becomes, in this convention,

$$g_{kl}(z) = \frac{\partial y^m}{\partial z^k} g_{mp}(y) \frac{\partial y^p}{\partial z^l}.$$

Sometimes we shall distinguish between an old and a new coordinate systems by using *primed indices*—i', j', and so on—for the new one. Thus, for example, formula 1.1.10 will be written as

$$g_{k'q'} = \frac{\partial y^k}{\partial z^{k'}} g_{kq} \frac{\partial y^q}{\partial z^{q'}} \quad \text{or} \quad g_{k'q'} = \frac{\partial y^k}{\partial z^{k'}} \frac{\partial y^q}{\partial z^{q'}} g_{kq}.$$

Consider the special case when the initial coordinates $x^1, \ldots, x^n$ are *Cartesian* and we transform them into new regular curvilinear coordinates $y^1, \ldots, y^n$. Formula 1.1.10 becomes

$$g_{kl}(y) = \frac{\partial x^m}{\partial y^k} g_{mp}(x) \frac{\partial x^p}{\partial y^l}.$$

But $g_{mp}(x) = \delta_{mp}$, where $\delta_{mp} = 0$ if $m \neq p$ and $\delta_{mp} = 1$ if $m = p$. Consequently,

$$g_{kl}(y) = \sum_m \frac{\partial x^m}{\partial y^k} \frac{\partial x^m}{\partial y^l}.$$

The functions g_{ij} give us the Euclidean metric in curvilinear coordinates. They have a clear geometric meaning. Consider an arbitrary point P in the domain C, and let $\gamma_1(t), \ldots, \gamma_n(t)$ be the coordinate lines through P of the curvilinear coordinate system $y^1, \ldots, y^n$ (Figure 1.17). Assume that P has coordinates $y_1 = c_1, \ldots, y_n = c_n$, so that γ_i has equation

$$y^i = t, \qquad y^j = c^j(P_0) \quad \text{for } j \neq i,$$

where t is the time parameter along γ_i.

The velocity vector $a_i = \dot{\gamma}_i$ of the coordinate curve γ_i has coordinates

$$a_i = \left(\frac{\partial x^1}{\partial y^i}, \ldots, \frac{\partial x^n}{\partial y^i} \right),$$

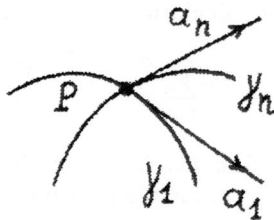

FIGURE 1.17.

where $x^1, \ldots, x^n$ are Cartesian coordinates in the domain C. Since

$$g_{ij}(y) = \sum_m \frac{\partial x^m}{\partial y^i} \frac{\partial x^m}{\partial y^j},$$

we obtain the simple and important formula

$$g_{ij}(y) = \langle a_i, a_j \rangle,$$

where a_i and a_j are the velocity vectors of the coordinate curves γ_i and γ_j. We recall that in this formula the scalar product is Euclidean, that is, it is a characteristic of the ambient space. Thus, we proved the following statement.

Lemma 1.1.11. *The functions $g_{ij}(y)$ representing the Euclidean metric with respect to a curvilinear coordinate system in a domain C are the scalar products of the velocity vectors of the corresponding coordinate lines γ_i and γ_j passing through the point under consideration.*

Thus, we see that under a coordinate transformation the matrix $G(y)$ transforms as the matrix of a quadratic form. In particular, if the initial coordinates are Cartesian, G is the identity matrix $(\delta_{ij}) = E$, so that in any other (curvilinear) coordinate system it can be written as $G(y) = AEA^T$, where A is the Jacobian matrix of the coordinate transformation.

Let's write explicit explicit formulas for the length of a smooth curve and for the matrix G in various curvilinear coordinate systems. These formulas are used in different problems of computer geometry and applications.

Polar coordinates in the plane. If r and φ are polar coordinates, the matrix G of the Euclidean metric has the form

$$G(r, \varphi) = \begin{pmatrix} 1 & 0 \\ 0 & r^2 \end{pmatrix}.$$

Hence the length of a curve $\gamma(t) = (r(t), \varphi(t))$ is

$$l(\gamma) = \int_a^b \left(\left(\frac{dr}{dt}\right)^2 + r^2 \left(\frac{d\varphi}{dt}\right)^2 \right)^{\frac{1}{2}} dt.$$

Cylindrical coordinates in three-dimensional space. In cylindrical coordinates r, φ, z, we have

$$G(r, \varphi, z) = \begin{pmatrix} 1 & 0 & 0 \\ 0 & r^2 & 0 \\ 0 & 0 & 1 \end{pmatrix}.$$

Thus the length of a curve $\gamma(t) = (r(t), \varphi(t), z(t))$ is

$$l(\gamma) = \int_a^b \left(\left(\frac{dr}{dt}\right)^2 + r^2 \left(\frac{d\varphi}{dt}\right)^2 + dz^2 \right)^{\frac{1}{2}} dt.$$

Spherical coordinates in three-dimensional space. In spherical coordinates r, θ, φ, we have

$$G(r, \theta, \varphi) = \begin{pmatrix} 1 & 0 & 0 \\ 0 & r^2 & 0 \\ 0 & 0 & r^2 \sin^2 \theta \end{pmatrix}.$$

Hence the length of a curve $\gamma(t) = (r(t), \theta(t), \varphi(t))$ is

$$l(\gamma) = \int_a^b \left(\left(\frac{dr}{dt}\right)^2 + r^2 \left(\frac{d\theta}{dt}\right)^2 + r^2 \sin^2 \theta \left(\frac{d\varphi}{dt}\right)^2 \right)^{\frac{1}{2}} dt.$$

It is sometimes convenient to deal with the element dl of arclength instead of the whole arclength. In the examples considered above these elementary differentials squared are of the form

$$dl^2 = dr^2 + r^2 \, d\varphi^2 \qquad \text{(polar coordinates on the plane)},$$
$$dl^2 = dr^2 + r^2 \, d\varphi^2 + dz^2 \qquad \text{(cylindrical coordinates in three-space)},$$
$$dl^2 = dr^2 + r^2 \, d\theta^2 + \sin^2 \theta \, d\varphi^2 \qquad \text{(spherical coordinates in three-space)}.$$

1.2. Metric Properties of Plane Curves: Frenet Formulas

Consider the Euclidean plane with Cartesian coordinates x and y. A smooth curve $r(t)$ in it is determined by its two components $x(t)$ and $y(t)$, which are smooth functions: $r(t) = (x(t), y(t))$. The velocity vector r' at time t is $v(t) = r'(t) = (x'(t), y'(t))$. (As usual, $f'(t)$ denotes the derivative df/dt of a function $f(t)$.) We are assuming that our curves are regular, so $v(t) \neq 0$ for all t. Recall that $|v(t)|$ denotes the length of the velocity vector in the Euclidean metric.

Let s denote the length of the arc going from a fixed point on the curve to another, variable, point. This length increases monotonically as the variable endpoint moves in the direction of increasing time parameter, so we may use s to reparametrize the curve. We call s the *natural parameter* of the curve. The vector-valued function

$$r(s) = (x(s), y(s))$$

is the *natural parametrization* of the curve, and we say that the curve is *parametrized by arclength* when we use this parametrization.

Lemma 1.2.1. *The velocity vector of a curve parametrized by arclength has constant length 1.*

This follows from the identity $dl = \left|\dfrac{dr}{dt}\right| dt$, because for $l = s$ we obtain

$$|v(s)| = \left|\frac{dr}{ds}(s)\right| = 1.$$

Thus, motion along a curve parametrized by arclength occurs at constant speed.

It is convenient to distinguish differentiation with respect to arclength s and with respect to an arbitrary parameter t. We will denote the former by a dot, so that $v(s) = \dot{r}(s)$; and we will continue to write $r'(t)$ for differentiation with respect to an arbitrary parameter t.

With each point on a curve we can associate, besides the velocity vector, the *acceleration vector*

$$v'(t) = \frac{dv}{dt}(t),$$

which depends smoothly on t. For a curve parametrized by arclength, it is easy to see that v' is always orthogonal to v:

Lemma 1.2.2. *Let $p(t)$ be a vector-valued function such that $|p(t)| = 1$ for all t. Then $p'(t)$ is always orthogonal to $p(t)$.*

Indeed, $\langle p(t), p(t) \rangle = 1$; after differentiation with respect to t we obtain $\langle p', p \rangle = 0$ for all t, i.e., the vectors p' and p are orthogonal.

Thus, each point of a smooth curve $r(s)$ parametrized by arclength has two associated orthogonal vectors: the velocity $v(s) = \dot{r}(s)$, and the acceleration $\dot{v}(s) = \ddot{r}(s)$ (Figure 1.18).

The acceleration need not have unit length: indeed, it may be zero. In order not to have to treat this case separately, it is convenient to define the *unit normal vector* to the curve for a given value of s as the unit vector $n(s)$ such that $e(s) = (v(s), n(s))$ is a positively oriented frame. (A *frame* is an ordered orthonormal basis of the ambient space, here the plane $\mathbb{R}^2$. When a frame is ordered so that it can be made to coincide with the standard frame (x, y) by means of a rigid motion, we say that it is *positively oriented*.) In other words, the vector $n(s)$ is obtained from $v(s)$ by a counterclockwise 90° rotation. If $\dot{v}(s) \neq 0$, we naturally have

$$n(s) = \pm \frac{\dot{v}(s)}{|\dot{v}(s)|},$$

the positive sign being chosen if and only if the curve's velocity vector is turning counterclockwise. We call $e(s)$ the Frenet frame of the curve at the

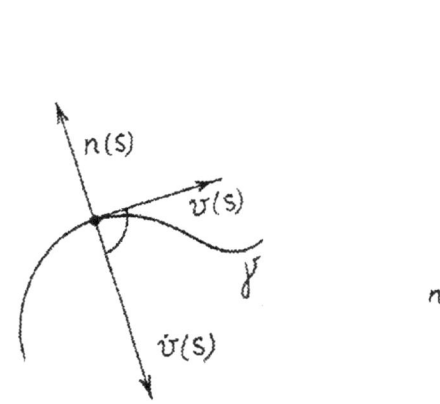

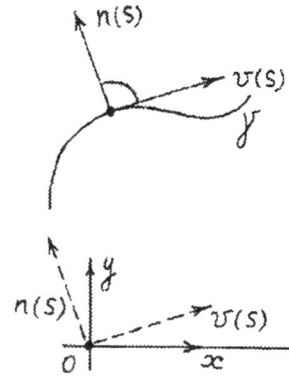

FIGURE 1.18.   FIGURE 1.19.

given value of s. As s changes, we obtain along the curve a smooth field of
frames $e(s) = (v(s), n(s))$.

Giving a positively oriented frame is the same as giving a rotation of a
plane about the origin (rotate the standard frame (x, y) to match the given
frame: see Figure 1.19). Therefore the Frenet frame field along a curve defines
a smooth map from the parameter interval into the group of orthogonal
matrices, that is, the group of rotations of the plane. Put another way, the
original curve in the plane gives rise to an associated curve in the orthogonal
group, whose points are represented as orthogonal 2×2 matrices.

Definition 1.2.3. Let $r(s)$ be a smooth plane curve parametrized by arclength.
The *signed curvature* of the curve at parameter value s is the real number
$k(s)$ such that $\dot{v}(s) = k(s)n(s)$. The *unsigned curvature* (or simply *curvature*
is $|k(s)|$, and the *curvature radius* is $R(s) = 1/|k(s)|$.

It is easy to see from the definition that unsigned curvature is the mag-
nitude of the acceleration of a point moving along the curve (with constant
speed $|\dot{r}(s)| = 1$); thus we have, in terms of the components $x(s)$ and $y(s)$ of
the vector $r(s)$:

$$|k(s)| = (\ddot{x}^2 + \ddot{y}^2)^{1/2}.$$

Example 1.2.4.

(a) The curvature of a straight line is equal to zero.
(b) The curvature of a circle of radius r is equal to $1/r$.

In many specific problems, the curve is not given as a function of arclength, and it may be difficult or impossible to reparametrize it explicitly
by arclength. It is useful, therefore, to have formulas for the curvature in
terms of an arbitrary parameter t.

Proposition 1.2.5. *Let $r(t)$ be a curve parametrized by an arbitrary parameter t. Then the curvature for a given value of the parameter is*

$$k = \frac{x''y' - y''x'}{(x'^2 + y'^2)^{3/2}},$$

where x', x'', y', y'' denote derivatives with respect to t.

Consider the motion of the frame $e(s) = (v(s), n(s))$ when the arclength parameter s varies. It turns out that the derivatives of the frame vectors satisfy simple relations, called the *Frenet formulas*.

Theorem 1.2.6. *For a plane curve parametrized by arclength,*

$$\dot{v}(s) = k(s)n(s), \qquad \dot{n}(s) = -k(s)v(s).$$

You can find the proof in [Dubrovin et al. 1992][Dubrovin et al. 1985] [Dubrovin et al. 1990][Mishchenko and Fomenko 1988], for example.

We can also write the Frenet formulas in matrix form. Writing the frame $e(n)$ as a column vector, each of whose entries is a vector in the plane, we have

$$\begin{pmatrix} \dot{v} \\ \dot{n} \end{pmatrix} = X \begin{pmatrix} v \\ n \end{pmatrix}, \qquad \text{where} \quad X = \begin{pmatrix} 0 & k \\ -k & 0 \end{pmatrix}$$

is a skew-symmetric matrix. This can be written also as $\dot{e}(s) = X(s)e(s)$, where Xe denotes the result of the action of the matrix X on the frame e.

We now give an extremely useful geometrical interpretation of the Frenet formulas (Figure 1.20). Consider the Frenet frame $e = (v, n)$ at a point s, and again at a point $s + \Delta s$ infinitesimally distant from s. After translating the frame $e(s + \Delta s)$ back so it's based at s, we can compare the two frames $e(s)$ and $e(s + \Delta s)$ at s. Now, $e(s + \Delta s)$ must be obtained from $e(s)$ by some rotation, whose (infinitesimal) angle we call $\Delta \varphi$. That is, $e(s)$ and $e(s + \Delta s)$ are related by an orthogonal transformation:

$$e(s + \Delta s) = g(\Delta s)e(s), \qquad \text{where} \quad g(\Delta \varphi) = \begin{pmatrix} \cos \Delta \varphi & \sin \Delta \varphi \\ -\sin \Delta \varphi & \cos \Delta \varphi \end{pmatrix}$$

Expanding the functions $\cos \Delta \varphi$ and $\sin \Delta \varphi$ in the small increment $\Delta \varphi$ and neglecting terms of the second order and higher in $\Delta \varphi$, we get

$$g(\Delta \varphi) = \begin{pmatrix} 1 & 0 \\ 0 & 1 \end{pmatrix} + \begin{pmatrix} 0 & \Delta \varphi \\ -\Delta \varphi & 0 \end{pmatrix} + \cdots,$$

that is,

$$e(s + \Delta \varphi) = e(s) + \begin{pmatrix} 0 & \Delta \varphi \\ -\Delta \varphi & 0 \end{pmatrix} e(s) + \cdots.$$

It follows from these simple calculations that

$$\dot{e}(s) = \begin{pmatrix} 0 & \dot{\varphi} \\ -\dot{\varphi} & 0 \end{pmatrix} e(s),$$

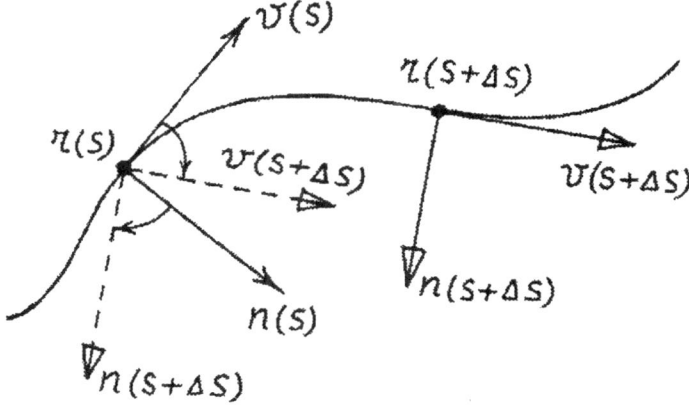

FIGURE 1.20.

where $\dot\varphi = \dot\varphi(s)$ is the derivative of the angle of rotation of the Frenet frame when the point moves along the curve. On the other hand, we know that

$$\dot e(s) = X(s)e(s).$$

The comparison of these two formulas give us the relation $\dot\varphi = k$, where k is the signed curvature. We have proved the following important theorem.

Theorem 1.2.7. *The curvature $k(s)$ of a smooth plane curve is the derivative of the rotation angle of the Frenet frame, when the point moves along the curve with constant velocity. In symbols,*

$$k(s) = \frac{d\varphi}{ds}(s).$$

Thus, if $|k|$ is large, the Frenet frame is turning very fast at this point; if $k = 0$, it is not turning at all.

The following important result, whose proof can be found in [Fomenko 1987a], for example, says that we can recover a plane curve from its curvature function.

Proposition 1.2.8. *Given a smooth function $k(s)$, defined for $a \le s \le b$, there exists a smooth plane curve $r(s)$ having s as its arclength parameter and $k(s)$ as its signed curvature. Any two such curves can be made to coincide by means of a rigid motion; that is, the solution is unique up to a rigid motion.*

A *rigid motion* of the plane (or of three-space) can be expressed as a rotation followed by a translation.

It is easy to prove that *if the curvature of a plane curve is identically zero, the curve is a straight line.*

1.3. Metric Properties of Curves in Three-Space

Consider a smooth regular curve in three-dimensional space, given by its Cartesian coordinate functions: $r(t) = \big(x(t), y(t), z(t)\big)$. Just as in the two-dimensional case, we can take the velocity $v(t) = r'(t)$, which is everywhere nonzero by the regularity assumption. If, moreover, the acceleration vector $v'(t)$ and the velocity $v(t)$ are *linearly independent* for every t—that is, if $v'(t)$ is never a scalar multiple of $v(t)$, for any value of t—we say that the curve is *biregular*. In this case we can define a Frenet frame that varies smoothly with the parameter.

For simplicity, assume that the curve is parametrized by arclength. Then the velocity vector $v(s) = \dot{r}(s)$ has length 1, and the acceleration $\dot{v}(s)$ is orthogonal to it, by Lemma 1.2.2. The biregularity condition then simply says that $\dot{v}(s) = \ddot{r}(s)$ is nonzero. Therefore we can define the *normal vector*

$$n(s) = \frac{\dot{v}(s)}{|\dot{v}(s)|},$$

which also has length 1. (This is also called the *principal normal.*) Now define a third unit vector $b(s)$, orthogonal to both $v(s)$ and $n(s)$, and such that the frame

$$e(s) = \big(v(s), n(s), b(s)\big)$$

is positively oriented (that is, it can be made to match the standard frame (x, y, z) by applying a rigid motion). In other words, set $b(s) = v(s) \times n(s)$, where $\times$ is the cross product of vectors in three-space. See Figure 1.21. We call $b(s)$ the curve's *binormal vector*, and $e(s)$ its *Frenet frame* (for the given value of s).

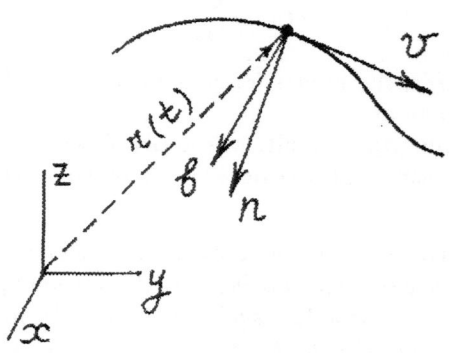

FIGURE 1.21.

Definition 1.3.1. The *curvature* of the curve at s is $k(s) = |\dot{v}(s)|$, that is, the magnitude of the acceleration vector (for a curve parametrized by arclength). Obviously, $\dot{v}(s) = k(s)n(s)$. The *radius of curvature* is $R(s) = 1/k(s)$.

Theorem 1.3.2. *The velocity, normal, and binormal vectors of a biregular smooth curve parametrized by arclength are connected by the relations*

$$\dot{v}(s) = k(s)n(s), \quad \dot{n}(s) = -k(s)v(s) + \tau(s)b(s), \quad \dot{b}(s) = -\tau(s)n(s),$$

where $\tau(s)$ is a real-valued function of s. In matrix form,

$$\begin{pmatrix} \dot{v} \\ \dot{n} \\ \dot{b} \end{pmatrix} = \begin{pmatrix} 0 & k & 0 \\ -k & 0 & \tau \\ 0 & -\tau & 0 \end{pmatrix} \begin{pmatrix} v \\ n \\ b \end{pmatrix}, \qquad or \qquad \dot{e} = Xe,$$

where the skew-symmetric 3×3 matrix X is defined by

$$X = \begin{pmatrix} 0 & k & 0 \\ -k & 0 & \tau \\ 0 & -\tau & 0 \end{pmatrix}$$

and Xe denotes the result of its action on the Frenet frame e.

Definition 1.3.3. The function $\tau = \tau(s)$ is called the *torsion* of the curve. The relations in Theorem 1.3.2 are called the *Frenet formulas*, and the matrix $X = X(s)$ is the *Frenet matrix*.

For a curve contained in a plane in space, the binormal vector is constant: It does not change as we move along the curve. In particular, the torsion is zero. In fact, zero torsion characterizes plane curves: *A curve in three-space lies in a plane if and only if its torsion is identically zero.*

We turn to the geometric interpretation of torsion (Figure 1.22). Consider the Frenet frame $e = (v, n, b)$ at s. and again at $s + \Delta s$, where Δs is infinitesimal. Translate $e(s + \Delta s)$ back so it's based at s, and compare $e(s + \Delta s)$ with $e(s)$. More precisely, project the frame $e(s + \Delta s)$ onto the plane spanned by $n(s)$ and $b(s)$. called the *normal plane*. The Frenet formulas say that the the projection of $v(s + \Delta s)$ is zero to first-order approximation, and that the variation in $n(s)$ and $b(s)$ is given by

$$\frac{dn}{ds} = \tau b. \qquad \frac{db}{ds} = -\tau n, \qquad \text{that is,} \qquad \begin{pmatrix} \dot{n} \\ \dot{b} \end{pmatrix} = \begin{pmatrix} 0 & \tau \\ -\tau & 0 \end{pmatrix} \begin{pmatrix} n \\ b \end{pmatrix}.$$

But this means that the motion of the projections (n, b) in the normal plane is a rotation with angular velocity τ. In other words, n and b are being rotated about v. with angular velocity τ. This explains why τ is called the torsion.

Consider. for example, the familiar *circular helix*, which winds round a cylinder (Figure 1.23). An arclength parametrization for this curve is

$$x = R \cos s, \quad y = R \sin s, \quad z = s.$$

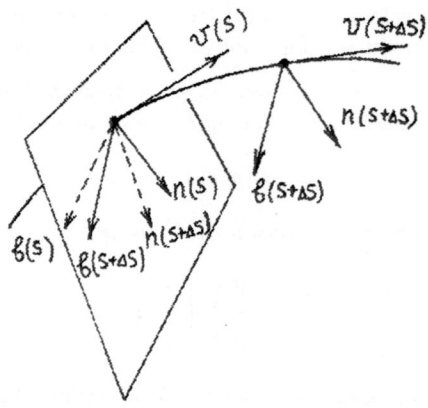

FIGURE 1.22.

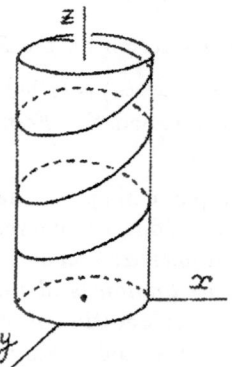

FIGURE 1.23.

It is easy to see that its curvature and torsion are constant (check it!).

It is an important fact that the curvature and torsion of a space curve completely determine the curve, so long as the curvature does not vanish. More precisely, we have the following three-dimensional counterpart to Statement 1.2.8:

Theorem 1.3.4. *Given smooth functions $k(s) > 0$ and $\tau(s)$, defined for $a \leq s \leq b$, there exists a smooth curve $r(s)$ in three-space having s as its arclength parameter, $k(s)$ as its curvature, and $\tau(s)$ as its torsion. Such a curve is unique up to a rigid motion.*

See the proof in [Struik 1950], for example.

2. The Notion of a Riemannian Metric

2.1. Riemannian Metrics on Domains of $\mathbb{R}^n$

Definition and simplest properties. In the preceding chapter we associated with each curvilinear coordinate system $x = (x^1, \ldots, x^n)$ on a domain C a smooth functional matrix $G(x)$, and these matrices, at each point, behave as a quadratic form under a change-of-coordinate transformations. The matrices $G(x)$ arose from the formula for the element of length of a curve, and we saw that the length of a curve can be calculated in a curvilinear coordinate system using the $G(x)$. The property of transforming at each point like a quadratic form turns out to be of major importance.

Definition 2.1.1. Let C be a domain in $\mathbb{R}^n$. Giving C a *Riemannian metric* means associating to every regular coordinate system $x = (x^1, \ldots, x^n)$ a functional matrix $G(x) = (g_{ij}(x))$, satisfying the following conditions:

(1) the functions $g_{ij}(x^1, \ldots, x^n)$ are smooth;
(2) the matrices $G(x)$ are symmetric, that is, $g_{ij}(x) = g_{ji}(x)$;
(3) the matrices $G(x)$ are positive definite, that is, $vG(x)v^T > 0$ for any nonzero vector $v \in \mathbb{R}^n$; and
(4) under a coordinate transformation $x \to y$ the matrix $G(x)$ is transformed by the rule

$$G(y) = AG(x)A^t,$$

where A is the Jacobian matrix of the change of coordinates $x \to y$. In other words, the matrix $G(x)$ must transform like the matrix of a quadratic form.

Clearly, if we fix *one* regular coordinate system $x = (x^1, \ldots, x^n)$ and define a functional matrix $G(x)$ satisfying conditions (1)–(3), there is a unique way to define a functional matrix $G(y)$ with respect to any other coordinate system in such a way that condition (4) is satisfied. Therefore we can define a Riemannian metric by giving a functional matrix in a particular coordinate system, so long as it satisfies conditions (1)–(3).

We show explicitly the action of the transformation rule. Assume that the coordinate transformation is $x \to x'$. Then

$$g_{i'j'} = \frac{\partial x^i}{\partial x^{i'}} \frac{\partial x^j}{\partial x^{j'}} g_{ij},$$

or. in matrix form, $G' = AGA^t$. (Recall the Einstein summation convention from the preceding chapter.)

Length of a curve. Suppose a Riemannian metric $G(x) = (g_{ij}(x))$ is given on a domain C. and let $\gamma(t) = (x^1(t), \dots, x^n(t))$ be a smooth curve given in the same curvilinear coordinate system.

Definition 2.1.2. The *length* of the curve γ from the point $\gamma(a)$ to the point $\gamma(b)$, with respect to the Riemannian metric $G(x)$, is

$$l(r) = \int_a^b \left(g_{ij}(x) \frac{dx^i}{dt} \frac{dx^j}{dt} \right)^{\frac{1}{2}} dt = \int_a^b \left(g_{ij}(x) x^{i'} x^{j'} \right)^{\frac{1}{2}} dt.$$

Here $'$ means differentiation with respect to the time parameter t.

If two smooth curves $\gamma_1(t)$ and $\gamma_2(t)$ intersect at a point P, the *angle* φ between them with respect to the Riemannian metric is defined by

$$\cos \varphi = \frac{g_{ij}(x) x_1^{i'}(t) x_2^{j'}(t)}{\left(g_{ij}(x) x_1^{i'}(t) x_1^{j'}(t) \right)^{\frac{1}{2}} \left(g_{ij}(x) x_2^{i'}(t) x_2^{j'}(t) \right)^{\frac{1}{2}}}.$$

The definition of a Riemannian metric can be formulated in more invariant terms, without the explicit use of coordinates. Indeed, a Riemannian metric defines at each point of a domain a *bilinear form* $\langle \, , \, \rangle_G$ on the set of all tangent vectors to smooth trajectories passing through that point. To explain this in more detail, we take two regular curves $\gamma(t)$ and $\alpha(t)$ such that $P = \gamma(0) = \alpha(0)$ (Figure 2.1). We consider their tangent vectors (velocity vectors) $\xi = \dot\gamma(0)$ and $\eta = \dot\alpha(0)$, with coordinates $(\xi^1, \dots, \xi^n)$ and $(\eta^1, \dots, \eta^n)$, in the coordinate system $x = (x^1, \dots, x^n)$. Then we may consider the bilinear form

$$\langle \xi, \eta \rangle_G = g_{ij}(x) \xi^i \eta^j,$$

which associates a number to any pair of vectors ξ and η.

Lemma 2.1.3. *The smooth map* $(\xi, \eta) \to \langle \xi, \eta \rangle_{G(x)}$ *defines a positive definite bilinear form that depends smoothly on the point x.*

See the proof, for example, in [Dubrovin et al. 1992; Dubrovin et al. 1985: Dubrovin et al. 1990; Mishchenko and Fomenko 1988].

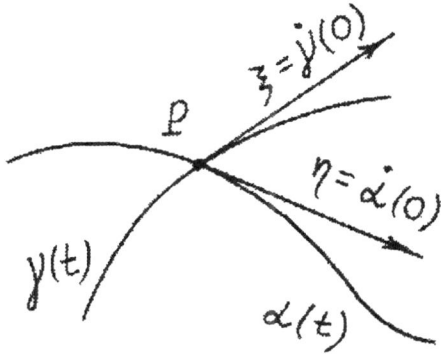

FIGURE 2.1.

Invariant definition of a Riemannian metric. We therefore arrive at the following coordinate-free definition of a Riemannian metric:

Definition 2.1.4. Let C be a domain in $\mathbb{R}^n$. Giving C a *Riemannian metric* means associating to each point of C a positive definite, symmetric bilinear form (scalar product) on vectors tangent to smooth curves through that point. This form should depend smoothly on the point.

Lemma 2.1.3 shows that this definition is equivalent to Definition 2.1.1. The lemma implies, in particular, that the length of a smooth curve with respect to a Riemannian metric is well-defined, that is, it does not depend on the coordinate system used in Definition 2.1.2.

The standard Euclidean scalar product provides an example of a Riemannian metric, called the *Euclidean metric*. If $x^1, \ldots, x^n$ are Cartesian coordinates in $\mathbb{R}^n$, then $G(x) = (\delta_{ij})$ is the identity matrix. In any other curvilinear coordinate system y obtained from x by a regular transformation, we have

$$G(y) = AG(x)A^t = AA^t,$$

where A is the Jacobian matrix of this transformation.

Definition 2.1.5. A Riemannian metric G defined in a domain C is called *Euclidean* if there exists a coordinate system y in C such that $G(y)$ is the identity matrix.

There exist *non-Euclidean* Riemannian metrics, that is, metrics that do not satisfy Definition 2.1.5. In fact, almost any metric defined at random will be non-Euclidean (except in dimension 1). At the present moment in our exposition we are unable to *prove* that some particular metric is non-Euclidean, since we cannot exclude the possibility that it will take the form $\sum_{i=1}^{n}(dx^i)^2$

in *some* coordinate system. It is intuitively clear that to rule out this possibility one has to find *invariants* that are preserved under regular coordinate transformations. Once these invariants are calculated for two metrics and shown to be distinct, we will know that the two metrics are not the same. Such invariants do exist and we shall define them soon.

The Euclidean metric in various coordinate systems.   As we see, the standard Euclidean metric loses its simple "Euclidean form" when written in an arbitrary curvilinear coordinate system y. We now write down its expression $G(y)$ in the specific curvilinear coordinate systems considered in the preceding chapter. Actually what the formulas below express is the squared element of length,

$$ds^2 = g_{ij}(y)\, dy^i dy^j,$$

which conveys the same information as the matrix $G(y) = (g_{ij}(y))$ but is shorter to write down when the matrix has many zero entries.

(1) Euclidean metric in polar coordinates (r, φ) on the plane:

$$ds^2 = dr^2 + r^2 d\varphi^2.$$

(2) Euclidean metric in cylindrical coordinates (r, φ, z) in three-space:

$$ds^2 = dr^2 + r^2\, d\varphi^2 + dz^2.$$

(3) Euclidean metric in spherical coordinates (r, θ, φ) in three-space:

$$ds^2 = dr^2 + r^2(d\theta^2 + \sin^2\theta\, d\varphi^2).$$

2.2.   Important Example: The Standard Two-Sphere

We start with the Euclidean plane $\mathbb{R}^2$, with Cartesian coordinates x and y and metric

$$ds^2 = dx^2 + dy^2.$$

The associated bilinear form is the standard scalar product on the plane:

$$\langle \xi, \eta \rangle = \xi^1\eta^1 + \xi^2\eta^2.$$

Now consider the standard two-dimensional sphere of radius R in $\mathbb{R}^3$. This is, by definition, the set of points that are at distance R from the origin $(0, 0, 0)$; we denote it by S^2. Suppose that a smooth curve $\gamma(t)$ lies on the two-sphere, and that we need to calculate the length of γ. Let x, y, z denote the Cartesian coordinates of $\mathbb{R}^3$. We may consider the curve $\gamma(t) = (x(t), y(t), z(t))$ as part of ambient three-space, which has the Euclidean metric

$$ds^2 = dx^2 + dy^2 + dz^2,$$

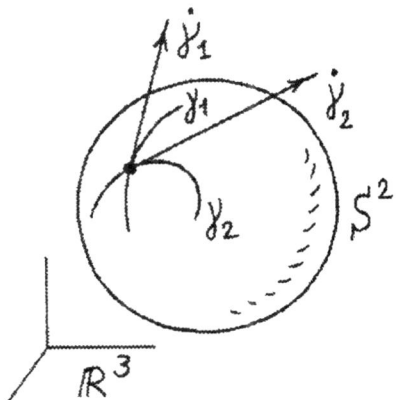

FIGURE 2.2.

and calculate the curve's length using Definition 2.1.2.

In such calculations we use the properties of the Euclidean metric only at points in the vicinity of S^2. In fact, we could restrict the Euclidean metric from $\mathbb{R}^3$ to S^2, and express it there in terms of coordinates on S^2. Since the sphere is given in $\mathbb{R}^3$ by a single equation, the position of a point on it is determined by two parameters (one fewer than in $\mathbb{R}^3$). This is especially obvious when we work in spherical coordinates (r, θ, φ) in $\mathbb{R}^3$: the sphere of radius R is defined by the single equation $r = R = \text{const}$, and is parametrized by the remaining two coordinates.

We now obtain an explicit expression for the scalar product of two vectors tangent to curves lying on S^2. These vectors are tangent to the sphere itself (Figure 2.2). Let $\gamma_1(t) = \big(R, \theta_1(t), \varphi_1(t)\big)$ and $\gamma_2(t) = \big(R, \theta_2(t), \varphi_2(t)\big)$. Then the velocity vectors are

$$\gamma_1'(t) = \big(0, \theta_1', \varphi_1'\big) \quad \text{and} \quad \gamma_2'(t) = \big(0, \theta_2', \varphi_2'\big).$$

The scalar product of the tangent vectors is

$$\langle \gamma_1', \gamma_2' \rangle = R^2 \theta_1' \theta_2' + R^2 \sin^2 \theta \, \varphi_1' \varphi_2'.$$

Thus, we can interpret this value as the scalar product of two tangent vectors $(\theta_1' . \varphi_1')$ and (θ_2', φ_2') relative to a new bilinear form

$$R^2 \theta_1' \theta_2' + R^2 \sin^2 \theta \, \varphi_1' \varphi_2',$$

corresponding to the metric

$$R^2 \, d\theta^2 + R^2 \sin^2 \theta \, d\varphi^2.$$

Note that this quadratic form can be obtained from the Euclidean metric

$$dr^2 + R^2 \big(d\theta^2 + \sin^2 \theta \, d\varphi^2\big)$$

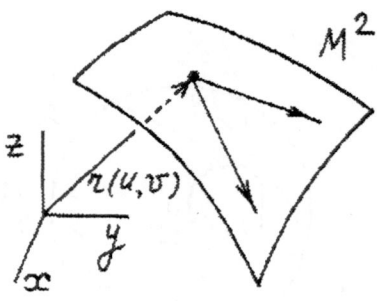

FIGURE 2.3.

in $\mathbb{R}^3$ by substituting the equation $r = R$ that determines the two-sphere. The Riemannian metric $R^2 (d\theta^2 + \sin^2 \theta \, d\varphi^2)$ thus obtained on S^2 is said to be *induced* by the ambient Euclidean metric of three-space.

This concrete example will be generalized below, and will illustrate the general notion of an *induced Riemannian metric*—roughly speaking, a metric obtained by restriction to a lower-dimensional space. To formalize this, we first introduce the concept of a *surface*, something to which the ambient Riemannian metric can be restricted.

2.3. First Definition of Surfaces

One definition of a surface in Euclidean space is as a sets of points parametrized by a vector-valued function of several variables, satisfying a natural condition called *nondegeneracy* or *regularity*. To formalize this, consider a smooth map $r : U \to \mathbb{R}^3$, where U is a domain in the Euclidean plane. We can write

$$r(u, v) = \big(x(u, v), \, y(u, v), \, z(u, v)\big),$$

where the pair of parameters (u, v) varies in U. We are interested in the set of points in $\mathbb{R}^3$ obtained as u and v vary, as shown in Figure 2.3. Now consider the vectors

$$r_u = \left(\frac{\partial x}{\partial u}, \frac{\partial y}{\partial u}, \frac{\partial z}{\partial u}\right) \quad \text{and} \quad r_v = \left(\frac{\partial x}{\partial v}, \frac{\partial y}{\partial v}, \frac{\partial z}{\partial v}\right).$$

These vectors are tangent to the *coordinate curves* obtained when we vary the parameter u with fixed v and when we vary v with fixed u. *Nondegeneracy* or *regularity* means that r_u and r_v define a plane, for any value of (u, v); this is the same as saying that r_u and r_v are linearly independent for any (u, v). In particular, both should always be nonzero.

Definition 2.3.1. Suppose that U is a domain in $\mathbb{R}^2$, and that $r : U \to \mathbb{R}^3$ is a smooth map. The image of r is called a smooth *parametrized surface* if the vectors r_u and r_v are linearly independent for any (u, v).

Given a parametrized surface, there is an important quadratic form on it, induced by the ambient Riemannian metric $ds^2 = \sum g_{ij}(p) \, dp^i \, dp^j$, where $p^1, \ldots, p^n$ are coordinates in $\mathbb{R}^3$. It is defined by regarding $p = r(u, v)$ as a point of the surface, and expressing the dependence of this point on u and v in the formula for ds^2. We obtain, as a result, the quadratic form

$$dl^2(u, v) = \sum g_{ij}(p(u, v)) \, dp^i(u, v) \, dp^j(u, v).$$

Definition 2.3.2. The quadratic form $dl^2(u, v)$ is called the Riemannian metric *induced* on the parametrized surface $r(u, v)$ by the metric g_{ij} of three-space.

For example, suppose that the ambient metric in $\mathbb{R}^3$ is the Euclidean one,

$$ds^2 = dx^2 + dy^2 + dz^2.$$

Then the induced Riemannian metric is

$$dl^2 = dx(u, v)^2 + dy(u, v)^2 + dz(u, v)^2,$$

or, using the chain rule,

2.3.3.

$$dl^2 = (x_u^2 + y_u^2 + z_u^2) \, du^2 + 2(x_u x_v + y_u y_v + z_u z_v) \, du \, dv + (x_v^2 + y_v^2 + z_v^2) \, dv^2$$
$$= \langle r_u, r_u \rangle \, du^2 + 2 \langle r_u, r_v \rangle \, du \, dv + \langle r_v, r_v \rangle \, dv^2.$$

Consider again a sphere S^2 of radius R in $\mathbb{R}^3$, parametrized by spherical coordinates θ and φ on S^2. (This sphere, as a whole, is not a parametrized surface in the sense of Definition 2.3.1, because the regularity condition is only satisfied for $0 < \theta < 2\pi$, so the poles cannot be included in the parametrization. However, we can think of a piece of the sphere as a parametrized surface.) In Section 2.2 we endowed S^2 with the metric induced by the ambient Euclidean metric of $\mathbb{R}^3$, and obtained earlier an explicit expression for this metric with respect to the spherical coordinates θ and φ on S^2, namely

$$dl^2 = R^2(d\theta^2 + \sin^2\theta d\varphi^2).$$

We now see that this formula is a particular case of 2.3.3.

Spherical coordinates are not the only useful coordinate system on a sphere. We discuss another one, based on the *stereographic projection* of the sphere S^2 onto the plane $\mathbb{R}^2$ (Figure 2.4). Place the center of the sphere of radius R at the origin 0 and consider the coordinate plane $\mathbb{R}^2(x, y)$ through 0. Let N and S denote the north and south poles on the sphere, and P an arbitrary point on S^2, different from N. Connect N to P, and extend the segment NP to the point Q where it intersects the plane $\mathbb{R}^2(x, y)$. Now map P to Q. This gives a map $h : S^2 \to \mathbb{R}^2$ called *stereographic projection*. The construction implies that h is defined at all points of the sphere, except at the north pole N. We often say that h maps the north pole to the "point at infinity", which is not part of the plane, but can be "adjoined" to it.

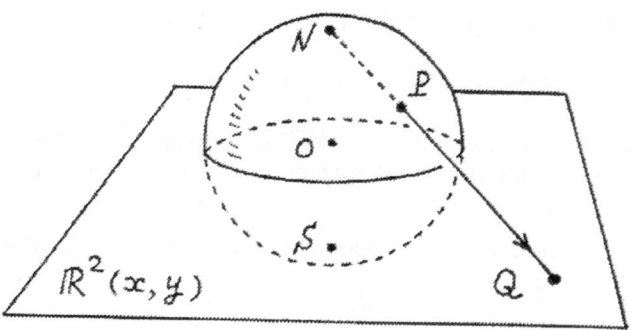

FIGURE 2.4.

Introducing coordinates both on the sphere and on the plane, we can write the map h in analytic form. Consider, for example, spherical coordinates θ, φ on S^2, and polar coordinates r, φ on $\mathbb{R}^2$. Because of the way in which we placed the sphere and the plane in $\mathbb{R}^3$, both sets of coordinates are restrictions of the spherical coordinates r, θ, φ in $\mathbb{R}^3$. In particular, stereographic projection h is easily seen to preserve the coordinate φ, and it is sufficient, in order to write h explicitly, to relate r (on the plane) and θ (on the sphere).

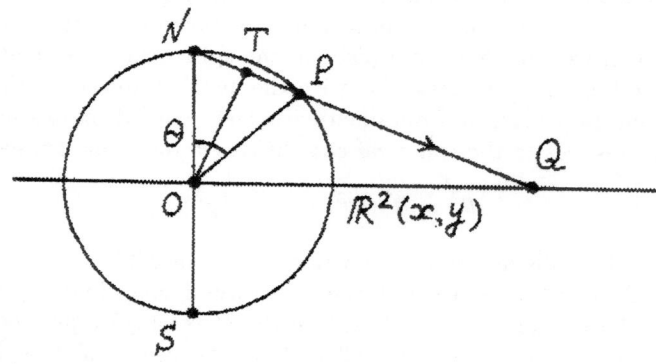

FIGURE 2.5.

Figure 2.5 shows how to do this. We take the section of S^2 by the plane containing the points P, Q, and N. Since the angle ONT is equal to $\frac{1}{2}(\pi - \theta)$, we obtain from the triangle ONQ that $r = R\cot(\theta/2)$. Thus, the final equations for the coordinate transformation are

$$\varphi = \varphi \quad \text{and} \quad r = R\cot\frac{\theta}{2}.$$

The Jacobian of this transformation is $(R/2)\sin^2(\theta/2)$; consequently, the transformation is *regular* at all points except at the north pole. Hence, we can introduce on the sphere S^2 coordinates induced by polar coordinates on a Euclidean plane. The induced Riemannian metric on the sphere in these coordinates takes the form

$$dl^2 = \frac{R^2}{(R^2 + r^2)^2}\,(dr^2 + r^2\,d\varphi^2).$$

Note that this Riemannian metric differs from the Euclidean metric on the plane, expressed in polar coordinates as $dr^2 + r^2 d\varphi^2$, only by the a multiplicative factor

$$\frac{4R^2}{(R^2 + r^2)^2}.$$

Metrics that differ only by a multiplicative scalar function are called *conformally equivalent*.

Definition 2.3.4. A Riemannian metric $g_{ij}(y)$ is called *conformal* if it is conformally equivalent to the Euclidean metric, that is, if there exists a coordinate system $x = (x^1, \ldots, x^n)$ such that

$$g_{ij}(x)\,dx^i\,dx^j = \lambda(x)\sum_i (dx^i)^2,$$

where $\lambda(x)$ is some smooth positive function.

3. Local Theory of Surfaces

3.1. Embedded Surfaces

As a definition of a surface, Definition 2.3.1 is not entirely satisfactory, because it does not cover many objects that we would naturally like to call surfaces, such as a sphere. Indeed, there are three common ways in which objects that we consider surfaces in $\mathbb{R}^3$ can arise:

Graph representation. As the *graph* of a smooth function $f(x, y)$ of two variables, that is, the set of points of the form $(x, y, f(x, y))$. This is illustrated in Figure 3.1.

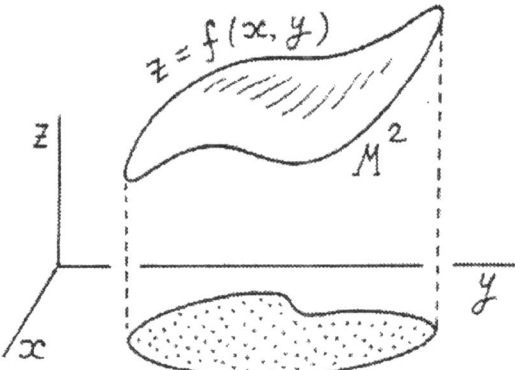

FIGURE 3.1.

Implicit representation. As a *zero-set*, that is, the set of all points (x, y, z) that satisfy the equation $F(x, y, z) = 0$, where F is a smooth function that is assumed to be regular at all points satisfying $F = 0$. (A point in the domain of F is called *regular*, and we say that F is regular at that point, if at least one of the partial derivatives F_x, F_y, F_z is nonzero there; otherwise the point is called *singular*. Thus, regular is the same as nonsingular. This is a particular case of Definition 3.4.2 below.) See Figure 3.2.

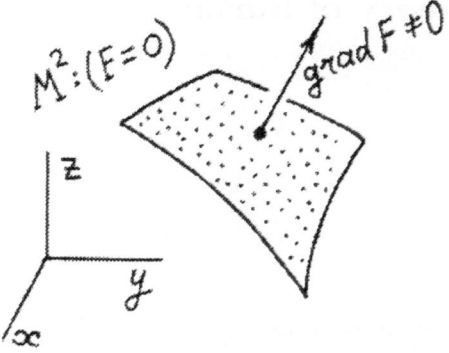

FIGURE 3.2.

Parametric representation. As a parametrized surface, that is, a set of points of the form

$$r(u, v) = \big(x(u, v),\ y(u, v),\ z(u, v)\big),$$

where u and v range over some domain in the plane, r is a smooth map, and the vectors r_u and r_v are always linearly independent. This is precisely Definition 2.3.1. See Figure 3.3.

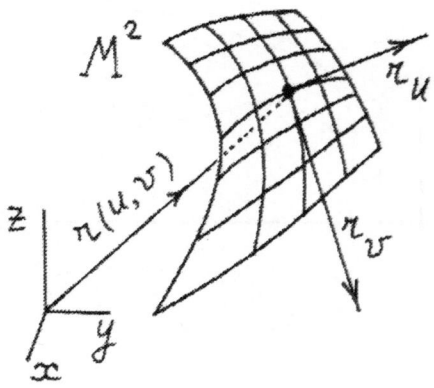

FIGURE 3.3.

Relating the three representations. It is clear that a graph can be regarded as a particular case of an implicitly defined surface, if we take $F(x, y, z) = f(x, y) - z$. A graph can also be regarded as a parametrized surface, with $x = u$ and $v = y$. Conversely, one can show using the *implicit function* *theorem* says that an implicitly or parametrically defined surface is *locally* a graph.

To state this more precisely, suppose that $P \in \mathbb{R}^3$ is any point on some implicitly defined surface. Then there is some domain containing P inside which we can express one of the coordinates of points on the surface as a function of the other two coordinates. Thus, inside that domain, the surface is the graph of some smooth function $z = f(x, y)$, or $x = f(y, z)$. or $y = f(x, z)$. (The representation $z = f(x, y)$ is possible if the partial derivative in z of the defining function is nonzero at P; otherwise we may have to resort to one of the other two possibilities. Since, by our definition, some partial derivative is nonzero at any point on the surface, at least one of the three possibilities is always available.)

Similarly, for any $(u_0, v_0) \in \mathbb{R}^2$ in the domain of a parametrically defined surface, there is some (perhaps smaller) domain containing (u_0, v_0). inside which u and v can be written as functions of two of the coordinates x, y, z. Substituting into the equation of the third coordinate, we get a local expression of the surface as a graph $z = f(x, y)$, or $x = f(y, z)$, or $y = f(x, z)$.

Definition of an embedded surface. Now we are prepared to generalize our surfaces to $\mathbb{R}^n$. for an arbitrary dimension n. We start by introducing the notion of a *local embedded surface* in $\mathbb{R}^n$, by which we mean essentially a parametrized surface where the parametrization is a homeomorphism.

Take a domain (open set) U in $\mathbb{R}^2$, with Cartesian coordinates u and v. Consider a smooth vector-valued function $r : U \to \mathbb{R}^n$, with components $x^1, \ldots, x^n$:

$$r(u, v) = (x^1(u, v), \ldots, x^n(u, v)).$$

Since the x^i are smooth functions, we can form the partial derivatives with respect to u and v:

$$r_u = \left(\frac{\partial x^1}{\partial u}, \ldots, \frac{\partial x^n}{\partial u} \right); \quad r_v = \left(\frac{\partial x^1}{\partial v}, \ldots, \frac{\partial x^n}{\partial v} \right).$$

Definition 3.1.1. The image $r(U)$ in $\mathbb{R}^n$ is called a (smooth, regular) *embedded local surface* in $\mathbb{R}^n$ if r is a homeomorphism as a map from U to $r(U)$, and if r_u and r_v are linearly independent for any $(u, v) \in U$.

Definition 3.1.2. A set $M \subset \mathbb{R}^n$ is called a (smooth) *embedded surface* if, for any point P of M, there is an open ball $B \subset \mathbb{R}^n$ containing P and such that the intersection $M \cap B$ is a local embedded surface. A smooth embedded surface is also called an *embedded two-manifold*.

The standard two-sphere in $\mathbb{R}^3$, minus the north pole, is a local embedded surface; the role of r may be played by stereographic projection. We essentially showed this on page 31. when we introduced stereographic projection. although we didn't check there that r is a homeomorphism. Likewise, the sphere minus the south pole is a local embedded surface. Therefore the whole sphere is an embedded surface, since any point of it lies in one set or the other.

In fact, an implicitly defined surface in $\mathbb{R}^3$ is always an embedded surface according to this definition, because near any point it looks like a graph, which is a local embedded surface. See page 37.

3.2. The Tangent Plane to a Surface

Consider an embedded local surface M in $\mathbb{R}^3$, given by the map $r = r(u, v)$. By assumption, for any given u, v, the two vectors

$$r_u = (x_u, y_u, z_u) \quad \text{and} \quad r_v = (x_v, y_v, z_v)$$

are linearly independent, and so determine a plane. We denote this plane by $T_P M$, where $P = r(u, v)$.

Definition 3.2.1. The tangent vector (velocity vector) at P of any smooth curve contained in M and going through P is called a *tangent vector* to M at P. The *tangent plane* to M at P is the set of all tangent vectors to M at P.

The reader should check that r_u and r_v are tangent vectors at P (think of the coordinate curves); that any linear combination of r_u and r_v is also a tangent vector at P; and that any tangent vector at P can be expressed as a linear combination of r_u and r_v. Therefore the tangent plane really is a plane—in fact, none other than $T_P M$.

Next we consider implicitly defined surfaces. Let M be defined by the equation $F(x, y, z) = 0$, where F is everywhere regular on M. We define tangent vectors and the tangent plane at a point $P \in M$ exactly as in Definition 3.2.1.

Lemma 3.2.2. *The tangent plane to M at P is given by the equation*

$$F_x(P)x + F_y(P)y + F_z(P)z = 0,$$

where F_x, F_y, and F_z are the partial derivatives of F at P. In other words, a vector (x, y, z) is tangent to the surface at P if and only if $F_x x + F_y y + F_z z = 0$.

Proof. Consider a smooth curve $\gamma = \big(\xi(t), \eta(t), \zeta(t)\big)$ on M with $\gamma(0) = P$. Because $F = 0$ on M, we have

$$F\big(\xi(t), \eta(t), \zeta(t)\big) = 0 \quad \text{for all } t.$$

Differentiating with respect to t we obtain $F_x \xi'(t) + F_y \eta'(t) + F_z \zeta'(t) = 0$. Substituting $t = 0$, this becomes $F_x(P)\xi'(0) + F_y(P)\eta'(0) + F_z(P)\zeta'(0) = 0$. But $\big(\xi'(0), \eta'(0), \zeta'(0)\big)$ is the tangent vector to γ at P. Since γ was arbitrary, this shows that any tangent vector at P satisfies the equation in the lemma.

Conversely, if (ξ, η, ζ) is a vector satisfying the equation in the lemma, we want to find a curve through P such that (ξ, η, ζ) is the tangent vector to the curve at P. We use the local expression of M as a graph (see page 37); for instance, if $F_z(P) \neq 0$ we write $z = f(x, y)$. We make our curve go through $P = (x_0, y_0, f(x_0, y_0))$ at $t = 0$, and vary the x- and y-coordinates linearly:

$$\gamma(t) = (x_0 + t\xi, \ y_0 + t\eta, \ f(x_0 + t\xi, y_0 + t\eta)).$$

So the velocity vector is $\dot{\gamma}(0) = (\xi, \eta, f_x\xi + f_y\eta)$. Now, it is well-known (and can easily be checked by fixing $y = y_0$) that $f_x = -F_x/F_z$; similarly $f_y = -F_y/F_z$. Therefore the z-component of $\dot{\gamma}(0)$ is $-(F_x\xi + F_y\eta)/F_z$, which equals ζ because $F_x\xi + F_y\eta + F_z\zeta = 0$ by assumption and $F_z \neq 0$. $\square$

The equation $F_x x + F_y y + F_z z = 0$ has an interesting interpretation in terms of the gradient $\operatorname{grad} F$. The gradient of a function is really a *covector*, not a vector. But because we are working in Cartesian coordinates x, y, z, we can identify $\operatorname{grad} F$ with the vector (F_x, F_y, F_z), and we shall do so. (*Important remark:* In general curvilinear coordinates this is no longer valid.) The preceding discussion says that

$$\operatorname{grad} F = (F_x(P), F_y(P), F_z(P))$$

is orthogonal to every tangent vector to M at P, and therefore is orthogonal to $T_P M$. See Figure 3.4.

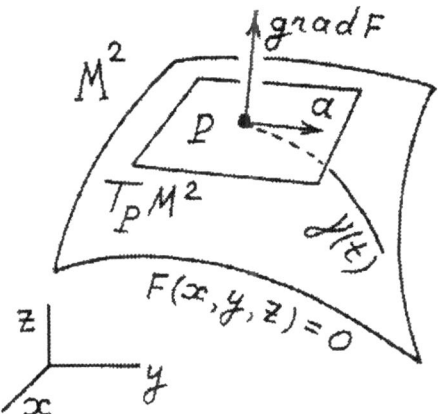

FIGURE 3.4.

Since grad F is normal to the surface and nonzero, we can define the *unit normal* to M at P as

$$n(P) = \frac{\operatorname{grad} F}{|\operatorname{grad} F|}$$

$$= \left(\frac{F_x}{(F_x^2 + F_y^2 + F_z^2)^{\frac{1}{2}}}, \frac{F_y}{(F_x^2 + F_y^2 + F_z^2)^{\frac{1}{2}}}, \frac{F_z}{(F_x^2 + F_y^2 + F_z^2)^{\frac{1}{2}}} \right).$$

This is very useful for the geometric description of surfaces in three-space.

3.3. Higher-Dimensional Manifolds in $\mathbb{R}^n$

Definitions 3.1.1 and 3.1.2 formalized the notion of surface that we will be primarily concerned with. These definitions can be easily extended to k-dimensional objects, called k-*manifolds*, for $k > 2$. It is sufficient to replace a domain $U \in \mathbb{R}^2$ by a domain $U \in \mathbb{R}^k$, with coordinates $u^1, \ldots, u^k$, and require that the vectors $r_{u^1}, \ldots, r_{u^k}$ form a linearly independent set.

Definition 3.3.1. A subset $M \subset \mathbb{R}^n$ is a k-*dimensional embedded manifold*, or simply k-*manifold*, if it has the following property. For any point $P \in M$, there is a domain $U \subset \mathbb{R}^k$ and a smooth map $r : U \to M$ satisfying these conditions:

(1) $r(U)$ is an open set in M (that is, the intersection of M with some domain of $\mathbb{R}^n$). and it contains P;
(2) r is a homeomorphism when considered as a map from U to $r(U)$;
(3) the partial derivatives $r_{u^1}, \ldots, r_{u^k}$, where $u^1, \ldots, u^k$ are Cartesian coordinates on U, form a linearly independent set of vectors in $\mathbb{R}^n$.

We say that the $u^1, \ldots, u^k$ are a set of *local coordinates* around P, and that r is a *coordinate chart*.

Remark. The word *embedded* means that the object in question is included in a larger space (here $\mathbb{R}^n$) in a nice way (see Section 3.4). It is possible to define the notion of manifolds abstractly, so that they are not embedded anywhere. This will be done in Chapter 5. Until then, we will simply say k-*manifold* to refer to a k-dimensional embedded manifold.

When we want to make practical use of a k-manifold, or prove facts about it, or define things on it, we typically need to employ local coordinates.

In later chapters another notion will be necessary, that of *manifolds with boundary*. A manifold with boundary M is not a manifold in the sense of Definition 3.3.1. because it includes points beyond which M does not extend. A formal definition will be given in Chapter 5, but the concept is probably familiar, at least in the case of surfaces, and will be used informally in the next chapter. Two familiar examples of surfaces with boundary are the closed unit disk $\{(x, y) \in \mathbb{R}^2 : x^2 + y^2 \leq 1\}$, and the cylinder

$$S^1 \times [0, 1] = \{(x, y, z) \in \mathbb{R}^3 : x^2 + y^2 = 1 \text{ and } 0 \leq z \leq 1\}$$

In the first case the boundary is the unit circle S^1. In the second case the boundary consists of two circles, $S^1 \times \{0\}$ and $S^1 \times \{1\}$. If we replace the inequalities $\leq$ by $<$ in these formulas, we get surfaces in the sense of Definition 3.3.1, and unless we use the words "with boundary" that's the definition we will have in mind when we talk about a surface or manifold.

3.4. The Differential of a Map. Embeddings and Immersions

Recall also that tangent vectors to M at P are the linear combinations of the vectors $r_{u^1}, \ldots, r_{u^k}$. We say that a tangent vector $a \in T_P M$ has *components* $a^1, \ldots, a^k$ in *local coordinates* if $a = \sum a^k r_{u^k}$. This expression is unique, since the r_{u^i} form a basis of the tangent space. We usually write $a = (a^1, \ldots, a^k)$.

Definition 3.4.1. Let $f : M \to N$ be a map from a smooth m-manifold M into a smooth n-manifold N. We say that f is *smooth* if it is smooth when expressed in local coordinates, by which we mean the following. Suppose $P \in M$. Taking local coordinates $x^1, \ldots, x^m$ in a neighborhood of P and local coordinates $y^1, \ldots, y^n$ in a neighborhood of $f(P)$, we can write the map f in the form

$$y^1 = y^1(x^1, \ldots, x^m), \quad \ldots, \quad y^n = y^n(x^1, \ldots, x^m).$$

Saying that f is smooth at P is saying that the functions $y^1, \ldots, y^n$ are smooth at P. If f is smooth at all points, it is a smooth map between manifolds.

Assume from now on that f is smooth. The *differential* of f at P is a linear map $df_P : T_P M \to T_{f(P)} N$ from the tangent space to P into the tangent space of $f(P)$, defined in local coordinates by the Jacobian matrix of f. Recall that the Jacobian matrix is the matrix of partial derivatives

$$df = \begin{pmatrix} \dfrac{\partial y^1}{\partial x^1} & \cdots & \dfrac{\partial y^n}{\partial x^1} \\ \vdots & & \vdots \\ \dfrac{\partial y^1}{\partial x^m} & \cdots & \dfrac{\partial y^n}{\partial x^m} \end{pmatrix}$$

(note that we use the same symbol for this matrix and the differential of the map). If $a = (a^1, \ldots, a^m) \in T_P M$ is a tangent vector to M at P, its image

$df(a)$ under df is the vector $b \in T_{f(P)}N$ having components

$$b^i = \sum_j \frac{\partial y^i}{\partial x^j} a^j.$$

This formula defines the linear map df explicitly.

Definition 3.4.2. Let $f : M \to N$ be a smooth map. A point P of M is called a *regular point* of f if the differential

$$df_P : T_P M \to T_Q N$$

is *surjective*, that is, if its image is all of $T_{f(P)}N$. In particular, for regular points to exist we must have $\dim M \geq \dim N$. A point $Q \in N$ is called a *regular value* of f if all points of the inverse image $f^{-1}(Q)$ are regular points of f. In particular, any point in the complement of the image of M is automatically a regular value of f, since the condition is vacuously satisfied. A point or value that is not regular is called *singular* or *critical*.

Saying that the differential is onto at P is saying that the Jacobian matrix at P has rank n (recall that the rank of a matrix is the number of linearly independent columns, that is, the dimension of the image of the corresponding linear map).

The following theorem is very important.

Theorem 3.4.3. *Let $f : M \to N$ be a smooth map of smooth manifolds, and let $Q \in f(M)$ be a regular value of f. Then the inverse image $f^{-1}(Q)$ is a smooth manifold of dimension $\dim f^{-1}(Q) = \dim M - \dim N = m - n$.*

This, too, follows from the implicit function theorem. A proof is given in [Guillemin 1974, p. 21]; see also [Dubrovin et al. 1992; Dubrovin et al. 1985; Dubrovin et al. 1990; Mishchenko and Fomenko 1988].

Definition 3.4.4. A smooth map $f : M \to N$ is called an *immersion* if at any point P from M the differential

$$df_P : T_P M \to T_{f(P)}N$$

is *injective*, that is, no nonzero vector maps to zero. If, moreover, f is a homeomorphism when considered as a map from M to $f(M)$, we say that f is *embedding*.

If we have two smooth manifolds in $\mathbb{R}^n$ contained in one another, the smaller one is called a *submanifold* of the larger one. If $f : M \to N$ is a smooth map, Theorem 3.4.3 says that the inverse image of a regular value is a submanifold of M, and Definition 3.4.4 says that $f(M)$ is a submanifold of N if f is an embedding.

Figure 3.5(a) shows an immersion of the circle S^1 into $\mathbb{R}^2$; Figure 3.5(b) shows an embedding of S^1 into $\mathbb{R}^2$.

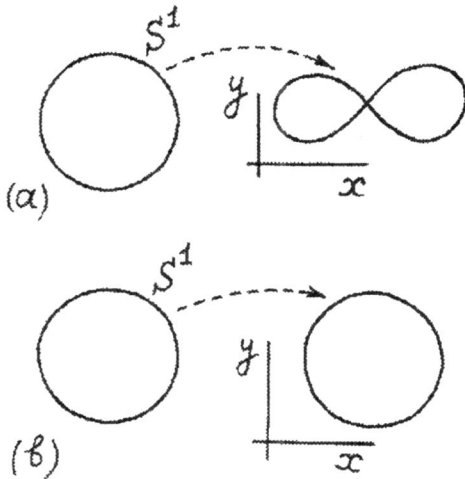

FIGURE 3.5.

3.5. The First Fundamental Form. Isothermal Coordinates. Area

We now go back to surfaces in Euclidean space. Consider in $\mathbb{R}^3$, with coordinates x, y, z, an embedded surface M given locally by

$$r(u, v) = \big(x(u, v), y(u, v), z(u, v)\big),$$

where r_u and r_v are of course linearly independent for any (u, v).

Definition 3.5.1. The Riemannian metric induced on M by the ambient Euclidean metric is called the *first fundamental form* of M.

According to Equation 2.3.3, the first fundamental form is

$$ds^2 = E\,du^2 + 2F\,du\,dv + G\,dv^2,$$

where

$$E = \langle r_u, r_u \rangle = x_u^2 + y_u^2 + z_u^2,$$
$$F = \langle r_u, r_v \rangle = x_u x_v + y_u y_v + z_u z_v,$$
$$G = \langle r_v, r_v \rangle = x_v^2 + y_u^2 + z_v^2.$$

We can write the same expression as

$$ds^2 = g_{ij}\,dx^i\,dx^j,$$

where $g_{11} = E$, $g_{12} = g_{21} = F$, $g_{22} = G$, $x^1 = u$, $x^2 = v$.

If the surface is given as a graph $z = f(x, y)$, we can set $u = x$ and $v = y$, and the formulas can be written more simply in local coordinates x, y as

$$g_{11} = 1 + f_x^2, \quad g_{12} = g_{21} = f_x f_y, \quad g_{22} = 1 + f_y^2.$$

If the surface is given implicitly by the equation $F(x, y, z) = 0$, we can write the induced metric as follows. Assume that F_z, say, is nonzero at the point of interest, and write the surface locally as the graph of a smooth function $z = f(x, y)$, just as in the proof of Lemma 3.2.2. Using the formulas $f_x = -F_x/F_z$ and $f_y = -F_y/F_z$, we obtain

$$g_{11} = 1 + \frac{F_x^2}{F_z^2}, \quad g_{12} = g_{21} = \frac{F_x F_y}{F_z^2}, \quad g_{22} = 1 + \frac{F_y^2}{F_z^2}$$

in local coordinates x, y. These formulas should be modified in the obvious way if we must choose instead y, z or x, z as local coordinates.

Isothermal or conformal coordinates on a surface. Consider an embedded surface $M \subset \mathbb{R}^3$ with local coordinates denoted $u = x^1$ and $v = x^2$. We will be considering on M an *arbitrary* smooth Riemannian metric ds^2, not necessarily the one induced by the Euclidean metric of $\mathbb{R}^3$. Saying that the Riemannian metric is smooth means that it has the expression $ds^2 = g_{ij} \, dx^i dx^j$, where the g_{ij} are smooth functions of x^1 and x^2.

Definition 3.5.2. The local coordinates are called *conformal* or *isothermal* for the given Riemannian metric if

$$g_{11} = g_{22} = \lambda, \quad g_{12} = g_{21} = 0,$$

where λ is a positive smooth function. Equivalently, we can write

$$ds^2 = \lambda(u, v)(du^2 + dv^2).$$

The following theorem says that isothermal coordinates exist around any point of a surface. It follows from the existence of solutions for the so-called *Beltrami's equation*; a proof can be found in [Chern 1955]. See also [Dubrovin et al. 1992; Dubrovin et al. 1985; Dubrovin et al. 1990].

Theorem 3.5.3. *Let M be a surface in $\mathbb{R}^3$, with a Riemannian metric ds^2. For any point $P \in M$, we can introduce isothermal local coordinates in a neighborhood of P; that is, we can choose a parametrization $r(u, v)$ of this neighborhood in such a way that*

$$ds^2 = \lambda(u, v)(du^2 + dv^2).$$

Note that the coordinates u and v serve, generally speaking, only for a certain neighborhood of the given point.

In isothermal coordinates u, v, the coordinate curves $u = $ constant and $v = $ constant are orthogonal with respect to the given Riemannian metric. Indeed, the scalar product of the tangent vectors r_u and r_v is $g_{12} = g_{21}$, which is zero by assumption.

Area of a surface. We continue to consider an arbitrary smooth Riemannian metric on M, with expression $ds^2 = g_{ij}\, dx^i\, dx^j$. We form the smooth function $g = \det(g_{ij})$, which is everywhere positive because of the definition of a Riemannian metric (the matrix (g_{ij}) is positive definite). Let U be a region on M bounded by some smooth or piecewise smooth curve γ on the surface M. We assume that the closure of U is a *compact set*, and that it is entirely contained in a single coordinate neighborhood, as illustrated in Figure 3.6. (The *closure* of a subset U of a topological space X is the smallest closed subset of X containing U.)

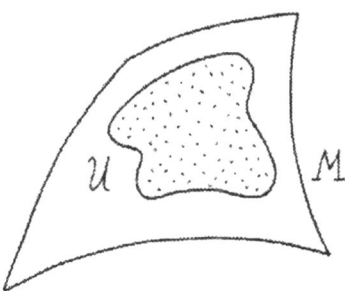

FIGURE 3.6.

Definition 3.5.4. The *area* of U with respect to the metric ds^2 is

$$\sigma(U) = \iint \sqrt{g}\, dx^1\, dx^2,$$

where the integral is taken over the inverse image of U in the (x^1, x^2)-plane.

We introduce the useful abbreviated expression

$$d\sigma = \sqrt{g}\, dx^1\, dx^2,$$

called the *element of area* or *differential of area* of the surface (or of the Riemannian metric g_{ij}).

The next theorem gives explicit local formulas for the area of a surface in $\mathbb{R}^3$ in graph, implicit, or parametric representation, *with respect to the Riemannian metric induced from the Euclidean metric of* $\mathbb{R}^3$. These formulas follow easily from the corresponding expressions for g_{ij} given at the beginning of this section (page 43).

Theorem 3.5.5. (a) *For a surface given as a graph* $z = f(x, y)$, *the area of a region U on the surface that projects to a region V of the (x, y)-plane is*

$$\sigma(U) = \iint_V \sqrt{1 + f_x^2 + f_y^2}\, dx\, dy.$$

(b) *For a surface given by $F(x, y, z) = 0$, the area of a region that has a one-to-one orthogonal projection onto the region V of the (x, y)-plane is*

$$\sigma(U) = \iint_V \frac{\sqrt{F_x^2 + F_y^2 + F_z^2}}{|F_z|} \, dx \, dy,$$

provided that $F_z \neq 0$ for all points in U.

(c) *For a surface given parametrically by $r = r(u, v)$, the area of the image U of a region V of the coordinate plane is*

$$\sigma(U) = \iint_V |r_u \times r_v| \, du \, dv = \iint_V \sqrt{EG - F^2} \, du \, dv.$$

Here $\times$ denotes the cross product in Euclidean space, and E, F, G are the coefficients of the first fundamental form.

Figure 3.7 illustrates the three cases.

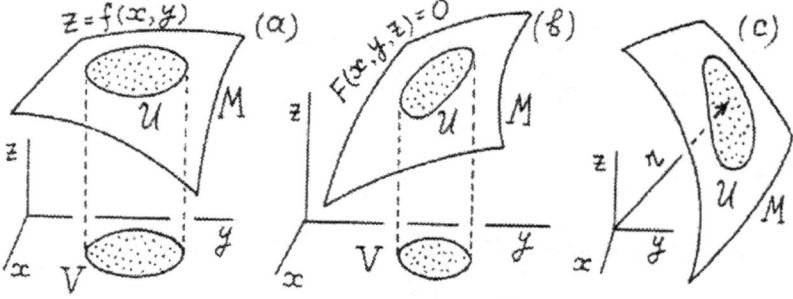

FIGURE 3.7.

3.6. The Second Fundamental Form. Mean and Gaussian Curvatures

Mean and Gaussian curvatures of the surface. Consider an embedded surface M in $\mathbb{R}^3$, and suppose that at the point $P \in M$ the tangent plane is orthogonal to the z-axis, as shown in Figure 3.8. In fact, any point can be made to satisfy this condition after an orthogonal change of coordinates: Just define the z-direction to be normal to the surface, and choose mutually perpendicular directions in the tangent plane to be the x- and y-directions. Now M may be given locally, in the neighborhood of P, by an equation $z = f(x, y)$. By means of a translation we can also assume that $P = (0, 0, 0)$.

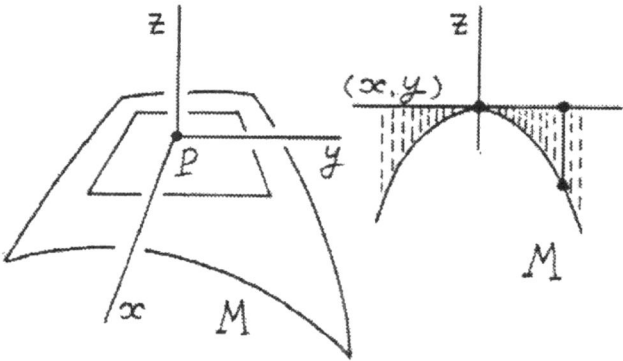

FIGURE 3.8.

Because the (x, y)-plane is tangent to the surface at P, the partial derivatives f_x and f_y of f vanish at $(0, 0)$. Now consider the partial derivatives of second order. We have $d^2 f = f_{xx}\, dx^2 + 2f_{xy}\, dx\, dy + f_{yy}\, dy^2$. The matrix of this quadratic form,

$$H_P f = \begin{pmatrix} f_{xx} & f_{xy} \\ f_{yx} & f_{yy} \end{pmatrix},$$

is called the *Hessian* of f.

Definition 3.6.1. Still in the same notation, the *principal curvatures* of the surface at P are the eigenvalues λ_1 and λ_2 of $H_P f$, which are real because the Hessian is a symmetric matrix. The *mean curvature* H of the surface at P is the trace of $H_P f$, which equals

$$H = f_{xx} + f_{yy} = \lambda_1 + \lambda_2.$$

(*Remark:* Some other texts define H as half this value.) The *Gaussian curvature* K of the surface at P is the determinant of $H_P f$, which equals

$$K = f_{xx} f_{yy} - f_{xy}^2 = \lambda_1 \lambda_2.$$

The *Theorema Egregium* of Gauss states that what we now call the Gaussian curvature is an *intrinsic invariant*, that is, it depends only on the internal metric properties of the surface. To appreciate the import of this statement, imagine slicing off a segment of a beach ball. The plastic piece you get can be deformed *isometrically* in $\mathbb{R}^3$, that is, without subjecting the plastic to stretching or compression. (An *isometry* is a map that preserves distances.) Gauss's theorem says that the product of the principal curvatures at a point does not change at all during this distortion. The mean curvature H, by contrast, is not invariant under isometries; it depends on the concrete embedding of the surface in three-space.

We formulated our definition of principal, mean, and Gaussian curvatures in terms of special local coordinates on the surface. Sometimes this is inconvenient. so now we will approach the same notions in terms of arbitrary local coordinates.

The second fundamental form.   Let the surface M be given in parametric form by $r(u, v)$. The unit normal vector m to M at the point of interest P is given by

$$m = \frac{r_u \times r_v}{|r_u \times r_v|},$$

because m is normal to the vectors r_u and r_v that determine the tangent plane $T_P M$. Now let $(u(t), v(t))$ be a smooth curve in the parameter plane, mapped by r to a curve $\gamma(t) = r(u(t), v(t))$ in M that goes through P at $t = 0$, say. Differentiating the expression of γ we obtain $\gamma' = r_u u' + r_v v'$, and consequently

$$\gamma'' = (r_{uu} u'^2 + 2r_{uv} u' v' + r_{vv} v'^2) + (r_u u'' + r_v v'').$$

We are interested in the component of the acceleration $\gamma''(0)$ normal to the surface. because intuitively this measures how much the surface is warped, as indicated in Figure 3.9. Taking the scalar product of the preceding equality with m and using the orthogonality of m with r_u and r_v, we get

$$\langle \gamma'', m \rangle = \langle r_{uu}, m \rangle u'^2 + 2\langle r_{uv}, m \rangle u' v' + \langle r_{vv}, m \rangle v'^2,$$

where all derivatives are evaluated at $t = 0$. This is a quadratic form in u' and v', whose coefficients are traditionally denoted by $L = \langle r_{uu}, m \rangle$, $M = \langle r_{uv}, m \rangle$, $N = \langle r_{vv}, m \rangle$. We have proved the following result:

Lemma 3.6.2. *The normal component of the acceleration vector of a curve* $\gamma(t) = r(u(t), v(t))$ *at* P *is a quadratic form in the components* u' *and* v' *of the velocity of the curve at* P. *This is called the second fundamental form of the surface at* P. *and is given by the matrix*

$$Q = \begin{pmatrix} L & M \\ M & N \end{pmatrix},$$

where $L, M. N$ *are the normal components of the vectors* r_{uu}, r_{uv}, r_{vv}.

This matrix is symmetric but not necessarily nondegenerate. A surface may even have $Q = 0$, as in the case of a plane parametrized linearly, where all derivatives $r_{uu}. r_{uv}, r_{vv}$ are identically zero.

Note that the coefficients L, M, N of the second fundamental form depend only on the point P on the surface and not on the curve $\gamma(t)$.

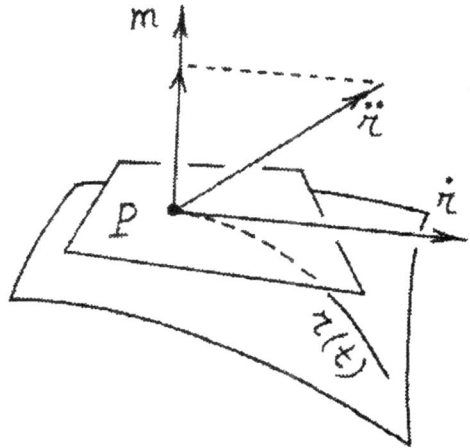

FIGURE 3.9.

Curvature of curves on a surface.   Now suppose that the smooth curve γ is parametrized by arclength; as usual, we denote the parameter by s instead of t. It follows from the Serret–Frenet formulas (see also **Figure 3.10**) that

$$\ddot{r} = \frac{d^2 r}{ds^2} = kn,$$

where n is the principal normal to the curve $\gamma(s)$ and $k(s)$ is its curvature. Consequently, we obtain

$$\langle \ddot{r}, m \rangle = k \langle n, m \rangle = k \cos \theta,$$

where θ is the angle between the unit vectors m and n (**Figure 3.10**).
    Hence,

$$k \cos \theta \, ds^2 = \langle \ddot{\gamma}, m \rangle \, ds^2 = L \, du^2 + 2M \, du \, dv + N \, dv^2.$$

Comparing this formula with the expression derived earlier for ds^2, we get

$$k \cos \theta = \frac{L \, du^2 + 2M \, du \, dv + N \, dv^2}{E \, du^2 + 2F \, du \, dv + G \, dv^2} = \frac{q_{ij} \, dx^i \, dx^j}{g_{ij} \, dx^i \, dx^j},$$

where on the right-hand side we have set $q_{11} = L$, $q_{12} = q_{21} = M$, $q_{22} = N$, $x^1 = u$, $x^2 = v$ (the g_{ij} were defined earlier).
    Note that we can now write for *any* curve $\gamma(t) = r\big(x^1(t), x^2(t)\big)$ (not just one parametrized by arclength)

$$k \cos \theta = \frac{q_{ij} x^{i\prime} x^{j\prime}}{g_{ij} x^{i\prime} x^{j\prime}}.$$

The preceding discussion has shown the following result:

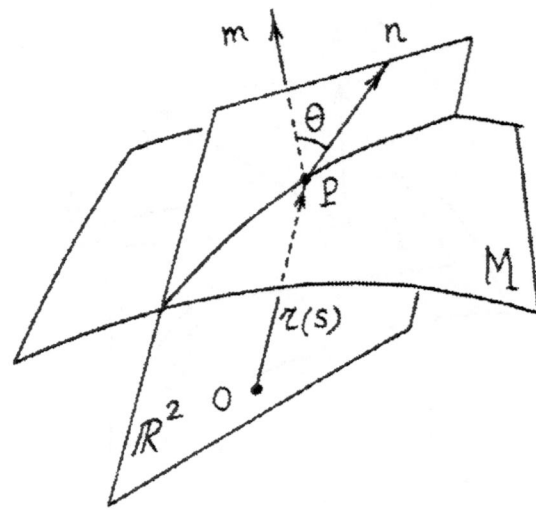

FIGURE 3.10.

Theorem 3.6.3. *The curvature at a point P of a curve $\gamma(t)$ on a surface in $\mathbb{R}^3$, multiplied by the cosine of the angle between the normal to the surface at P and the principal normal to the curve at P, is the ratio between the second and first fundamental forms, evaluated at P in the direction of the curve.*

An important particular case is when the curve is obtained by intersecting the surface with a plane normal to the surface at P. Then $n = \pm m$, so that $\cos\theta = \pm 1$.

Theorem 3.6.4. *If the curve is a cross-section of the surface by a plane normal to the surface of P, the corresponding normal curvature of the curve at P is given by*

$$\pm k = \frac{q_{ij}x^{i\prime}x^{j\prime}}{g_{ij}x^{i\prime}x^{j\prime}},$$

where the minus sign corresponds to the case when m and n have opposite directions.

Recall that here the $x^{i\prime}$ denote the components of the velocity vector for the curve at the point in question.

Thus we have associated with each point P of a surface M in $\mathbb{R}^3$ two quadratic forms

$$ds^2 = g_{ij}dx^idx^j \quad \text{and} \quad \langle \ddot{r}, m\rangle \, dt^2 = q_{ij}dx^idx^j,$$

the first of which is positive definite. These forms have matrices

$$G = \begin{pmatrix} g_{11} & g_{12} \\ g_{21} & g_{22} \end{pmatrix}, \quad Q = \begin{pmatrix} q_{11} & q_{12} \\ q_{21} & q_{22} \end{pmatrix}.$$

Eigenvalues of the fundamental forms. Gaussian and mean curvatures. We consider the equation $\det(Q - \lambda G) = 0$ in λ. Since G is positive definite, and therefore invertible, this equation is equivalent to $\det(G^{-1}Q - \lambda I) = 0$, where I is the identity matrix, so the roots are the eigenvalues of $G^{-1}Q$. According to the theory of quadratic forms, these eigenvalues are real because G and Q are symmetric and G is positive definite.

Definition 3.6.5. The roots λ_1 and λ_2 of $\det(Q - \lambda G) = 0$ are called the *principal curvatures* of the surface at the point P. They are the eigenvalues of $G^{-1}Q$. The product $K = \lambda_1 \lambda_2$ is called the *Gaussian curvature* of the surface at P, and the sum $H = \lambda_1 + \lambda_2$ is called the *mean curvature*. Thus,

$$K = \det(G^{-1}Q) = \frac{\det Q}{\det G}, \quad H = \operatorname{trace}(G^{-1}Q).$$

Clearly, a vector $e_1 \in T_P M$ satisfies $(Q - \lambda_1 G)e_1 = 0$ if and only if it is a λ_1-eigenvector of $G^{-1}Q$. The space of these vectors determines the *principal direction* at P associated with λ_1. The principal direction associated with λ_2 is defined likewise. If the principal curvatures λ_1 and λ_2 are the same, Q and G are proportional, and all directions are principal. We then call P an *umbilic*, or *umbilical point*.

We now compare this general definition of the Gaussian and mean curvatures with Definition 3.6.1, which applied to surfaces in the special form of Figure 3.8. Recall that, in that special case, M is given by $z = f(x, y)$ with $f(0,0) = 0$, $f_x = 0$ and $f_y = 0$ at the point $P = (0, 0, 0)$, so that $T_P M$ is the xy-plane. Putting $x = u$ and $y = v$, we get at the point P the values $g_{11} = 1$, $g_{22} = 1$, and $g_{12} = g_{21} = 0$, so that G is the identity matrix I. We also have at P the equalities

$$L = q_{11} = \langle r_{uu}, m \rangle = f_{xx},$$
$$M = q_{12} = \langle r_{uv}, m \rangle = f_{xy},,$$
$$N = q_{22} = \langle r_{vv}, m \rangle = f_{yy}$$

if we assume that the unit normal m is in the direction of the positive z-axis. Consequently, the second fundamental form at P is

$$L\,dx^2 + 2M\,dx\,dy + N\,dy^2 = f_{xx}\,dx^2 + 2f_{xy}\,dx\,dy + f_{yy}\,dy^2 = d^2 f.$$

Because $G = I$, the principal curvatures are the eigenvalues λ_1 and λ_2 of

$$\begin{pmatrix} f_{xx} & f_{xy} \\ f_{yx} & f_{yy} \end{pmatrix},$$

which is the Hessian of the function f.

Thus, the Gaussian curvature K is simply $f_{xx}f_{yy} - f_{xy}^2$ (the determinant of the Hessian) and the mean curvature is $f_{xx}+f_{yy}$ (the trace of the Hessian). We see that Definition 3.6.1 agrees with Definition 3.6.5.

A direct calculation shows that, for a surface M given in $\mathbb{R}^3$ as the graph $z = f(x, y)$, the mean curvature H is given at any point by

3.6.6.
$$H = \operatorname{div}\left(\frac{f_x}{\sqrt{1 + f_x^2 + f_y^2}}, \frac{f_y}{\sqrt{1 + f_x^2 + f_y^2}}\right).$$

(Recall that, in Cartesian coordinates, the divergence of a plane vector field $v = (v^1, v^2)$ is the function $\operatorname{div} v = \partial v^1/\partial x^1 + \partial v^2/\partial x^2$.)

Now we calculate the Gaussian and mean curvatures for the standard sphere of radius R. It is clear that all normal cross-sections are the circles of radius R on the sphere, called *great circles*. Thus, the normal curvature is everywhere and in all directions just the constant R^{-1}. Hence the two principal curvatures are both equal to R^{-1} everywhere, so that the Gaussian curvature is $K = R^{-2}$ and the mean curvature is $H = 2R^{-1}$.

A surface whose Gaussian curvature is positive at every point is *strictly convex*.

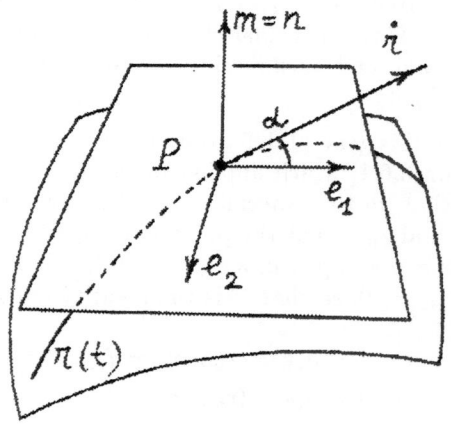

FIGURE 3.11.

Euler's formula. Consider a basis e_1, e_2 in $T_P M$ consisting of eigenvectors of $G^{-1}Q$. It is an important fact about quadratic forms that, if $\lambda_1 \neq \lambda_2$, such a basis orthogonalizes G and Q simultaneously; in other words, if A is the matrix with columns e_1, e_2, the matrices $G' = A^t G A$ and $Q' = A^t Q A$ are diagonal. Evaluating both forms at e_1 and e_2 shows that

$$Q' = G' \begin{pmatrix} \lambda_1 & 0 \\ 0 & \lambda_2 \end{pmatrix}.$$

Note that, by construction, the off-diagonal entry of G' is the scalar product (according to G) of e_1 and e_2, so the principal directions are orthogonal with respect to this scalar product, and therefore also with respect to the Euclidean scalar product in $\mathbb{R}^3$. This can also be arranged if $\lambda_1 = \lambda_2$, for then all vectors in $T_P M$ are eigenvectors of $G^{-1}Q$, and we are free to pick e_1 and e_2 orthogonal. In any case G' and Q' are diagonal, and related as above.

Now arrange the surface as in Definition 3.6.1, with the normal m at P pointing toward the positive z-axis. Rotate the xy-plane so that e_1 becomes a multiple of the unit x-vector and e_2 a multiple of the unit y-vector. In the local coordinates x and y, we know that G is the identity matrix at P; the preceding discussion shows that then

$$Q = \begin{pmatrix} \lambda_1 & 0 \\ 0 & \lambda_2 \end{pmatrix}.$$

Thus, the first fundamental form at P is $dx^2 + dy^2$, and the second fundamental form is $\lambda_1 dx^2 + \lambda_2 dy^2$.

Consider the cross-section of M by a normal plane at P (a plane containing the normal vector). It follows from Theorem 3.6.4 that the curvature k of the cross-section at P is

3.6.7. $$k = \frac{\lambda_1 x'^2 + \lambda_2 y'^2}{x'^2 + y'^2} = \lambda_1 \cos^2 \alpha + \lambda_2 \sin^2 \alpha,$$

where α is defined by

$$\cos^2 \alpha = \frac{x'^2}{x'^2 + y'^2}, \quad \sin^2 \alpha = \frac{y'^2}{x'^2 + y'^2}$$

and depends only on the plane chosen for the cross-section. Equation 3.6.7 is called *Euler's formula*, and k is called the *normal curvature* of the surface (in the direction of the plane). The dependence of the curvature $k = k(\alpha)$ on the angle α is shown in Figure 3.12. It is clear that λ_1 and λ_2 are the minimal and maximal values for $k(\alpha)$.

Since the manipulations that led to this formula can be applied to any point P, and since the formula does not involve a particular coordinate system, we see that it holds at any point on any surface. Thus we have proved:

Theorem 3.6.8. *The normal curvature at a point on a surface is given by* $k = \lambda_1 \cos^2 \alpha + \lambda_2 \sin^2 \alpha$, *where λ_1 and λ_2 are the principal curvatures at that point, and α is the angle between the tangent vector to the normal section and the principal direction corresponding to λ_1. If $\lambda_1 > \lambda_2$, then λ_1 and λ_2 are respectively the largest and smallest values of the normal curvature at the point.*

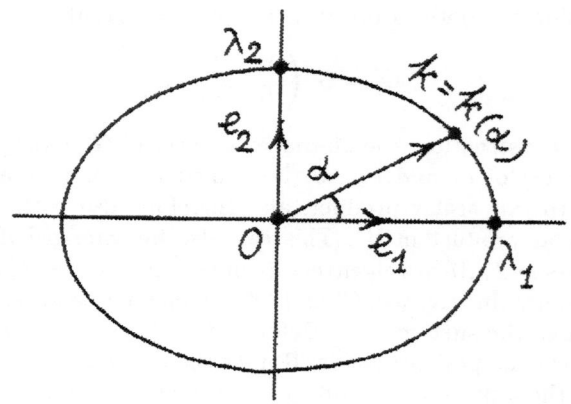

FIGURE 3.12.

The geometric meaning of the Gaussian curvature. We once more assume that the surface is given (in some domain) by $z = f(x, y)$, and that the normal at $P = (0, 0, 0)$ points up. We have seen that the Gaussian curvature at P is

$$K = f_{xx} f_{yy} - f_{xy}^2.$$

We discuss the geometric interpretation of the sign of K. We need to consider three cases, illustrated in the three parts of Figure 3.13.

(a) $K > 0$, $\lambda_1 > 0$, $\lambda_2 > 0$. According to Euler's formula $k = \lambda_1 \cos^2 \alpha + \lambda_2 \sin^2 \alpha$, all normal curvatures K are *positive*, that is, at the point P the surface bends upwards in all directions. Hence $f(x, y)$ has a local *minimum* at P.

(b) $K > 0$, $\lambda_1 < 0$, $\lambda_2 < 0$. By the same token, $f(x, y)$ has a local *maximum* at P.

(c) $K < 0$, so that λ_1 and λ_2 have different signs. Some cross-sections curve up, and some curve down. The function f is *saddle-shaped* at P.

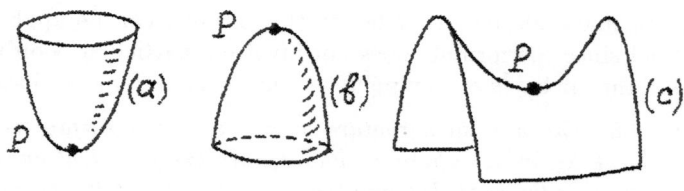

FIGURE 3.13.

In coordinate-free language, we have this result, illustrated in Figure 3.14:

Theorem 3.6.9. *If the Gaussian curvature K is positive at P, there is a neighborhood P throughout which the surface lies on one side of the tangent plane $T_P M$. If K is negative, the surface intersects $T_P M$, and the intersection is homeomorphic to two smooth segments that cross at P.*

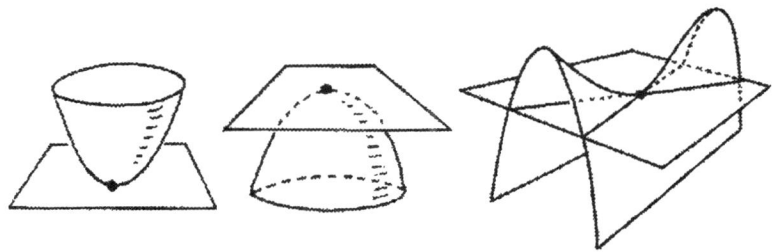

FIGURE 3.14.

3.7. Surfaces of Constant Mean Curvature

The Poisson–Laplace Theorem.   The mean curvature $H = \lambda_1 + \lambda_2$ occurs naturally in many physical problems. As an example, we present the *Poisson–Laplace theorem.*

Theorem 3.7.1. *Suppose a smooth surface M in $\mathbb{R}^3$ is the interface (separation surface) between two fluid media that are in equilibrium. Let p_1 and p_2 be the pressures in the media. Then the mean curvature H of the surface M is equal to*

$$H = h(p_1 - p_2),$$

where the constant h is determined by the physical properties of the media and $p_1 - p_2$ is the pressure difference (see Figure 3.15). In particular, if p_1 and p_2 are uniform throughout the respective media, the interface is a surface of constant mean curvature (that is, the mean curvature is independent of the point on the surface).

Soap films.   We apply this theorem to well-known physical objects: soap bubbles and soap films. They arise, for example, when we dip a wire contour into a soap solution and pull it out slowly. We discuss two cases, shown in the two parts of Figure 3.16:

(a) a closed soap bubble, that is, a soap film without boundary, and

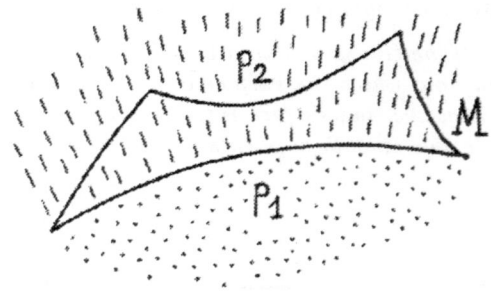

FIGURE 3.15.

(b) a soap film bounded by a wire contour.

In case (a) the soap film separates two media with different pressures. As a model, we can take a soap bubble made by blowing through a tube (Figure 3.17). Consequently, here we have

$$H = h(p_1 - p_2) > 0,$$

with an appropriate choice of direction for the normal to the surface. If we neglect gravity, Theorem 3.7.1 says that the mean curvature is constant throughout the whole surface of the bubble. A difficult theorem proved by Aleksandrov in 1955 says that any embedded surface in $\mathbb{R}^3$ that has constant mean curvature and is homeomorphic to a sphere is in fact a *round sphere* (by which we mean a standard sphere of some radius; often we loosely call a sphere a space that is just homeomorphic to S^2). Details can be found in [Hopf 1983]. for example. Single soap bubbles in equilibrium, therefore, must be spherical.

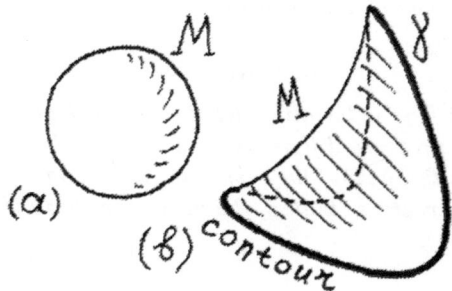

FIGURE 3.16.

Remark. Hopf [1983] strengthened Aleksandrov's theorem to show that any *immersion of the sphere* in $\mathbb{R}^3$ having constant mean curvature is a round

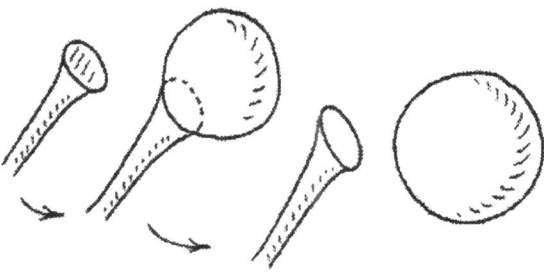

FIGURE 3.17.

sphere. That is, allowing self-intersections does not change the picture as far as surfaces modeled on the sphere are concerned. However, if we allow other topological types, there are other immersed surfaces of constant mean curvature $\mathbb{R}^3$, whose construction is far from trivial; this is currently an area of active research. The first examples of such surfaces were the *Wente tori* [Wente 1987].

Minimal surfaces.   Turning now to the case of Figure 3.16(b), we see that when the film has a boundary, it does not disconnect space, so the medium on both sides of the film is, in fact, the same, and the pressure must be the same as well. See Figure 3.18, where the arc represents a cross-section of the two-dimensional film. Therefore in this case $H = 0$: the surface has zero mean curvature.

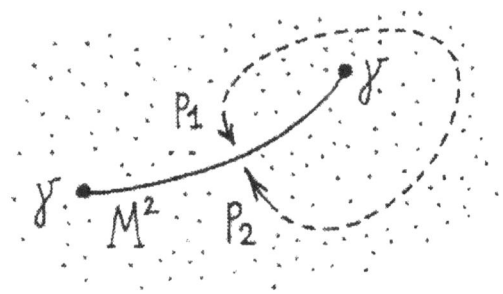

FIGURE 3.18.

Surfaces of zero mean curvature are also known as *minimal surfaces* because they are locally area-minimizing, in the following sense.

Consider all possible small enough perturbations of the surface M in $\mathbb{R}^3$ that do not move the boundary of the surface (Figure 3.19). We call a perturbation *small* if it is small in amplitude and localized inside a small region in space (that is, outside a certain small ball the surface remains unchanged).

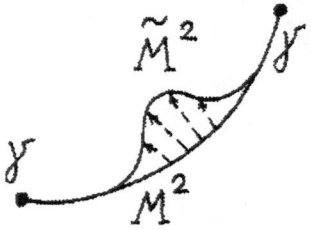

FIGURE 3.19.

Definition 3.7.2. A surface in three-space is called *locally minimal* if no small perturbation of it decreases its area.

Theorem 3.7.3. *A surface M in Euclidean three-space is locally minimal if and only if its mean curvature is identically zero.*

Corollary 3.7.4. *A soap film in $\mathbb{R}^3$ bounded by some wire contour is a locally minimal surface.*

Analytic expression for the local minimality of a surface. From the definition of mean curvature H we have

$$H = \frac{GL - 2MF + EN}{EG - F^2},$$

where $ds^2 = E\,du^2 + 2F\,du\,dv + G\,dv^2$ is the first quadratic form and $L\,du^2 + 2M\,du\,dv + N\,dv^2$ is the second. According to Theorem 3.5.3, we can (locally) choose conformal coordinates u and v; in such coordinates we have $E = G$ and $F = 0$. Consequently, in these coordinates the condition $H = 0$ is equivalent to the identity

$$L + N = 0.$$

From this it follows that

$$\frac{\partial^2 r}{\partial u^2} + \frac{\partial^2 r}{\partial v^2} = 0,$$

where $r = r(u, v)$ is the parametrization of the surface in $\mathbb{R}^3$. Thus, in conformal coordinates we have

$$\Delta r = 0,$$

where Δ is the *Laplace operator*. By definition, this means that r is a *harmonic* vector-valued function with respect to the conformal coordinates.

Hence, the equality $H = 0$ can be regarded as the *differential equation of the minimal surface*.

Expanding Equation 3.6.6 we see that, for a surface given locally as a graph $z = f(x, y)$, the minimality condition is

$$(1 + f_x^2)f_{yy} - 2f_x f_y f_{xy} + (1 + f_y^2)f_{xx} = 0.$$

Simplest examples of minimal surfaces.

Example 3.7.5. A plane in $\mathbb{R}^3$ is obviously a minimal surface.

Example 3.7.6. Consider in $\mathbb{R}^2$ the curve given by the equation

$$y = a \cosh \frac{x}{a} = \tfrac{1}{2}a(e^{x/a} + e^{-x/a}),$$

where a is a constant (Figure 3.20). This curve is called a *catenary*, because it is the shape taken by a chain (in Latin, "catena") of uniform weight hanging from two points, such as A and B in Figure 3.20. The surface of revolution formed by rotating the catenary about the x-axis, shown in Figure 3.21, is a minimal surface (prove it!), called a *catenoid*. An implicit equation for the catenoid in Cartesian coordinates is

$$a^2(x^2 + y^2) = \cosh^2(az),$$

where a is a constant and z is the axis of symmetry.

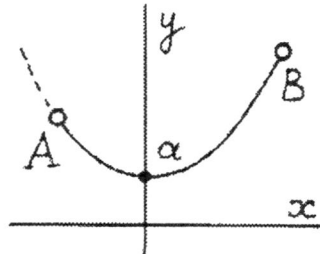

FIGURE 3.20.

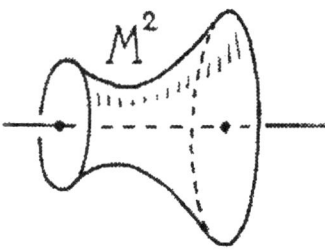

FIGURE 3.21.

Example 3.7.7. The *helicoid* is given in $\mathbb{R}^3$ by the equation

$$x \cos z = y \sin z;$$

locally we can write $z = \arctan x/y$. Geometrically, this surface is obtained when a straight line A that intersects orthogonally a vertical straight line B moves up B at constant speed, while also turning at constant speed. See Figure 3.22.

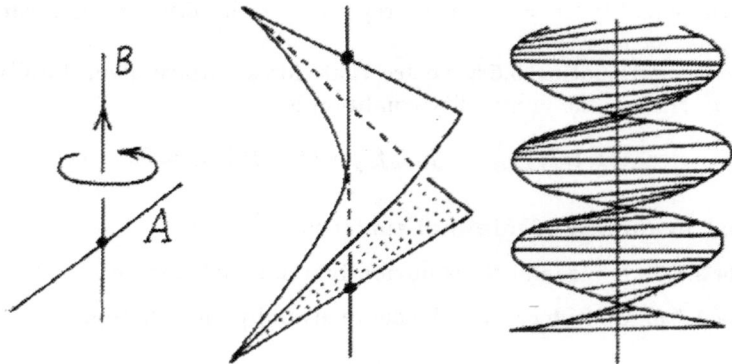

FIGURE 3.22.

Example 3.7.8. *Scherk's surface* is given by the equation

$$z = \frac{1}{a} \ln \frac{\cos(ay)}{\cos(ax)}, \quad \text{or} \quad e^{az} = \frac{\cos(ay)}{\cos(ax)},$$

where a is a constant.

Singular points of minimal surfaces. Plateau principles.   Soap bubbles in practice are typically not single embedded surfaces; they usually consist of pieces of embedded surfaces that join together along curves. For instance, when two spherical bubbles coalesce into a double bubble, we get three pieces of spheres, that join together along a circle, giving rise to the local situation shown in Figure 3.23.

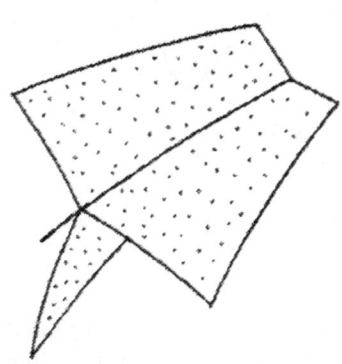

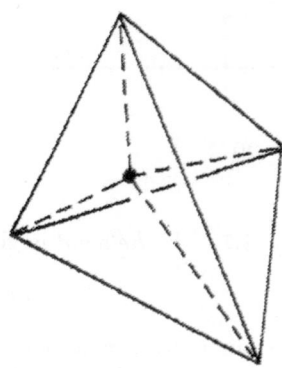

FIGURE 3.23.   FIGURE 3.24.

The points where the bubble is not an embedded surface are called *singular points*, and they form the bubble's *singular locus*. Often we talk about such configurations as *surfaces of constant mean curvature with singular points*.

The singular locus can be complicated. Singular points are not isolated: they form arcs, which can meet together and form complicated skeletons. The local structure of a minimal surface near such singular points is described by the Plateau principles:

Smoothness Principle. Energy-minimizing surface configurations, such as soap films, consist of fragments of smooth surfaces of constant mean curvature meeting at a singular locus.

Branching Principle. The singular locus can have only two local configurations: three smooth sheets meeting along a smooth curve at 120° angles (Figure 3.23), or six smooth sheets (along with four singular curves) converging in a vertex, the angles between the curves being equal to $\arcsin 2\sqrt{2}/3 \approx 109°28'$ (Figure 3.24).

Remark. Joseph Plateau (1801–1883) was a Belgian physicist, professor of physics and anatomy, who experimented extensively with soap films and other physical surfaces. The principles that bear his name were formulated as a result of physical experiments; although they were published over a century ago [Plateau 1873], it was not until much later that they were proved mathematically. See the *Scientific American* article [Almgren Jr. 1976].

It is convenient to visualize the situation of Figure 3.24 as follows: take a regular tetrahedron and place the sextuple point in its center; then connect this point with the vertices and span the four edges by six triangles.

From the mathematical point of view, there exist surfaces whose mean curvature vanishes almost everywhere (at all nonsingular points), but that have more than three sheets meeting at an edge: for example, two orthogonal planes meeting in a straight line (Figure 3.25). Nevertheless, attempts to construct a real soap film with more than three sheets meeting along a singular line fail. It turns out that any surface of zero curvature with self-intersections of multiplicity more than three is unstable.

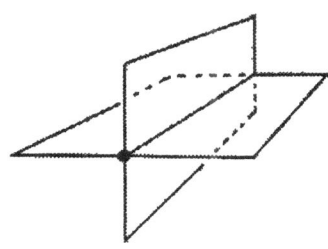

FIGURE 3.25.

An informal argument that makes this fact plausible is easy to give. Consider, for example, two planes meeting in a straight line. If the planes are not orthogonal, we can decrease the total area by splitting the singular line into two, as in Figure 3.26. But even when the planes are orthogonal, one can see that the total area can be decreased by taking a section by a plane orthogonal to the singular line, as shown in Figure 3.27. This gives two orthogonal lines

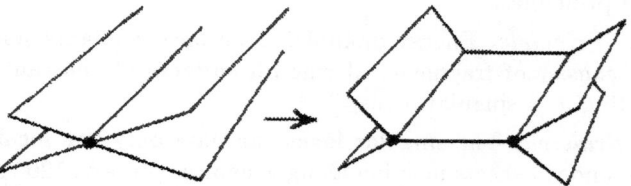

FIGURE 3.26.

in the plane. An easy calculation shows that the total lengths of the segments decreases if the quadruple point is split into two triple points, as shown in the figure. The same idea in one dimension higher serves to decrease the total area of the configuration in Figure 3.25. It also applies to configurations where more than four surfaces meet, as in Figure 3.26.

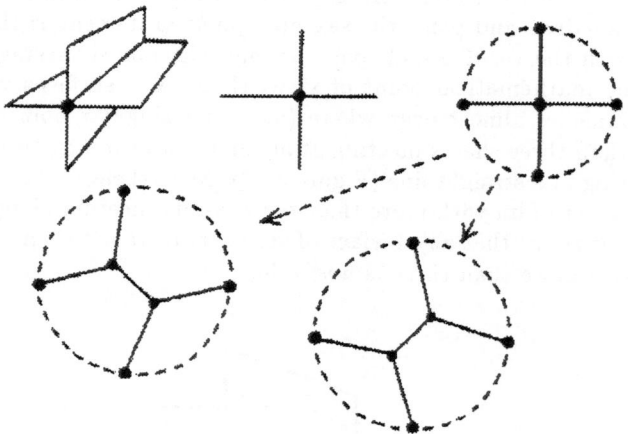

FIGURE 3.27.

Minimal surfaces in nature.   Many concrete problems in physics, chemistry, and biology can be reduced to the analysis of minimal surfaces. We

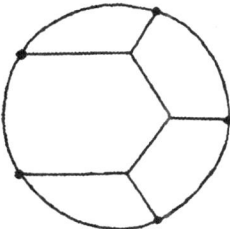

FIGURE 3.28.

illustrate with a biological example. Minimal films are realized as the most economical surfaces forming the tiny sea animals known as radiolarians.

Radiolarians consist of small blobs of protoplasm inside foam-like forms similar to soap films or bubbles. Since these organisms are fairly complex, the minimal films involved typically have many branch points and triple edges, on which the organism's fluids are mostly concentrated. The concentration of fluid leads to the deposit of salts along these edges, and this is what builds the animal's skeleton. After death, the soft tissues vanish gradually, and the hard skeleton remains. The skeleton therefore represents the singular locus of the animal's membranes (some of which have constant positive mean curvature). Three skeletons are drawn in Figure 3.29; their shapes are strikingly similar to those of certain actual soap film clusters, based on a regular tetrahedron, cube, and triangular prism, respectively (Figure 3.30). See also the books [Thompson 1917] and [Haeckel 1887].

(a)

FIGURE 3.29.

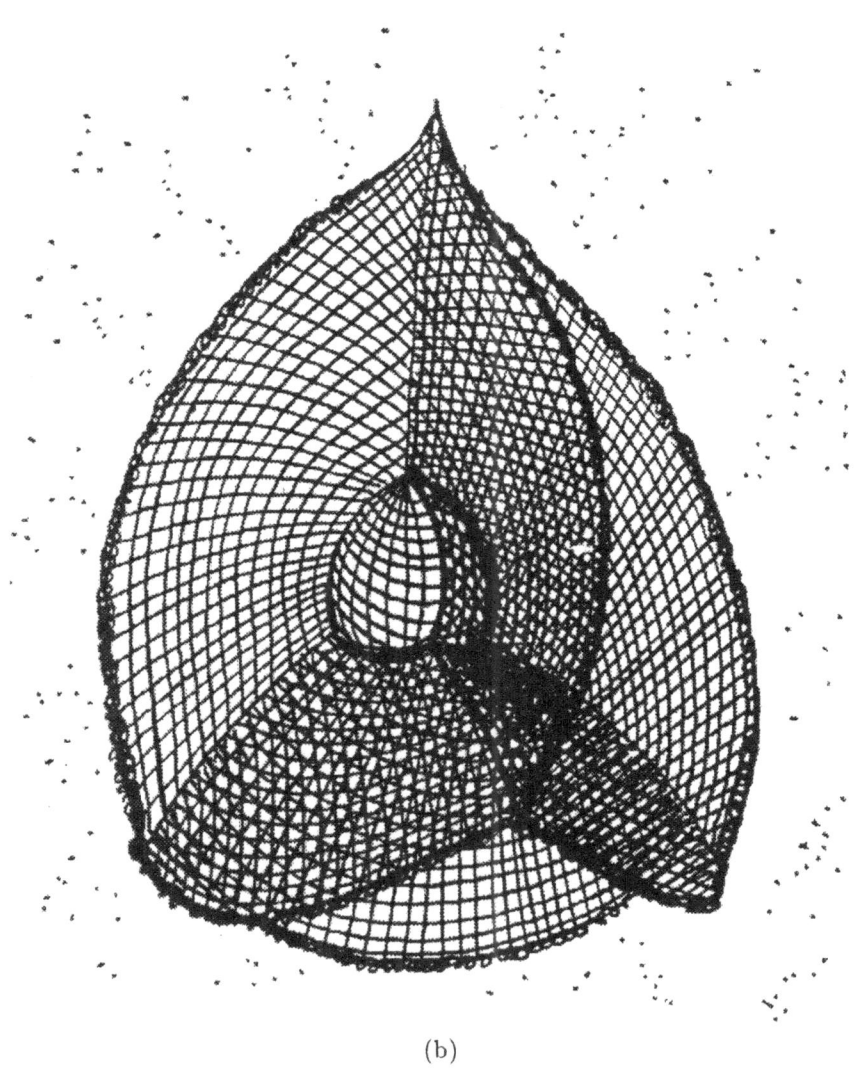

(b)

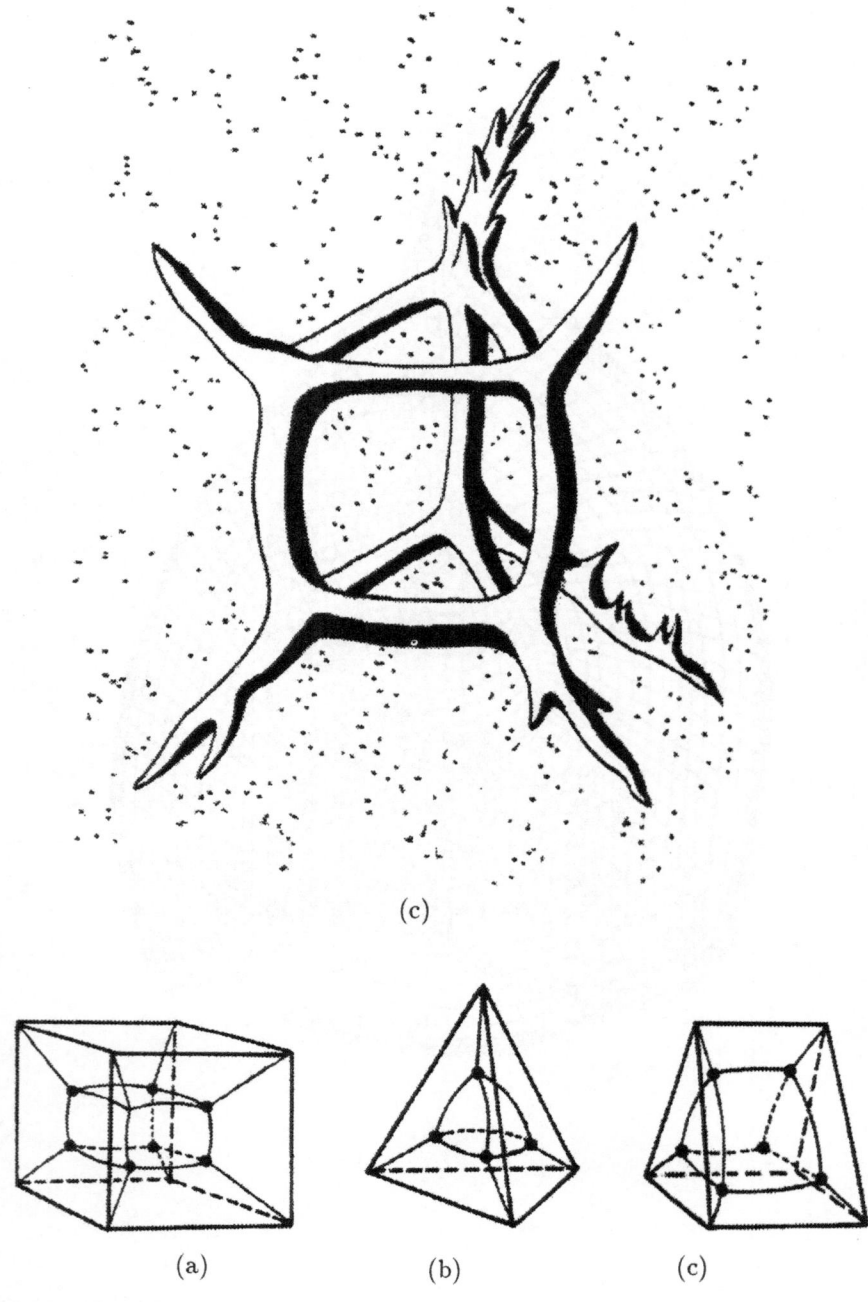

(c)

(a)   (b)   (c)

FIGURE 3.30.

4. The Classification of Surfaces

4.1. The Canonical Fundamental Polygon of a Surface

In this chapter we study embedded surfaces (two-manifolds) in $\mathbb{R}^n$ that are *compact* and *connected*. We recall the meaning of these concepts.

First, a subset of $\mathbb{R}^n$ is called *closed* if its complement is an open set (also known as a domain; see Chapter 1). Another, often useful, characterization, is this: If $X \subset \mathbb{R}^n$ is closed, and a sequence of points $x_1, \ldots, x_n, \ldots$ in X converges to a limit point in $\mathbb{R}^n$, this limit is also in X. Conversely, any $X \subset \mathbb{R}^n$ with this property is closed.

A subset of $\mathbb{R}^n$ is called *compact* if it is *closed* and *bounded* (a set is bounded if it is contained in some Euclidean ball of finite radius; see Figure 4.1).

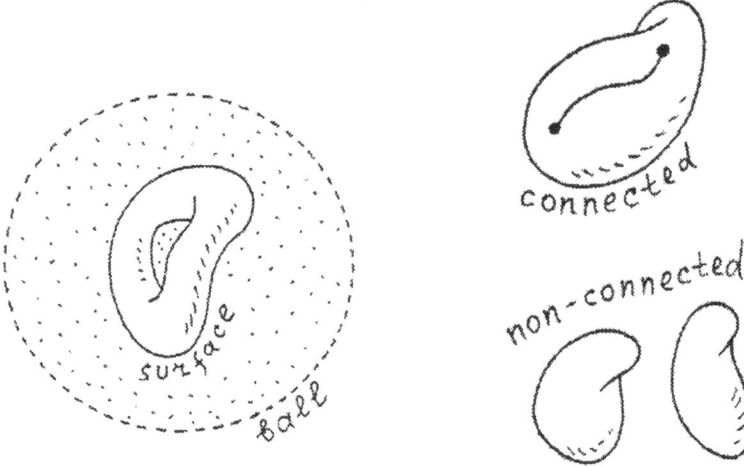

FIGURE 4.1.   FIGURE 4.2.

A sphere in $\mathbb{R}^3$ is a compact surface. The closed unit ball in $\mathbb{R}^3$, that is, the set of points at distance less at most 1 from the origin, is a compact subset of $\mathbb{R}^3$. The corresponding open ball, that is, the closed ball minus the bounding sphere, is not compact, because it is not closed: one can have a sequence of points in the ball converging to a point on the unit sphere, which is not in the open ball. A plane in $\mathbb{R}^3$ is not compact, because it is not bounded.

A surface is called *connected* (in full, *path-connected*) if any two points in it can be connected by a continuous curve that is contained in the surface. See Figure 4.2.

We are interested in studying the possible *topological types* of compact, connected surfaces. By this we mean that we consider all possible surfaces, and regard as equivalent those that are homeomorphic. The natural question arises: How to describe the topological types of surfaces? In other words, how can we classify all surfaces *up to homeomorphism*? The answer, given in Theorem 4.2.7, turns out to be very simple; it has been known for over a century, and has been proved in different ways.

The proof we give here is one of the earliest. It is intuitive and easy to visualize; moreover, its combinatorial approach can easily be used in computer geometry. This allows us to use the computer to recognize concrete surfaces given, for example, by some equation in three-space.

Remark. We could instead consider the classification of surfaces *up to diffeomorphism*. But it turns out that the answer is the same: there is no difference between homeomorphism types and diffeomorphism types of surfaces. This is a consequence of the following theorem:

Theorem 4.1.1. *If two smooth surfaces are homeomorphic, they are also diffeomorphic* (of course, the converse statement is also valid).

First step: Triangulating the surface. The tool that we will use to perform the classification of surfaces is *triangulations*. Roughly speaking, triangulating a surface means dividing it into (curvilinear) triangles, in such a way that every point either is in a triangle, or is in an edge between two triangles, or is a point where several triangles meet. To formalize this, we define a *triangle* on the surface M as the image of an embedding $f : \Delta \to M$, where Δ is a standard closed triangle in $\mathbb{R}^2$—say the set determined by the conditions $x \geq 0$, $y \geq 0$, and $x + y \leq 1$. By an *embedding* we mean that f is a *smooth map* defined in an open set of $\mathbb{R}^2$ containing Δ, that it is an *immersion*, and that it is a *homeomorphism* from Δ to $f(\Delta)$. These conditions ensure that $f(\Delta)$ really matches our intuitive notion of a triangle on the surface. The *edges* of $f(\Delta)$ are the images of the edges of Δ, and are therefore *arcs* (embedded curves), whose endpoints are the *vertices* of $f(\Delta)$.

Definition 4.1.2. Let M be a compact, connected surface. A *triangulation* of M consists of a finite number of points $P_i \in M$, called *vertices*, a finite number of arcs $\gamma_\alpha \subset M$, called *edges*, and a finite number of triangles $\Delta_p \subset M$, satisfying these conditions:

(a) any point of M belongs to some triangle Δ_p;

(b) each edge has as endpoints two distinct vertices, and contains no other vertex:

(c) each triangle has as edges three distinct arcs, and its interior intersects no arc and contains no vertex;

(d) any two triangles either do not intersect, or intersect at one common vertex, or intersect along one common edge.

Figure 4.3 shows examples of the three situations in (d). Figure 4.4 shows configurations that are forbidden by the conditions of the definition.

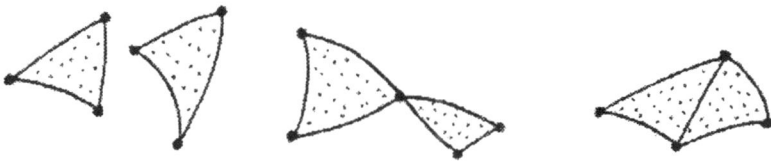

FIGURE 4.3.

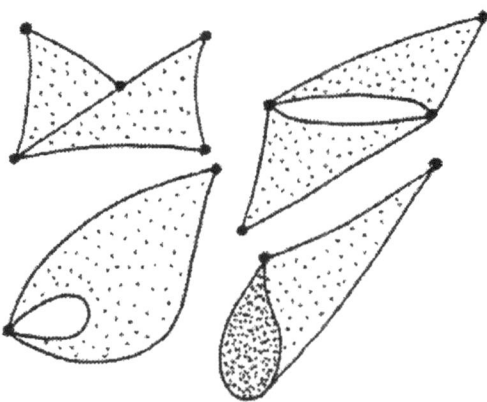

FIGURE 4.4.

Theorem 4.1.3. *Any smooth compact two-manifold admits a triangulation* (in fact, infinitely many).

See the proof in [Mishchenko and Fomenko 1988], for example.

Second step: Cutting up the surface. To carry out our classification, we will take an arbitrary compact, connected surface M and reduce it to a standard form. The first step in this process is to fix some triangulation of M. Then we mark each edge of the triangulation with a letter, without repetition. We also assign an arbitrary orientation to each edge; this corresponds to drawing an arrow along the edge, pointing one way or the other (Figure 4.5).

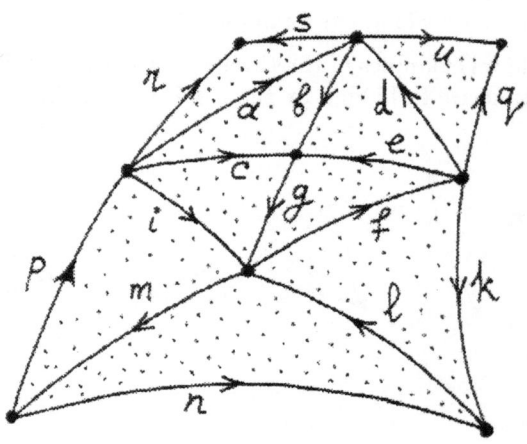

FIGURE 4.5.

Next we cut the surface along all the edges, decomposing it into a finite set of triangles, each side of which is marked with a letter and provided with an arrow, as shown in Figure 4.6. Each letter appears in the set of triangles *exactly twice*, because each edge of the triangulation was contained in exactly two triangles (this is a consequence of our definition of surface: no boundary edges are allowed).

Our aim is to glue again all these triangles in such a way as to obtain a plane polygon (more exactly, a polygonal region). We may consider all these triangles as plane triangles because only the marking of edges will be important, not the concrete form of triangle. We begin the procedure by choosing a triangle, say Δ_1. Consider any side of this triangle: it has a letter and an orientation. Since each letter appears in the entire collection of sides exactly twice, the other edge with the same letter can be found in another triangle, say Δ_2. It can't be the same triangle, by condition (c) of Definition 4.1.2. We now glue Δ_1 and Δ_2 along the common edge in such a way that the arrows point in the same direction, as shown in Figure 4.7.

(More precisely, we choose triangles Δ_1' and Δ_2' in the plane that share an edge and that map to Δ_1 and Δ_2, making the maps agree along the common edge. This gives a plane representation of the piece of M consisting of Δ_1 and Δ_2. From now on we will not mention explicitly the fact that we are building a map from a plane polygon to M, but simply think of the triangles $\Delta_1, \Delta_2, \ldots$ sometimes as part of M, and sometimes as lying in the plane.)

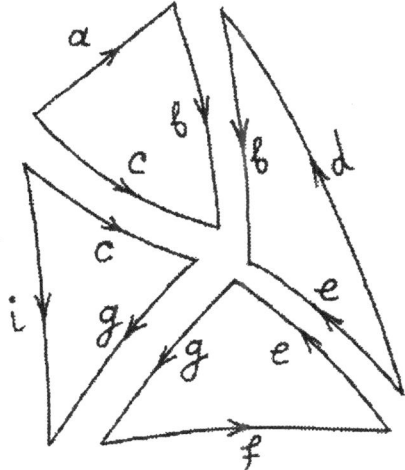

FIGURE 4.6.

We thus obtain a plane polygon W, whose edges are marked with letters and arrows. We erase the inner diagonal, coming from the edge where we glued (b in Figure 4.7).

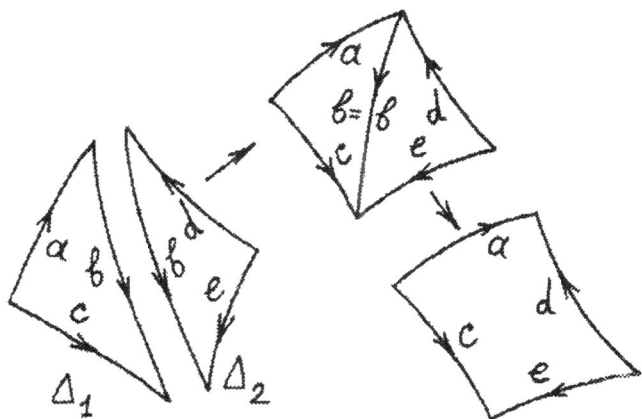

FIGURE 4.7.

Now we have two possibilities: either all letters on edges around W occur twice, or there is at least one letter without a match. In the first case, we claim that there are no other triangles in the triangulation. Indeed, suppose there were; then, because M is connected, we could draw an arc in M from a point in the interior of W to a point in the interior of a triangle outside W. This arc would have to go out of W at some point; but any curve getting out of W must do so through a vertex or an edge of W, and it is easy to show that, if W has no unpaired edges, no edge or vertex of W is shared by a triangle outside W.

In the remaining case, we choose an unpaired edge in W, and find the triangle where the same letter occurs again, say Δ_3. We glue this triangle to W along the edge in question, as shown in Figure 4.8. By the same reasoning of the preceding paragraph, we can continue this process until we have glued together all the triangles Δ_i. (Note that we can always find room to add triangles in the plane, because we can choose the triangles as narrow as necessary.)

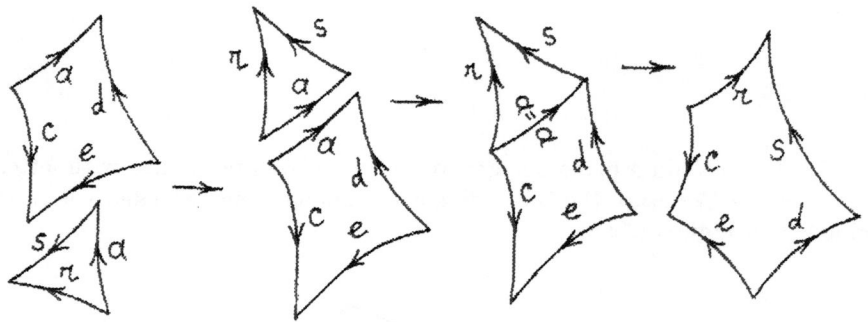

FIGURE 4.8.

The result is a plane polygon W with the following properties:

(1) The boundary ∂W consists of an even number of edges, each having a letter and an arrow; each letter appears in the boundary exactly twice.
(2) If we identify boundary edges of W having the same letter, respecting the arrows. we obtain a space homeomorphic to the initial surface M.

A polygon having these properties is said to *represent* M. Of course, W is not determined uniquely (even for a fixed triangulation).

Now we fix an orientation for W and go around its boundary, starting from an arbitrary vertex P, and writing successively all the letters we encounter along the way. If our direction of motion coincides with the arrow of the edge we're on, we write the letter to the power $+1$; otherwise, we write the letter to the power -1. When we come back to the initial point P we will have written some word

$$W = a_{i_1}^{\varepsilon_1} a_{i_2}^{\varepsilon_2} \ldots a_{i_N}^{\varepsilon_N},$$

where $\varepsilon_\alpha = \pm 1$. This word, or code, expresses W from the combinatorial point of view. Figure 4.9 shows an example.

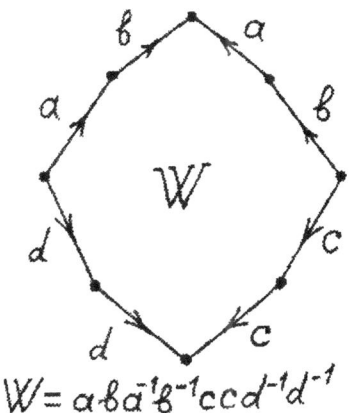

$$W = a\,b\,a^{-1}\,b^{-1}\,c\,c\,d^{-1}\,d^{-1}$$

FIGURE 4.9.

Summary. We have associated with a triangulated surface M a word W whose symbols are the edges of the triangulation. This word can be considered as a code for M. This encoding is not unique: infinitely many codes correspond to a single M.

Third step: Putting W in canonical form. We will simplify W in several steps, always preserving the property that it represents M.

In the following lemmas, the symbols a and b can represent either a letter assigned to an edge or its inverse.

Lemma 4.1.4. *Assume that W has the form*

$$W = \underline{\quad\quad} a a^{-1} \underline{\quad\quad},$$

where the dashes represent arbitrary subwords, not both of which are empty. Then the word obtained from W by deleting aa^{-1} also represents M.

The proof follows from Figure 4.10.

By repeated application of this lemma, we can assume that our words have no cancelable pairs aa^{-1} (unless the word is such a pair by itself, a case that we leave aside for now). We make this assumption from now on. If any of the subsequent constructions introduce cancelable pairs, we delete them before proceeding, without affecting the conclusions of the lemmas.

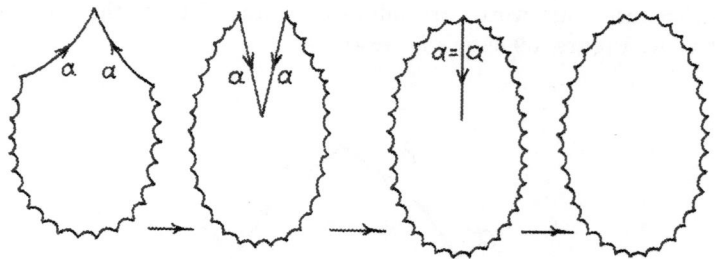

FIGURE 4.10.

Lemma 4.1.5. *The word W can be transformed to a word representing M and such that all the vertices of the new polygon become a single point under reconstruction* (by which we mean the gluing of all the edges, yielding M).

Proof. We separate the vertices of W into equivalence classes, each class consisting of all the vertices that become a particular vertex under reconstruction. If there is only one class of vertices, there is nothing to prove. So let W contain at least two classes, and consider one such class $\mathcal{P}$. Then there exists an edge a connecting a vertex P in class $\mathcal{P}$ to a vertex Q in a different class $\mathcal{Q}$. We take the edge c adjacent to a at P (Figure 4.11, left).

If $c = a$, we are in the case of the preceding lemma (the case of one arrow pointing into P and one pointing out is impossible, for this would mean that Q is in the same class of vertices as P). Using the lemma to make vertex P disappear, we reduce the number of vertices in class $\mathcal{P}$ by one.

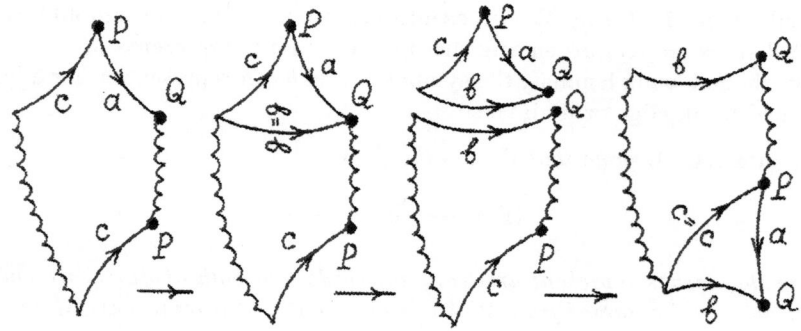

FIGURE 4.11.

If $c \neq a$, we transform W by the procedure shown in Figure 4.11. That is, we cut the polygon along the arc b, which corresponds to some arc in M (not necessarily part of the original triangulation), and then glue the two pieces back together along c and its match. As a result, class $\mathcal{P}$ loses one vertex, and class $\mathcal{Q}$ gains one vertex.

In either case the number of elements of $\mathcal{P}$ decreases. We can repeat the procedure (changing the class $\mathcal{Q}$ as needed) until $\mathcal{P}$ is exhausted, thus decreasing the number of classes by one. Continuing inductively, we see that we can reduce the number of classes to one. □

We assume from now on that the conclusion of the lemma is satisfied. The subsequent lemmas will not affect this property, since no new vertices will be created.

Lemma 4.1.6. *If W has the form*

$$W = XaYaZ,$$

where X, Y, Z are arbitrary subwords, the word

$$XY^{-1}aaZ$$

also represents M.

By Y^{-1} we mean Y written backwards with all letters replaced by their inverses. The proof that $XY^{-1}aaZ$ represents M is illustrated in Figure 4.12, where b is a new letter not otherwise present; at the end of the procedure we replace b by a.

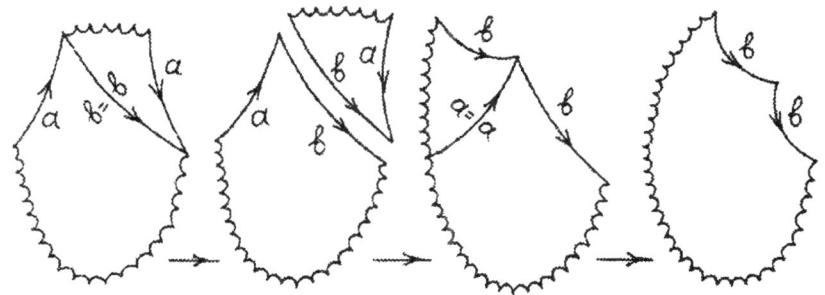

FIGURE 4.12.

Two equal letters brought together by this lemma remain together under further applications of this and subsequent steps. Thus, applying Lemma 4.1.6 repeatedly following each subsequent step, we can assume from now on that like letters are adjacent.

Lemma 4.1.7. *If W contains a pair a and a^{-1} separated by other letters, there exists another pair b and b^{-1} that alternate with a and a^{-1}, like this:*

$$W = -a-b-a^{-1}-b^{-1}- \qquad or \qquad W = -b^{-1}-a-b-a^{-1}-.$$

Proof. Assume, to the contrary, that for any b between a and a^{-1} its match b or b^{-1} appears again between a and a^{-1}. Glue W along the two copies of a, as in Figure 4.13. The result is a tube with two boundary components. By assumption, each edge lies in the same boundary component as its match; therefore reconstruction of the surface leads to at least two vertices in M, contrary to the assumption made after Lemma 4.1.5.

Therefore there is some b in the interval from a to a^{-1} whose match is outside that interval. The match cannot be b, by the assumption that like letters are adjacent. Therefore it must be b^{-1}, and the lemma is proved.   □

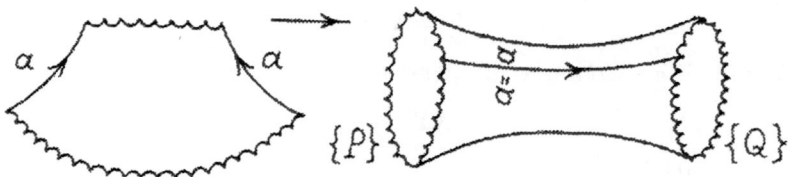

FIGURE 4.13.

Lemma 4.1.8. *If W has the form*

$$W = XbYaZb^{-1}Sa^{-1}T,$$

where X, Y, Z, S, T are arbitrary subwords, the word

$$Saba^{-1}b^{-1}ZYTX$$

also represents M.

The proof is illustrated in Figure 4.14, where c and d are new letters not otherwise present. At the end of the the procedure we replace c and d by a^{-1} and b^{-1}, respectively.

Using Lemma 4.1.7, we can apply Lemma 4.1.8 repeatedly until any pairs a, a^{-1} appear two at a time, in the form of *commutators* $aba^{-1}b^{-1}$. So far, then, we have transformed W into the product of commutators and squares aa. We can reduce further the case when there are both commutators and squares:

Lemma 4.1.9. *If W has the form*

$$W = Xaba^{-1}b^{-1}ccZ,$$

the word

$$XaabbccZ$$

also represents M.

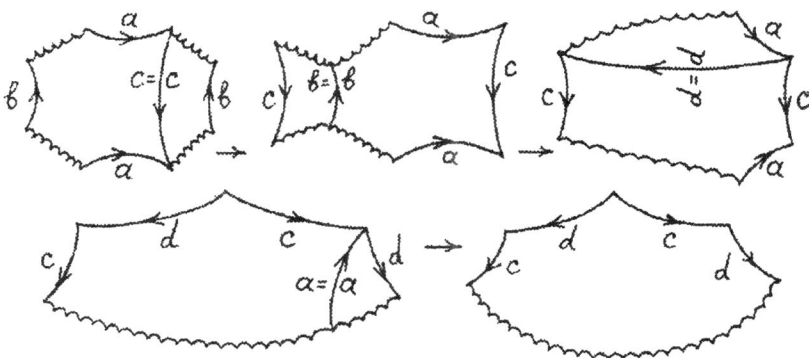

FIGURE 4.14.

Proof. We first transform W into $XabcbacZ$, by applying the procedure of Lemma 4.1.6 in reverse (the role of X is played by Xab, and that of Y by ba). Another application of Lemma 4.1.6 brings together the a's, and gives $Xb^{-1}c^{-1}b^{-1}aacZ$. Now we rename b^{-1} to b and apply Lemma 4.1.6 two more times, obtaining $XcbbaacZ$ and then $Xa^{-1}a^{-1}b^{-1}b^{-1}ccZ$. Finally rename a^{-1} to a and b^{-1} to b once more.   □

The upshot is that we can reduce W either to a product of commutators or to a product of squares. Recalling the case $W = aa^{-1}$, which we set aside after Lemma 4.1.4, we conclude that:

Lemma 4.1.10. *The manifold M is can be represented by a word of one of these forms:*

$$(1^*)\ W = aa^{-1}.$$
$$(1\ )\ W = a_1 b_1 a_1^{-1} b_1^{-1} \ldots a_g b_g a_g^{-1} b_g^{-1}.$$
$$(2\ )\ W = c_1 c_1 c_2 c_2 \ldots c_k c_k.$$

where g and k are positive numbers.

Definition 4.1.11. A plane polygon W representing M and having one the forms shown in Lemma 4.1.10 is called a *canonical fundamental polygon* of the given surface M. The corresponding system of cuts on the surface also is called *canonical* (Figure 4.15).

4.2. Reconstructing the Surface from the Canonical Polygon

What surfaces correspond to each type of word W?

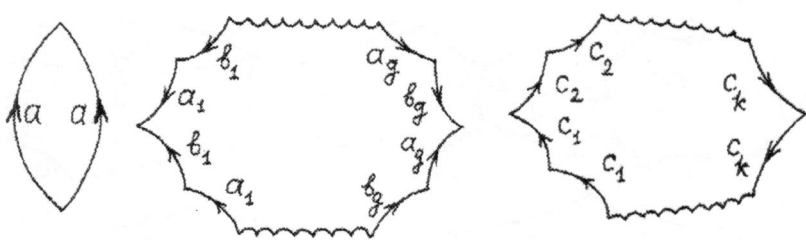

FIGURE 4.15.

Case 1*: the sphere. Figure 4.16 shows that a surface represented by $W = aa^{-1}$ is homeomorphic to a sphere. From now on we usually call any such surface a sphere, and denote it S^2.

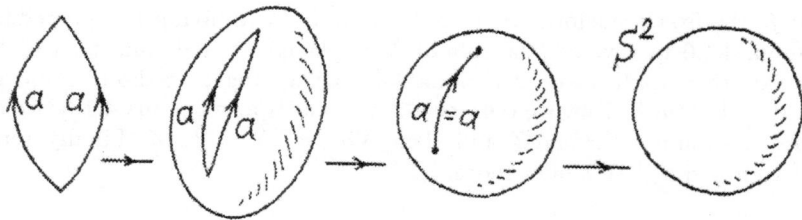

FIGURE 4.16.

Case 1: the torus and spheres with handles. Next consider a surface represented by $W = a_1 b_1 a_1^{-1} b_1^{-1} \ldots a_g b_g a_g^{-1} b_b^{-1}$. We start, for simplicity, with the case $g = 1$, so that

$$W = aba^{-1}b^{-1}.$$

As we can see from Figure 4.17, the surface is homeomorphic to a *torus*. We usually call any such surface a torus as well, and denote it T^2.

When the torus is drawn as in Figure 4.17, the cuts a and b are called a *parallel* and a *meridian*, respectively. Figure 4.18 shows two other ways of drawing a torus, including the canonical cuts. This figure shows that the torus can be thought of as a *sphere with one handle*.

Now consider the case of two commutators:

$$W = aba^{-1}b^{-1}cdc^{-1}d^{-1}.$$

The construction of the surface is demonstrated in Figure 4.19. The cuts a, b, c, d form the canonical system. Such a surface is called a *sphere with two handles*. Other representations of it are displayed in Figure 4.20.

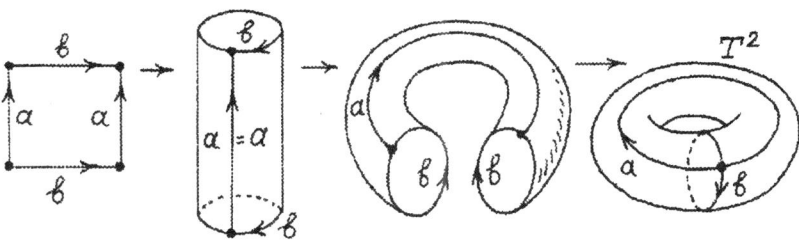

FIGURE 4.17.

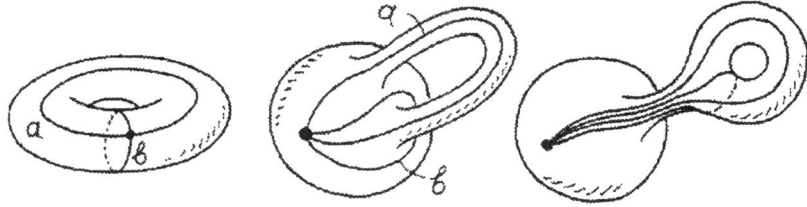

FIGURE 4.18.

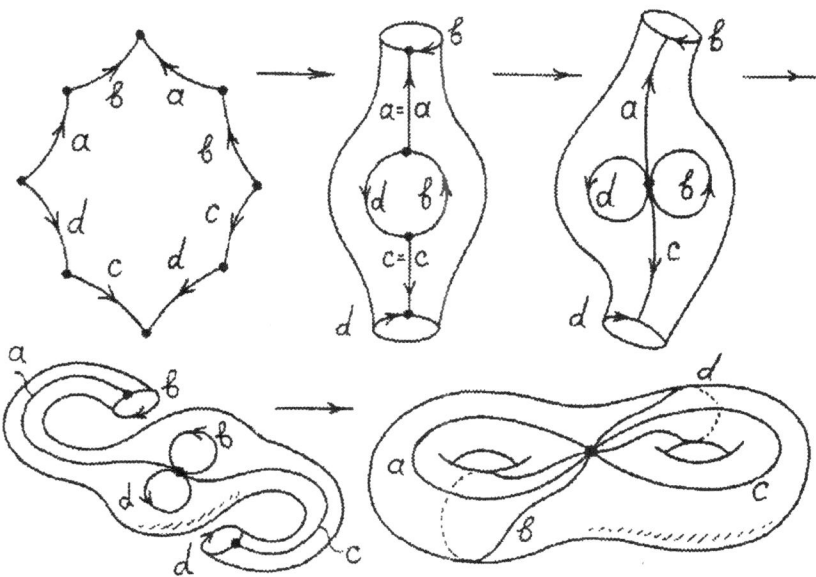

FIGURE 4.19.

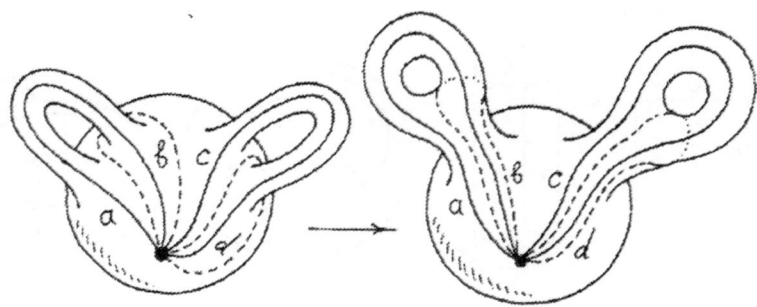

FIGURE 4.20.

Continuing in this way, we see that, for arbitrary $g > 0$:

Lemma 4.2.1. *A surface represented by the canonical fundamental polygon*

$$W = a_1 b_1 a_1^{-1} b_1^{-1} \ldots a_g b_g a_g^{-1} b_g^{-1}$$

is homeomorphic to a sphere with g handles.

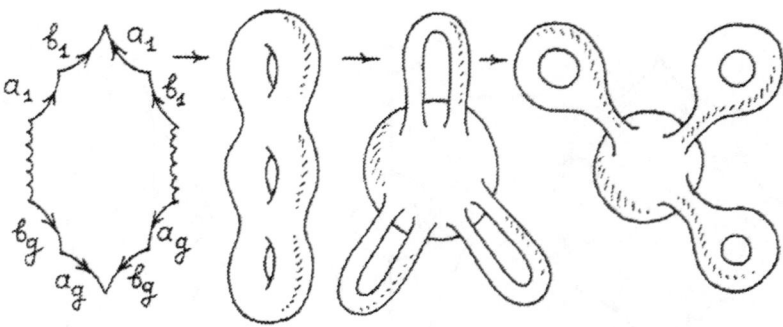

FIGURE 4.21.

See Figure 4.21. It is intuitively plausible, and can be proved (by using homology groups, for example: see Corollary 11.7.3), that g is determined by the homeomorphism type of the surface; that is, a sphere with g handles cannot be homeomorphic to one with h handles, for $g \neq h$. The number of handles is also called the *genus* of the surface, or of the homeomorphism type.

We can include the sphere S^2 in the sequence of spheres with handles, by putting $g = 0$. Thus, we can combine the two cases 1* and 1 into a single case: the sphere with g handles, where $g \geq 0$.

Connected sums. Before continuing the analysis of possible surfaces, we introduce a useful notion: the *connected sum* of a two surfaces M and P. This operation is shown in Figure 4.22. We consider in each surface an embedded closed disk, and remove the interior of the disk. This leaves the boundary curves of the disks, which are topologically circles. We then glue these two circles together using some homeomorphism. This gives a new surface K (without boundary), called the *connected sum* of M and P, and denoted

$$K = M \# P.$$

It can be shown that the homeomorphism type of K does not depend on the choices made in this construction.

Lemma 4.2.2. *The sphere with g handles is a connected sum of g copies of the torus T^2:*

$$\text{sphere with } g \text{ handles} = T^2 \# \cdots \# T^2 \quad (g \text{ copies}).$$

The proof follows from Figure 4.23.

The projective plane. We now consider the second type of fundamental regions,

$$W = c_1 c_1 c_2 c_2 \ldots c_k c_k.$$

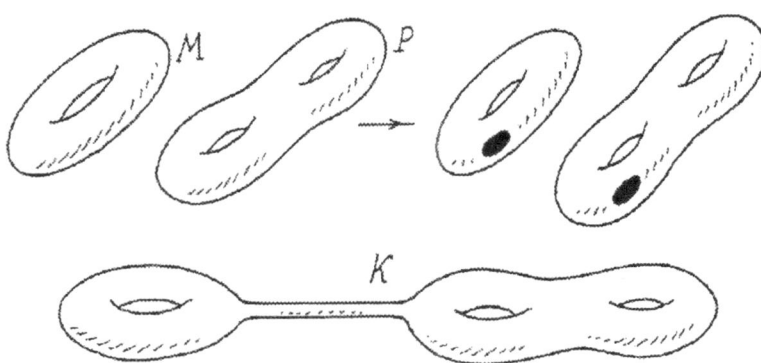

FIGURE 4.22.

FIGURE 4.23.

We start from the case $k = 1$, that is, $W = cc = c^2$.

Recall that the *projective plane* $\mathbb{RP}^2$ is defined as the space of lines in $\mathbb{R}^3$ that go through the origin. That is, each point of $\mathbb{RP}^2$ comes from one line of $\mathbb{R}^3$, and two points of $\mathbb{RP}^2$ are near if the angle between the corresponding lines is small.

Lemma 4.2.3. *The surface with fundamental polygon $W = c^2$ is homeomorphic to the projective plane $\mathbb{RP}^2$.*

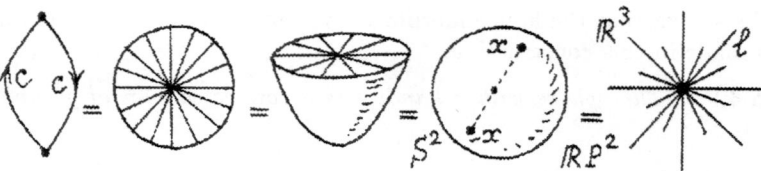

FIGURE 4.24.

Proof. The proof is illustrated in Figure 4.24. Gluing the edges of W together amounts to identifying opposite points on the boundary of a disk. The disk can be deformed into a hemisphere, so the resulting object is homeomorphic to a hemisphere with opposite points on its equator identified. This in turn is homeomorphic to a sphere with opposite points identified (that is, points x and $-x$, where x runs over the whole sphere). This last object is homeomorphic to the set of all straight lines in $\mathbb{R}^3$ passing through the origin, since each such line is uniquely determined by the two opposite points where it meets the unit sphere. By definition, this is the projective plane. $\square$

We can represent $\mathbb{RP}^2$ also as $W = abab$. The proof is illustrated in Figure 4.25.

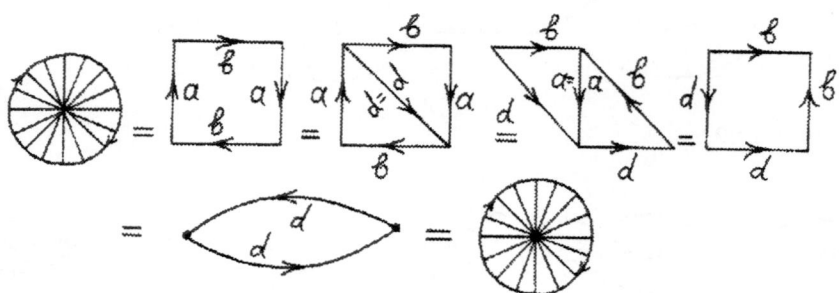

FIGURE 4.25.

We can use this new representation to realize the projective plane as a "surface" in $\mathbb{R}^3$. We use the quotation marks because it is impossible to embed $\mathbb{RP}^2$ in $\mathbb{R}^3$, that is, to realize it as a smooth surface without self-intersections. (Recall Definition 3.4.4.) The model that we suggest now gives a visual image of the projective plane, but is not an embedding or even an immersion. (An immersion will be shown following Lemma 4.4.1.)

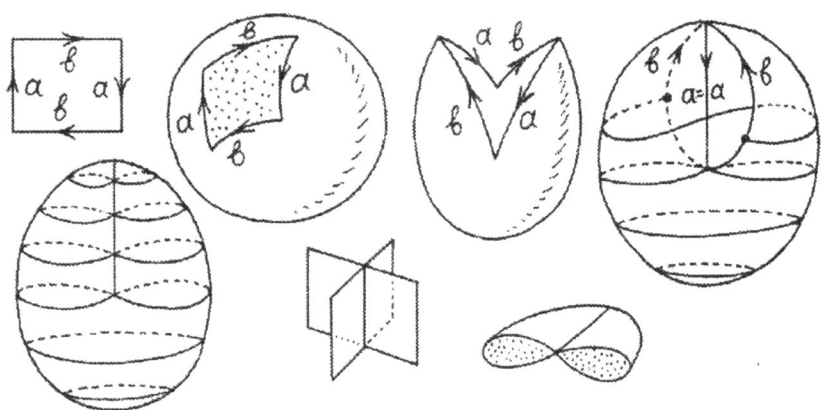

FIGURE 4.26.

The construction is shown in Figure 4.26. The result is a "surface" with a segment of self-intersections, and two singular at the ends of this segment. More exactly, we have a smooth map from $\mathbb{RP}^2$ to $\mathbb{R}^3$ that is regular (an immersion) except at two points—the points that map to the endpoints of the double line. These singular points cannot be eliminated. Figure 4.27 shows the cross-sections of the model by planes orthogonal to the singular segment, as we translate the plane vertically.

Recall that a *Möbius band* or *Möbius strip* μ is the surface with boundary represented in Figure 4.28. Its boundary is a circle.

Lemma 4.2.4. *The projective plane can be obtained by gluing a Möbius strip with a disk, by identification of their boundary circles. In symbols,*

$$\mathbb{RP}^2 = \mu + D^2.$$

The proof is illustrated in Figure 4.29, which decomposes the projective plane into a union of μ and D^2, with boundaries identified. The last step shown is simply the replacement of $a'ba''$ by c.

Here is an equivalent formulation of Lemma 4.2.4. First, remember that a sphere minus a small open disk is homeomorphic to a closed disk. Thus, if we take a sphere, remove an open disk, and glue in its stead a Möbius strip, we

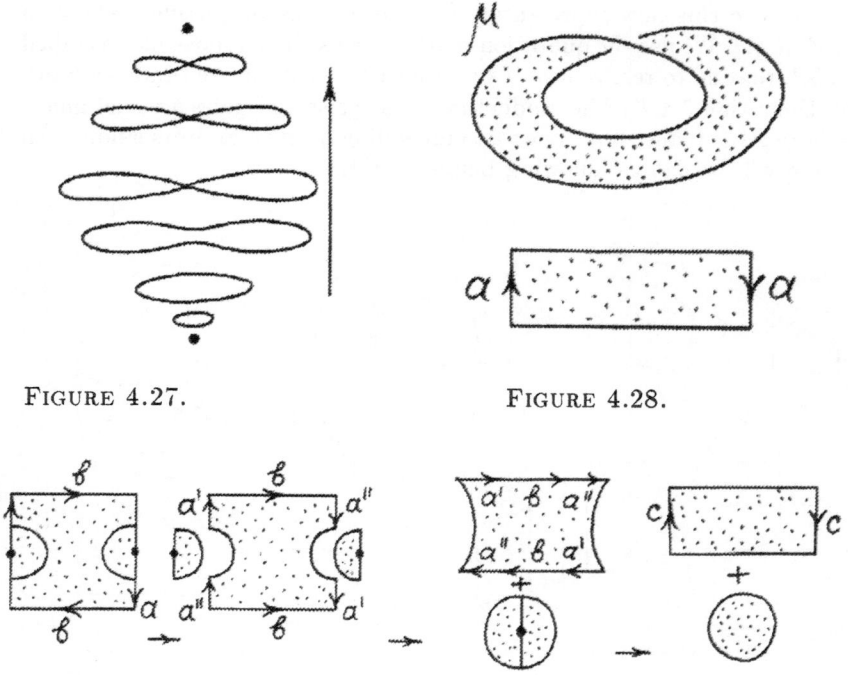

FIGURE 4.27.

FIGURE 4.28.

FIGURE 4.29.

get $\mathbb{RP}^2$. This is shown schematically in Figure 4.30; the wheel with spokes represents the Möbius strip, glued in place by means of some homeomorphism between its boundary circle and the boundary of the removed disk. We cannot actually embed the Möbius strip in three-space so that its boundary becomes a flat circle; if we could, the construction of Figure 4.30 would give us an embedding of the projective plane in $\mathbb{RP}^2$. We know that there is no such embedding.

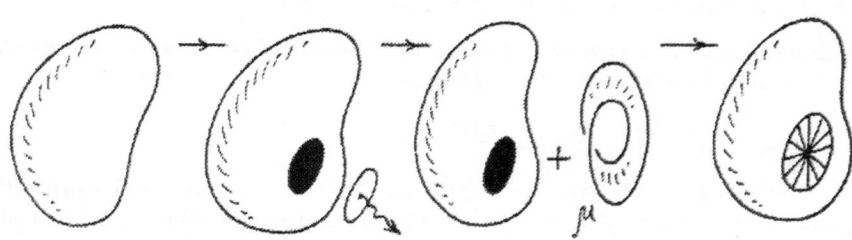

FIGURE 4.30.

Thus. the wheel with spokes in the figure is just a mnemonic for an abstract Möbius strip. The reason we use this diagram it is the following. In the "flat" model for the Möbius strip, shown in Figure 4.28, bottom, we see that the strip can be ruled by vertical line segments, connecting opposite points of the boundary circle. So, if we have a collection of segments joining antipodal points of a circle, in such a way that the segments vary continuously and don't intersect one another, we have a Möbius strip. This is what is conveyed by the wheel with spokes (or rather "nonintersecting diameters").

Now, the usual realization of a Möbius strip in three-space (Figure 4.28, top) is an embedding. but the boundary circle goes twice around the center line of the Möbius strip; in particular, it is not a flat curve. Can we map the Möbius strip into $\mathbb{R}^3$ *in any way* such that the boundary is a flat circle?

The answer follows from the equality $\mathbb{RP}^2 = \mu + D^2$ of Lemma 4.2.4. Consider the model of $\mathbb{RP}^2$ in Figure 4.26, and remove a disk; we should obtain a model of the Möbius band, because $\mathbb{RP}^2 - D^2 = \mu$. The removal of the disk can be conveniently carried out by cutting the model of $\mathbb{RP}^2$ with a horizontal plane. The result is seen in Figure 4.31. This "surface" is called a *crosscap*, and represents a Möbius strip in $\mathbb{R}^3$ with a flat boundary circle. It is not an embedding, or even an immersion!

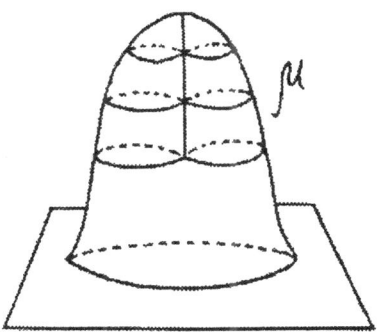

FIGURE 4.31.

Thus. when we simplify the boundary of a Möbius strip, we pay for it by a more complex representation of the Möbius strip itself.

The Klein bottle. Now consider the next fundamental polygon $W = a^2b^2$, from the series (2) of Lemma 4.1.10. It is convenient to represent W in the equivalent form $W = abab^{-1}$, which we can reach by applying Lemma 4.1.6 backwards. The corresponding transformation is shown in Figure 4.32.

Figure 4.33 shows the construction of the surface corresponding to W. The result is called a *Klein bottle*. It is shown here as an immersion in three-

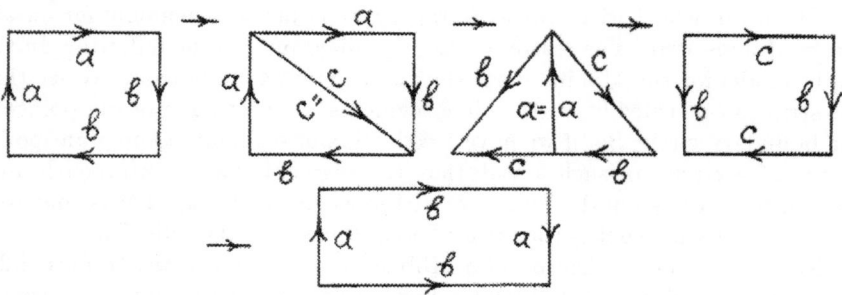

FIGURE 4.32.

space; it is not an embedding because of self-intersections! It can be proved that it is impossible to embed a Klein bottle in three-space.

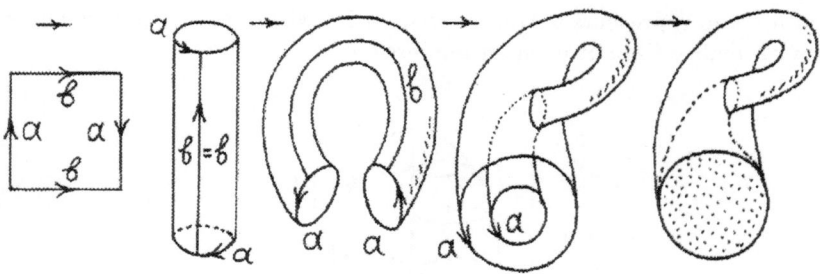

FIGURE 4.33.

Lemma 4.2.5. *The Klein bottle K has the following representations:*

(a) *as the result of gluing two Möbius strips along the boundary circles ($K = \mu + \mu$);*

(b) *as the result of removing two disks from the sphere and substituting Möbius strips or crosscaps;*

(c) *as the connected sum of two projective planes ($K = \mathbb{RP}^2 \# \mathbb{RP}^2$);*

(d) *as a cylinder (or annulus) where, for each of the two boundary circles, we identify opposite points together.*

Proof. The first statement follows from Figure 4.34. The remaining statements are reformulations of the first, because of the arguments demonstrated above for the case of projective plane; see also Figure 4.35. □

Figure 4.36 shows another representation of the Klein bottle, this time as a sphere with a handle turned inside out. Figure 4.37 shows yet another,

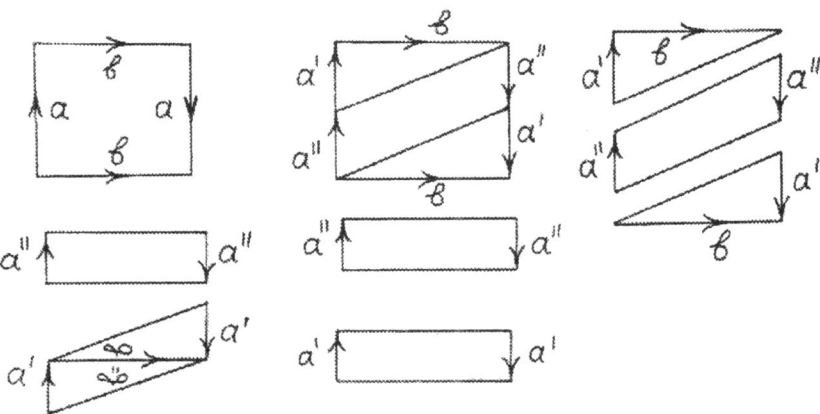

FIGURE 4.34.

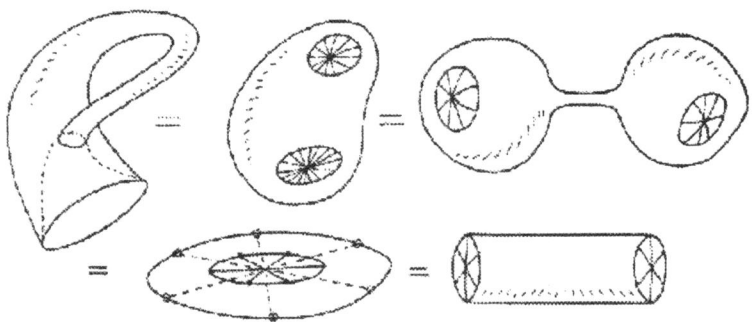

FIGURE 4.35.

based on Möbius bands. We cut the usual immersion of a Klein bottle with its plane of symmetry; it follows from the preceding discussion that we obtain two Möbius bands, immersed in three-space.

Spheres with many crosscaps.   Pursuing further the analysis that led to the expression of the projective plane and the Klein bottle as a sphere with one and two crosscaps. we obtain the following result (Figure 4.38).

Lemma 4.2.6. *The surface given by the canonical fundamental polygon*

$$W = c_1 c_1 c_2 c_2 \ldots c_k c_k$$

can be obtained by removing k disks from a sphere and replacing each with a Möbius band. or crosscap. It can also be obtained as the connected sum of k

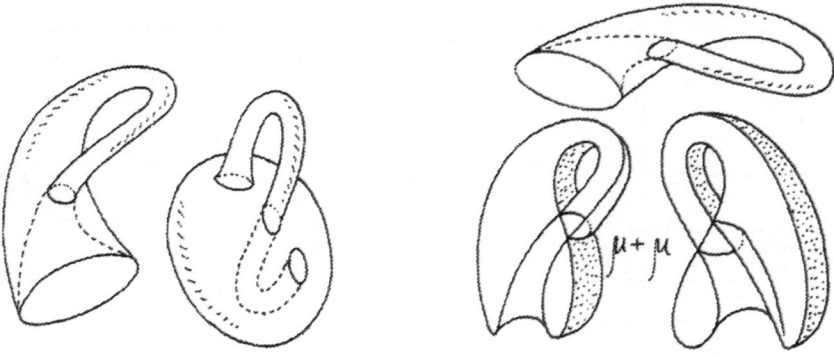

FIGURE 4.36.   FIGURE 4.37.

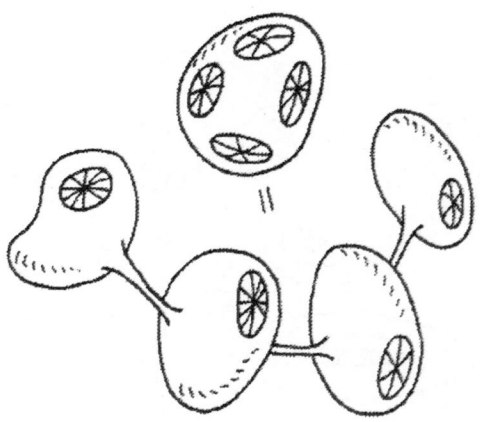

FIGURE 4.38.

copies of the projective plane $\mathbb{RP}^2$:

$$\text{sphere with } k \text{ crosscaps} = \mathbb{RP}^2 \# \cdots \# \mathbb{RP}^2 \quad (k \text{ copies}).$$

Just as in the case of Lemma 4.2.1, it can be proved that k is determined by the homeomorphism type of the surface; that is, a sphere with k crosscaps cannot be homeomorphic to one with h crosscaps, for $k \neq h$. The number of crosscaps is also called the *genus* of the surface. Moreover a surface with crosscaps cannot be homeomorphic to a surface of the first type; this will be proved in Lemma 4.3.2.

The classification theorem. All this put together implies the following result, the centerpiece of this section:

Theorem 4.2.7 (Classification Theorem for Surfaces).

(a) *Any smooth, compact, connected surface (without boundary) is homeomorphic (and diffeomorphic) either to a sphere with g handles, where $g \geq 0$, or to a sphere with k crosscaps. where $k > 0$.*

(b) *A surface of the first type cannot be homeomorphic to one of the second type, nor can surfaces with different values of g or k be homeomorphic.*

Remark. There exist other convenient representations of surfaces. For example, it can be shown using arguments much the same as the ones above that any smooth. compact, connected surface (without boundary) can be represented by a polygon of the form

$$W = a_1 a_2 \ldots a_{N-1} a_N a_1^{-1} a_2^{-1} \ldots a_{N-1}^{-1} a_N^{\varepsilon},$$

where $\varepsilon = -1$ if M is a sphere with handles (in which case N is even), and $\varepsilon = +1$ if M is a sphere with crosscaps (in which case N is arbitrary).

Transforming a handle and a crosscap into three crosscaps. We showed in Lemma 4.1.9 that a surface represented by a polygon containing both commutators and squares is also represented by one containing squares alone. The geometric interpretation of this fact is that a handle and a crosscap can combine to form three crosscaps. We now give a visual explanation of this fact, considering the simplest such surface: a sphere with one handle and one crosscap. We recall that a crosscap can be thought of as a Möbius strip plugged in place of a disk removed from the sphere.

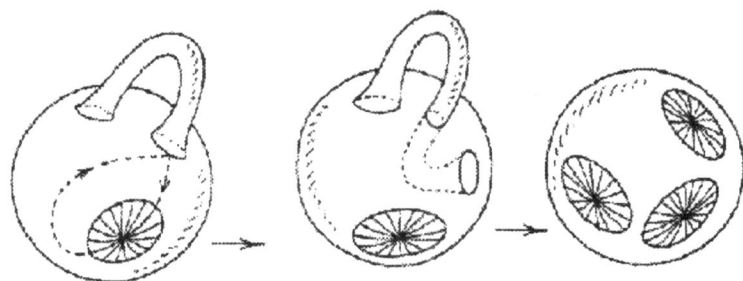

FIGURE 4.39.

Fix one end of the handle and move the other towards the Möbius strip, once around the center loop of the Möbius strip, and back. This is the first step in Figure 4.39, and it is shown in more detail in Figures 4.40 and 4.41. After this motion the handle is no longer outside the sphere: it is a "handle turned inside out" and differs from the ordinary position by the way in which the ends are glued to the sphere. We already know how a sphere with a handle turned inside out can be changed into a Klein bottle (Figure 4.36), and hence into a sphere with two Möbius strips or crosscaps (Figure 4.35). The same sequence of deformations takes us from the sphere with an inside-out handle and a crosscap to a sphere with three crosscaps; this is the second step in Figure 4.39.

4.3. Orientability

Consider a smooth connected surface M, and fix a point P on it. Consider an arbitrary *loop* γ on the surface, that is, a path $\gamma(t)$ that starts at P when $t = 0$ and returns to P when $t = 1$ (Figure 4.42). Consider also a basis (e_1, e_2) of the tangent plane $T_P M$ to the surface at P. Finally, consider any smooth *transport* of this basis along γ, that is, a deformation of this basis along γ in such a way that these vectors remain linearly independent and tangent to the surface throughout (see Figure 4.42, bottom). Formally, a smooth transport is a pair of vector-valued functions (v_1, v_2) defined on the unit interval $[0, 1]$, satisfying these conditions: $v_1(t)$ and $v_2(t)$ form a basis of $T_{\gamma(t)} M$ for each t; we have $e_1 = v_1(0)$ and $e_2 = v_2(0)$; and $v_1(t), v_2(t)$ vary smoothly with t.

When we return to P, we compare the orientation of the new basis $(e_1', e_2') = (v_1(1), v_2(1))$ at P with that of the initial basis (e_1, e_2). There are two possibilities: either the two bases have the same orientation, or they have opposite orientations.

Note that it doesn't make sense to ask whether a basis at some point $Q \neq P$ has the same orientation as the initial basis, because the vector spaces $T_P M$ and $T_Q M$ are different, and cannot be identified in any standard way. It is only when we get back to P that we can compare the orientation of the final basis with that of the initial one.

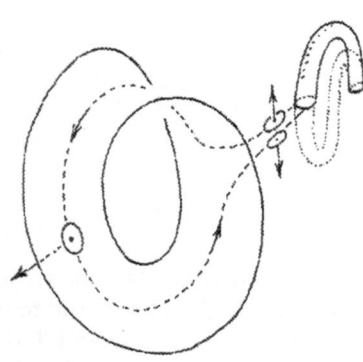

FIGURE 4.40.

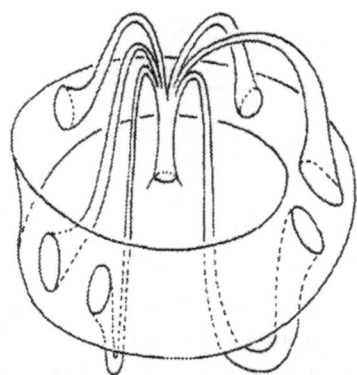

FIGURE 4.41.

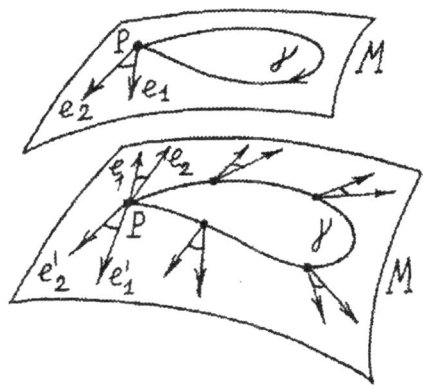

FIGURE 4.42.

All smooth (or even continuous) transports along γ must behave in the same way regarding whether or not the initial and final bases have the same orientation. The reason is the following. Suppose first that the two transports are such that the initial bases have the same orientation. Then it is possible to deform one into the other through a continuous family of transports (in technical terms, there is a homotopy between the two transports). In particular, the final basis of the first transport can be deformed into the final basis of the second. But if a basis *at a fixed point* varies continuously, its orientation cannot change; therefore the final bases of the two transports have the same orientation, like the initial bases. If we suppose instead that the two given transports have initial bases with opposite orientations, we reduce to the preceding case by switching the order of the vectors of one of the transports, thus reversing the orientation of all its bases.

We thus see that *reversing or preserving orientation under transport is a property of the loop γ itself.* Moreover, *this property cannot change if the loop varies continuously;* this is intuitively plausible, and a proof can be given along the following lines. We look at a small piece of γ at a time, so that we can pretend (using coordinate charts) that the surface is part of $\mathbb{R}^2$. In $\mathbb{R}^2$ it does make sense to compare bases at different points, since the tangent planes are canonically identified by translation. A basis that varies continuously in $\mathbb{R}^2$ cannot change orientation; therefore the orientation of the basis $(v_1(t), v_2(t))$ at the point $\gamma(t)$ does not change as the loop and the transport are deformed continuously

Definition 4.3.1. A surface M is called *orientable* if transport of bases along any loop γ in M preserves orientation. Conversely, M is called *nonorientable* if there exists some loop γ on M such that transport along γ reverses orientation.

Lemma 4.3.2. *Spheres with handles are orientable. Spheres with crosscaps are nonorientable.*

Proof. Consider the fundamental polygon for a sphere with handles. and a loop γ. If γ does not intersect the boundary of W, we are in $\mathbb{R}^2$, and by moving each vector parallel to itself we get a transport where the final basis coincides with the initial basis, so orientation is preserved. If γ does intersect the boundary of W, we can assume, after a small continuous deformation if necessary. that γ avoids the vertices of W and that all intersections with edges are transverse. Fixing a point in the interior of W and again changing γ continuously, we can assume that γ is composed of one or more loops based at P, each of which goes out of and back into W exactly once. For a loop of this sort we are in the situation of Figure 4.43, right. Parallel transport along γ brings the basis from the starting point to one edge a; the matching map between the two edges a preserves the orientation of the basis; and parallel transport again brings the basis back to P, with the same orientation.

The case of a sphere with Möbius strips attached is even easier: We only need to show that an orientation-reversing loop exists. Take any loop γ that crosses the boundary of the canonical fundamental polygon only once, in the interior of an edge a. Parallel transport along γ brings the basis from the starting point to a; the matching map between the two edges a this time *reverses* the orientation of the basis; and parallel transport again brings the basis back to P, with orientation opposite the initial one. Figure 4.44 illustrates the case of a Möbius strip. □

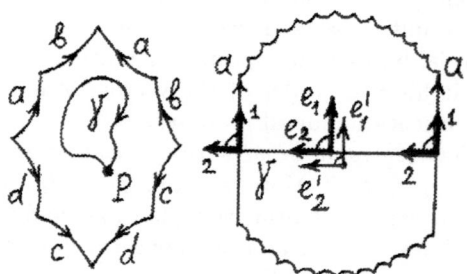

FIGURE 4.43.

It is clear that two diffeomorphic surfaces are either both orientable or both nonorientable. Thus, Lemma 4.3.2 and Theorem 4.1.1 together imply that a sphere with handles cannot be diffeomorphic to a sphere with crosscaps. as announced before Theorem 4.2.7.

The notion of orientability extends easily beyond two dimensions, to smooth n-dimensional manifolds.

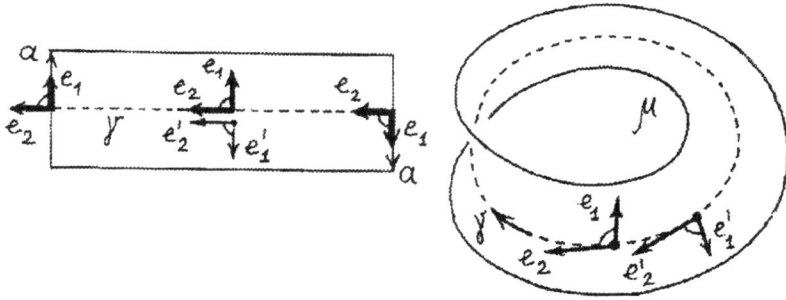

FIGURE 4.44.

4.4. Embedding and Immersing Surfaces in Three-Space

Lemma 4.4.1. *Any orientable surface can be embedded in $\mathbb{R}^3$. Any non-orientable surface can be immersed in $\mathbb{R}^3$.*

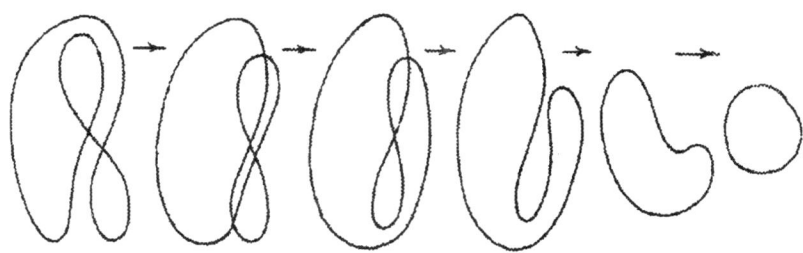

FIGURE 4.45.

Proof. Spheres with handles can be embedded in in $\mathbb{R}^3$ as in Figure 4.23, for example. As to nonorientable surfaces, they are connected sums of projective planes, so it is sufficient to prove that $\mathbb{RP}^2$ can be immersed in $\mathbb{R}^3$.

Consider the immersion of a Möbius strip in $\mathbb{R}^3$ as half of a Klein bottle, as shown in Figure 4.37, bottom. If we can attach a disk to the boundary of this Möbius strip while maintaining the immersion property, we will have an immersion $\mathbb{RP}^2 \to \mathbb{R}^3$.

To do this, imagine the boundary as a curve drawn on a horizontal plane, and move the plane up while at the same time deforming the curve as in Figure 4.45. The surface formed by this stack of closed curves is the immersion of a cylinder. The bottom edge matches the boundary of the Möbius strip, and the top edge is a small circle, which we can cap off with a standard disk, thus obtaining the immersed disk to be glued to the Möbius strip.   □

This immersion of $\mathbb{RP}^2$ in $\mathbb{R}^3$ can be put into a more symmetric position, shown in Figure 4.46. In this form it is called *Boy's surface*.

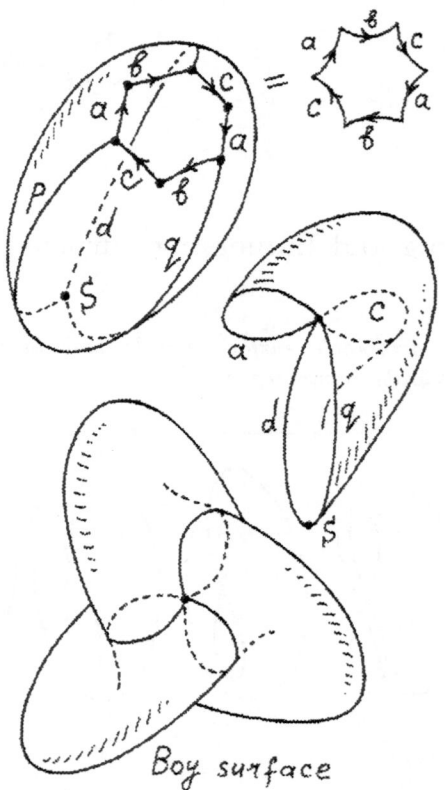

Boy surface

FIGURE 4.46.

5. Abstract Manifolds

In Chapter 3 we defined manifolds in $\mathbb{R}^n$ (see Definition 3.3.1), and we remarked that a more abstract definition is possible. so that a manifold does not have to exist in any $\mathbb{R}^n$. We introduce this definition in this chapter, and also a technical tool useful in the study of manifolds, called *partitions of unity*. The chapter closes with an application to the generation of topographic maps.

5.1. Definition of an Abstract Manifold

As we have seen, *coordinate systems* describing the position of a point in space are an important tool in the study of geometrical objects. Using coordinate systems, we can apply the methods of differential and integral calculus.

Even an object as simple as a circle or a sphere, however, cannot be described by a single coordinate system, which is why we introduced in Chapter 3 the notion of local coordinates. Thus, in cartography, the sphere (as the surface of the Earth) is reconstructed from several local charts, or maps (open disks). The domain of each chart can be projected separately onto part of the plane. The whole sphere can then be "constructed" from these maps by gluing them together according to certain rules. Thus, the sphere is obtained from several simpler objects by gluing them together along their common parts. This notion of gluing simple objects together was refined over the last decades of the nineteenth century and the first decades of the twentieth, and eventually led to the very general concept that we discuss here.

Before stating the definition. we recall that a topological space X is *Hausdorff* if it has the following property: if $x \in X$ and $y \in X$ are distinct points, there exist disjoint open sets U_x and U_y containing x and y, respectively. Most topological spaces arising from geometric constructions are Hausdorff: in particular. all metric spaces (sets with an appropriate notion of distance between any two points) are Hausdorff topological spaces. In fact, readers not very familiar with topology may substitute "metric space" for "topological space" in the following discussion.

Definition 5.1.1. A Hausdorff topological space M is called an *n-dimensional topological manifold* (or simply *topological n-manifold* or *n-manifold*) if any point of M has a neighborhood U in M homeomorphic to an open disk D^n in $\mathbb{R}^n$, and a countable collection of such neighborhoods covers M.

This definition can be restated in brief as follows: an n-dimensional manifold M^n is locally homeomorphic to an open disk D^n. Thus, if M is an n-dimensional manifold, we can find in M a countable system of open sets U_i and homeomorphisms

$$\varphi_i : U_i \to D^n$$

from U_i onto the disk D^n (Figure 5.1). The domains U_i must together cover the whole space M (that is, $M = \bigcup U_i$); they may intersect one another. Each open set U_i, together with the given homeomorphism $\varphi_i : U_i \to D^n$, is called a *chart* (in full, *coordinate chart*). Sometimes either U_i or φ_i is also referred to as a chart. A collection of charts defining the manifold M is called an *atlas for M* (or *on M*).

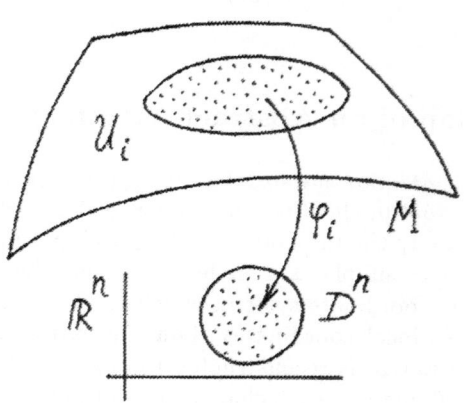

FIGURE 5.1.

Consider an atlas on a manifold M, and fix two charts (U_i, φ_i) and (U_j, φ_j), where $i \neq j$. Let $U_i \cap U_j$ be the intersection of U_i and U_j (Figure 5.2). Consider two copies D^n, one with coordinates $(x^1, \ldots, x^n)$ and the other with coordinates $(y^1, \ldots, y^n)$, corresponding to the image of the charts φ_i and φ_j (Figure 5.2).

Definition 5.1.2. The map $\varphi_{ij} : \varphi_i(U_i \cap U_j) \to \varphi_j(U_i \cap U_j)$ defined by

$$\varphi_{ij} = \varphi_j \varphi_i^{-1}$$

is called the *transition map*, or *gluing map*, associated with the charts (U_i, φ_i) and (U_j, φ_j).

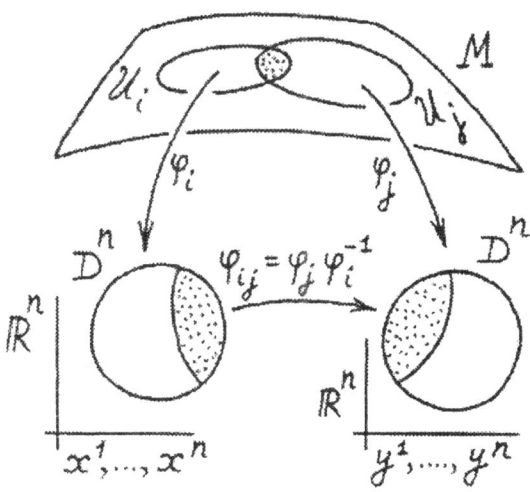

FIGURE 5.2.

Indeed, we are gluing the two local charts U_i and U_j along their common part by using the function φ_{ij}. It is clear that all these functions φ_{ij} are homeomorphisms.

The map φ_{ij} can be written in terms of the coordinates x^i and y^i as follows:

$$y^1 = y_{ij}^1(x^1, \ldots, x^n), \quad \ldots \quad y^n = y_{ij}^n(x^1, \ldots, x^n).$$

Of course, these functions will be different for different pairs of local charts U_i and U_j. This is why we use the pair of indices ij.

Definition 5.1.3. A topological manifold is called a *smooth manifold* if all transition maps φ_{ij} are smooth. In an analogous way we can define *analytic manifolds*, and so on.

We now compare the class of objects introduced in Definition 5.1.3 with the old manifolds of Chapter 3 (manifolds embedded in $\mathbb{R}^n$). It is obvious that any manifold according to the old definition is also a manifold according to Definition 5.1.3. In a sense, the converse is also true for smooth manifolds:

Theorem 5.1.4 (Whitney embedding theorem). *Any smooth manifold M^k can be realized as a smooth k-dimensional submanifold of $\mathbb{R}^{2k+1}$ (that is, as a manifold in the sense of Chapter 3).*

This is why we prefer most of the time to work with submanifolds of $\mathbb{R}^n$: We can get essentially the same objects, and the realization in $\mathbb{R}^n$ often makes it easier to work with the manifold because of the ambient Euclidean geometry. Also, many manifolds are naturally defined as subsets of some $\mathbb{R}^n$

anyway. In other cases it is more natural to define the manifold abstractly, by means of coordinate charts, and then an embedding into some $\mathbb{R}^n$ may not be easy to describe explicitly, although it must exist by Whitney's theorem. In these cases the abstract point of view is useful.

Definition 5.1.5. A Hausdorff topological space M is called an n-*dimensional topological manifold with boundary* (or simply n-*manifold with boundary*) if any point of M has a neighborhood U in M homeomorphic to an open subset of the closed halfspace

$$\mathbb{R}^{n-1} \times \{x_n \in \mathbb{R} : x_n \geq 0\} \subset \mathbb{R}^n,$$

and a countable collection of such neighborhoods covers M. The *boundary* of M. denoted ∂M, is the subset of M consisting of points for which the neighborhood U *cannot* be chosen as an open subset of the open halfspace

$$\mathbb{R}^{n-1} \times \{x_n \in \mathbb{R} : x_n < 0\}.$$

A topological manifold with boundary is called a *smooth manifold with boundary* if it has an atlas such that all transition maps φ_{ij} are smooth.

Thus, manifolds are particular cases of manifolds with boundary (the boundary being empty), but a manifold with boundary is not a manifold if the boundary is nonempty. If M is a manifold with boundary, $M - \partial M$ is a manifold. sometimes called the *interior* of M.

A compact manifold is often called closed, in the sense that it has no boundary and no "open edge"; but this terminology can lead to confusion and we will avoid it.

5.2. Partitions of Unity

This section is devoted to the concept of partitions of unity, which provide a technique for the approximation of continuous functions and the further development of manifold theory.

Definition 5.2.1. Let f be a function $f : M \to \mathbb{R}$. The *support* of f, denoted supp f, is the closure of the set $\{x : f(x) \neq 0\}$, that is, the smallest closed set outside of which f vanishes.

Lemma 5.2.2. *Let A and B be disjoint closed subsets of $\mathbb{R}^n$, with A bounded. There exists a smooth function φ on $\mathbb{R}^n$ taking the value 1 on A. the value 0 on B, and values $0 \leq \varphi(x) \leq 1$ elsewhere.*

Note that, because B is closed, the condition $\varphi = 0$ on B can also be written $B \cap \operatorname{supp} \varphi = \varnothing$.

Proof. We start by defining a function $f : \mathbb{R} \to \mathbb{R}$ (for fixed $a, b \in \mathbb{R}$) by the equation

$$f(x) = \begin{cases} e^{-1/(x-a)}\, e^{-1/(b-x)} & \text{if } a < x < b, \\ 0 & \text{otherwise.} \end{cases}$$

The graph of f is shown in Figure 5.3, and it is easy to check that f is smooth.

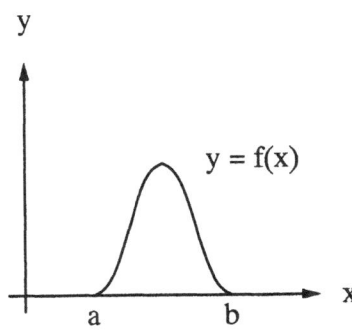

FIGURE 5.3.   FIGURE 5.4.

Next we define $F : \mathbb{R} \to \mathbb{R}$ by

$$F(x) = \frac{\displaystyle\int_x^b f(t)\, dt}{\displaystyle\int_a^b f(t)\, dt}.$$

It can be seen immediately that $F(x) = 0$ for $x \geq b$, that $F(x) = 1$ for $x \leq a$, and that $F(x)$ decreases monotonically from 1 to 0 for $a \leq x \leq b$. See Figure 5.4.

Now define a spherically symmetric function $\psi : \mathbb{R}^n \to \mathbb{R}$ that radially behaves like F:

$$\psi(x^1, \ldots, x^n) = F(r^2), \qquad \text{where } r^2 = (x^1)^2 + \cdots + (x^n)^2.$$

Clearly, $\psi(x) = 0$ for $r^2 \geq b$, $\psi(x) = 1$ for $r^2 \leq a$, and $\psi(x)$ decreases monotonically from 1 to 0 for $a \leq r^2 \leq b$. See Figure 5.5.

We know that A is compact, B is closed, and $A \cap B = \varnothing$. A simple argument involving the compactness property of A (every open cover of a compact set has a finite subcover) and the Hausdorff property of $\mathbb{R}^n$ shows that we can take finitely many open balls $D_1, \ldots, D_m$ such that $A \subset \bigcup_{i=1}^m D_i$ and $\bar{D}_i \cap B = \varnothing$ for all i, where $\bar{D}_i$ denotes the closure of D_i. Again because A is compact, we can find smaller balls D_i' concentric with D_i and such that $\bigcup_i^{m=1} D_i'$ still contains A. Now let ψ_i be the function ψ defined earlier, with

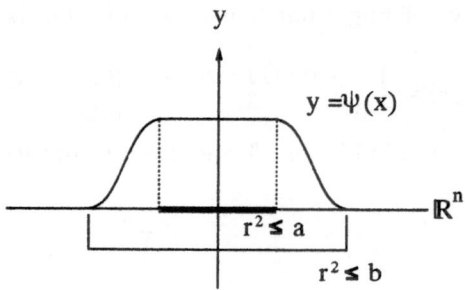

FIGURE 5.5.

a and b taken as the squared radii of D_i' and D_i, and translated so the center of symmetry is the center of D_i. In other words, $\psi_i(x)$ equals 1 inside D_i', it equals 0 outside D_i, and it satisfies $0 \leq \psi_i \leq 1$ everywhere.

The desired function $\varphi : \mathbb{R}^n \to \mathbb{R}$ can then be taken as

$$\varphi(x) = 1 - \prod_{i=1}^{m}(1 - \psi_i(x)).$$

It is clear that φ is smooth, that it takes the value 1 wherever some ψ_i does, that it takes the value 0 wherever all ψ_i do, and that $0 \leq \varphi(x) \leq 1$ everywhere. □

Lemma 5.2.3. *Let C be a compact subset of a smooth manifold M and let V be an open subset of M containing C. There exists a smooth function $\varphi : M \to \mathbb{R}$ such that $0 \leq \varphi(x) \leq 1$ everywhere, $\varphi(x) = 1$ on C, and $\varphi(x) = 0$ outside V.*

Proof. Let $(U_\alpha, \varphi_\alpha)$ be a chart of M, and let S_α be a compact subset of U_α. Since $\varphi(U_\alpha)$ is open in $\mathbb{R}^n$ and $\varphi(S_\alpha)$ is compact in $\mathbb{R}^n$, Lemma 5.2.2 says that there exists a smooth function f_α on $\mathbb{R}^n$ such that $f(x) = 1$ on $\varphi_\alpha(S_\alpha)$ and supp $f_\alpha \subset \varphi_\alpha(U_\alpha)$. Pull f_α back to a function F_α on M, that is, set

$$F_\alpha(P) = \begin{cases} f_\alpha(\varphi_\alpha(P)) & \text{for } P \in U_\alpha, \\ 0 & \text{for } P \notin U_\alpha. \end{cases}$$

One can see easily that F_α is smooth, $F_\alpha(P) = 1$ on S_α, and $F_\alpha(P) = 0$ outside U_α.

Now we turn to our attention to the compact set $C \subset V$. Since C is compact, there exists open sets $U_1, \ldots, U_m$ and compact sets $S_1, \ldots, S_m$ such that $S_\alpha \subset U_\alpha$. $C \subset \bigcup_{\alpha=1}^{m} S_\alpha$, and $\bigcup_{\alpha=1}^{m} S_\alpha \subset V$. For each $\alpha = 1, \ldots, m$ we can find a function F_α such that F_α is smooth, $F_\alpha = 1$ in S_α, and $F_\alpha = 0$ outside U_α. Hence the function $F = 1 - \prod_{\alpha=1}^{m}(1 - F_\alpha)$ is smooth, it equals 1 on C. and it vanishes outside $\bigcup_{\alpha=1}^{m} U_\alpha$, as desired. □

Definition 5.2.4. Let M be a smooth manifold, and let $\{U_\alpha\}_{1 \le \alpha \le N}$ be a finite family of open sets covering M. A family of functions $\varphi_\alpha : M \to \mathbb{R}$ is a *partition of unity subordinate to the cover* $\{U_\alpha\}$ if it satisfies the following conditions:

(1) supp $\varphi_\alpha \subset U_\alpha$ for all α;
(2) $0 \le \varphi_\alpha(x) \le 1$ for all $x \in M$ and all α; and
(3) $\sum_\alpha \varphi_\alpha(x) = 1$ for all $x \in M$.

Lemma 5.2.5 (Existence of a partition of unity). *If M is a compact smooth manifold and $\{U_\alpha\}_{1 \le \alpha \le m}$ is a finite cover of M, there exists a partition of unity subordinate to $\{U_\alpha\}$.*

Proof. There exist open sets V_α, for $1 \le \alpha \le m$, such that $\bar{V}_\alpha \subset U_\alpha$ and $\{V_\alpha\}$ also covers M. By applying Lemma 5.2.3 to each pair $(\bar{V}_\alpha, U_\alpha)$, we can find smooth functions ψ_α such that $0 \le \psi_\alpha(x) \le 1$ everywhere, $\psi_\alpha(x) = 1$ on $\bar{V}_\alpha$, and $\psi_\alpha(x) = 0$ outside U_α. The function $\psi = \sum_{\alpha=1}^{m} \psi_\alpha$ is smooth and positive on M. If we set $\varphi_\alpha = \psi_\alpha / \psi$, we have the desired partition of unity. $\square$

5.3. An Application to Visual Geometry: Mountain Guide Maps

Many ancient and medieval paintings were drawn having multiple viewpoints, each valid for one or more objects of interest. After the Renaissance, when the perspective view became dominant as an exact and hence scientific way of drawing, multiple-viewpoint pictures have declined, and have survived only in limited cases such as mountain guide maps, certain medical drawings, and in some schools of art, such as cubism.

When we try to understand how machines are configured, we draw them as seen from different sides. In human visual cognition, multiple-viewpoint pictures are natural and there is no reason for them to be rejected. One illustrative example is our visual memory: Most of us remember our homeland as we saw it when traveling around it, and therefore from many viewpoints.

This section is a step toward the science of multiple-viewpoint pictures. We conducted a case study to test the hypothesis that there is a way to model multiple-viewpoint pictures exactly ("exactly" in the sense that we can define the pictures without ambiguity, and hence can generate them automatically). We chose the popular case of mountain guide maps.

In a mountain guide map, mountain peaks, mountain passes, and lakes are highlighted to characterize land undulations. To represent such land features clearly, the map is often drawn with multiple viewpoints. For example, consider the difference between an ordinary perspective picture and a mountain guide map, illustrated in Figure 5.6. Part (a) of the figure shows an ordinary perspective picture. The lake is partially hidden by surrounding mountains, while the mountain skyline is seen clearly. Part (b) shows a mountain guide map where the viewpoint of the area containing the lake is changed, so that we can see the whole scene of the lake from a height as well as the mountain skyline as seen from the foot of the mountain. In this way, a mountain guide map can extract as much information as possible when projecting a three-dimensional land surface onto the plane.

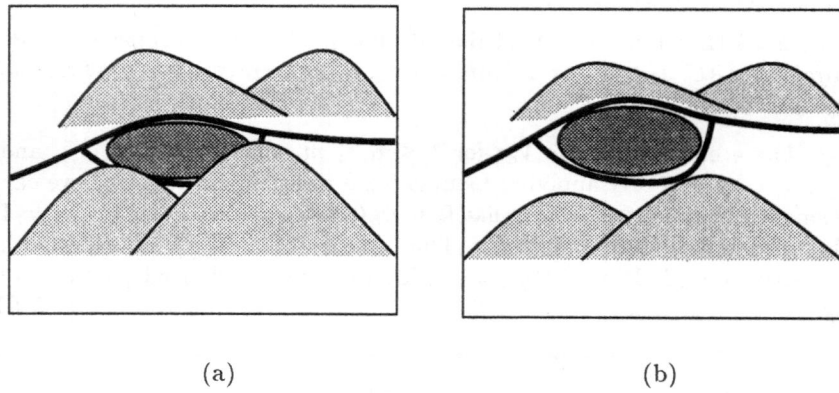

(a)   (b)

FIGURE 5.6.

So far, commercially available mountain guide maps have been drawn without appropriate modeling, and hence include ambiguous representations. Modeling the process of drawing a mountain guide map requires an understanding of the steps involved. The typical steps are these:

− Select areas that include one peak, pass, or lake each.
− Combine the areas together to construct the entire land surface.
− Project each area as seen from a viewpoint, usually a vista point of the area.

We call peaks, passes and lakes *characteristic points*, and the area including one characteristic point is called a *characteristic area*. Figure 5.7 illustrates the steps just described. In fact, the steps parallel the construction of a manifold.

We present here some visual examples of images for mountain guide maps, generated using the concept of manifolds [Kunii and Takahashi 1993; Takahashi and Kunii 1994]. The method of projecting a land surface as seen from multiple viewpoints is realized by blending the viewpoints using a partition of unity.

Figures 5.8 and 5.9 are the basic images for the mountain guide maps around Lake Ashinoko, a famous tourist area with a scenic crater lake in Japan. Here we see the influences on the images when the viewpoint or view direction of the area including the lake is changed. Figure 5.8(a) is the

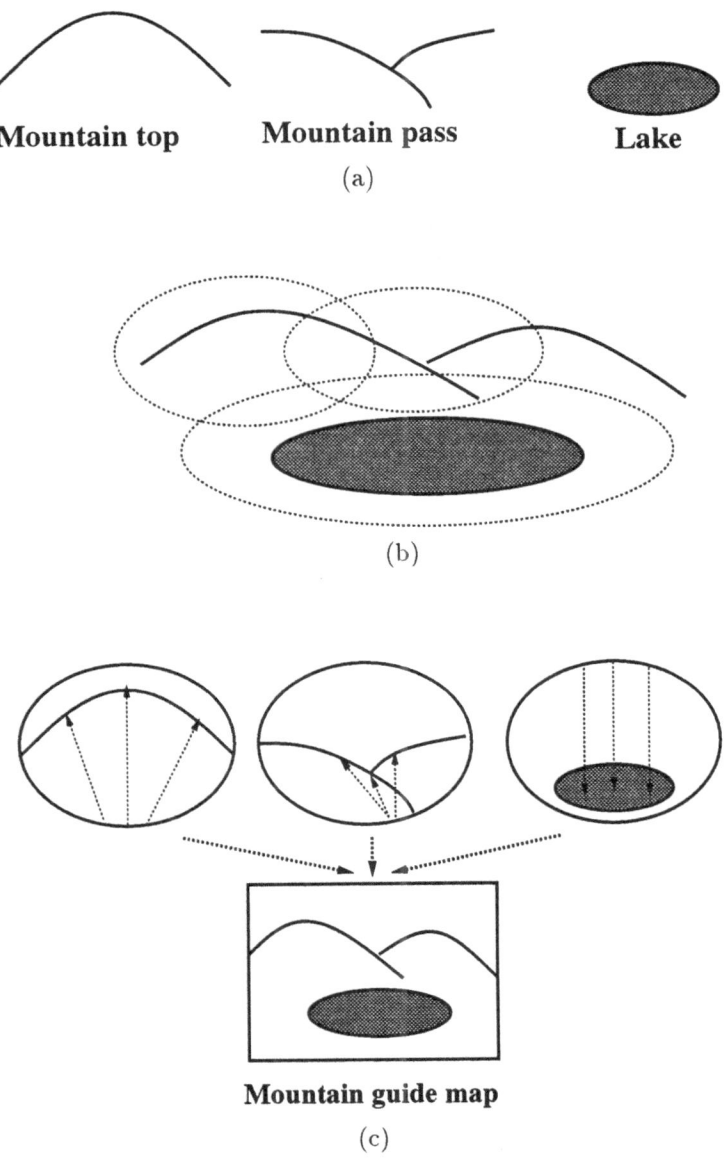

Mountain top   **Mountain pass**   **Lake**

(a)

(b)

Mountain guide map

(c)

FIGURE 5.7.

image under perspective with one viewpoint, that is, ordinary perspective. Figure 5.8(b) is the image under perspective with multiple viewpoints. The viewpoint of the area including the lake is raised to avoid the surrounding mountains. Figure 5.9(a) is the image under parallel projection with one view direction, that is, ordinary parallel projection. Figure 5.9(b) is the image under the parallel projection with multiple view directions. The whole lake is visible because it and the surrounding area are seen from a view direction different from the one used for other areas.

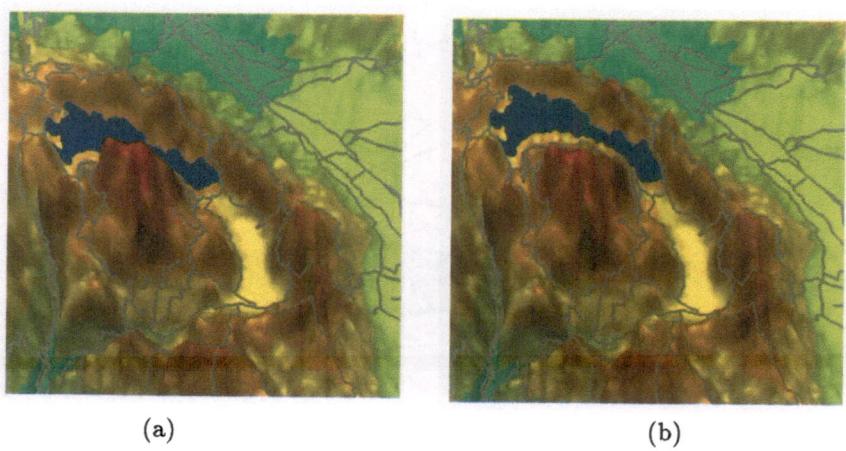

(a)   (b)

FIGURE 5.8.

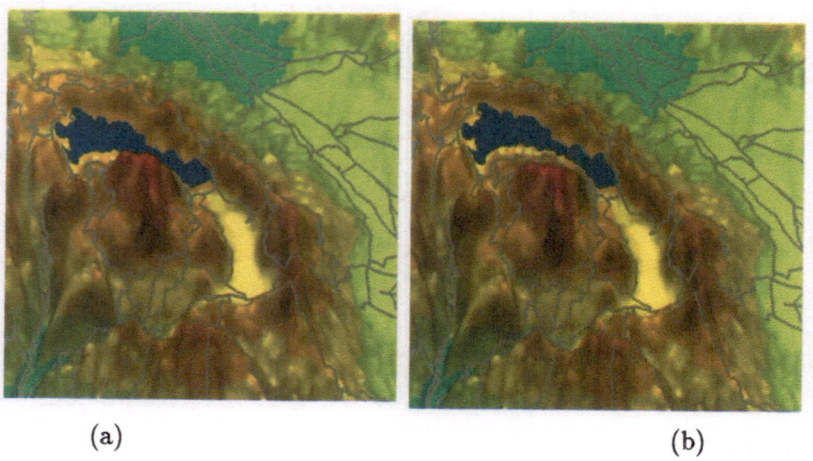

(a)   (b)

FIGURE 5.9.

6. Critical Points and Morse Theory

In this chapter we introduce Morse theory, a systematic way of studying certain features of smooth functions on manifolds. We will primarily consider surfaces and three-manifolds, because the main applications of Morse theory in computer geometry are concentrated in these dimensions.

6.1. Critical Points and Morse Functions

Consider a smooth function f on a smooth manifold M of dimension n, with local coordinates $(x^1, \ldots, x^n)$. We will be interested in the *critical points* of f; we recall from Chapter 3 what they are.

The differential df_x of $f : M \to \mathbb{R}$ at x (Definition 3.4.1) is a linear map from the tangent space $T_x M$ to the tangent space $T_{f(x)}\mathbb{R}$; this latter space can be identified with $\mathbb{R}$. By Definition 3.4.2, saying that x is a critical point of f is the same as saying that df_x is not surjective, and since $\mathbb{R}$ is one-dimensional this is the same as saying that $d_x f$ is the zero map.

In coordinates, this has the following translation: $x \in M$ is a critical point of f if all partial derivatives $\partial f / \partial x^i = f_{x^i}$ vanish at x.

Also according to Definition 3.4.2, the image of a critical point is a *critical value*.

When x is a critical point of f, it is possible to regard the second differential of f as a *symmetric bilinear form* on $T_x M$. Indeed, let

$$d^2 f = \left(\frac{\partial^2 f(x_0)}{\partial x^i \partial x^j} \right) = (f_{x^i x^j})$$

be the matrix of second partial derivatives of f. Then, under a regular coordinate transformation $x^i = x^i(y^1, \ldots, y^n)$, where $1 \le i \le n$, the matrix $d^2 f$ transforms as follows, *if all the first derivatives f_{x_i} vanish:*

6.1.1.
$$f_{x^i x^j} = \sum_{p,q} \frac{\partial y^p}{\partial x^i} \frac{\partial y^q}{\partial x^j} f_{y^p y^q}.$$

That is, $d^2 f$ transforms as the matrix of a quadratic form. This can also be expressed in coordinate-invariant language by regarding $d^2 f$ as a symmetric bilinear form $d^2 f : T_x M \times T_x M \rightarrow \mathbb{R}$. Explicitly, this bilinear form, when applied to tangent vectors a and b, yields

$$d^2 f(a, b) = \sum_{i,j} a^i f_{x^i x^j} b^j,$$

where $(a^1, \ldots, a^n)$ and $(b^1, \ldots, b^n)$ are the expressions of a and b in the coordinate system $(x^1, \ldots, x^n)$. When x is not a critical point, by contrast, Equation 6.1.1 does not hold, and we cannot regard the matrix $d^2 f$ as an invariant bilinear form.

Let B be a symmetric bilinear form on a finite-dimensional vector space V, and let Q be the corresponding quadratic form. The *index* of B (or of Q) is the dimension of a maximal linear subspace W where B is *negative definite* (that is, such that $B(w, w) = Q(w) < 0$ for all nonzero $w \in W$). If B is brought to diagonal form (recall that every symmetric bilinear form can be diagonalized by a suitable change of coordinates), we see immediately that the index equals the number of negative eigenvalues. The *nullity* of B (or of Q) is the dimension of the nullspace of B (that is, the set of $w \in V$ such that $B(v, w) = 0$ for all $v \in V$). The nullity equals the multiplicity of the zero eigenvalue. A form is *degenerate* or *singular* if its nullity is greater than zero, that is, if its determinant vanishes.

Definition 6.1.2. A critical point x of f is called *nondegenerate* if $d^2 f$ is nondegenerate at that point. This is equivalent to the condition $\det d^2 f \neq 0$ at x. The *index* of x is the index of $d^2 f$ at x. The *nullity* of x is the nullity of $d^2 f$ at x.

By the preceding discussion, these definitions do not depend on the choice of a local coordinate system. In this book we will deal mostly with nondegenerate critical points.

Definition 6.1.3. A smooth function on a smooth manifold is called a *Morse function* if all its critical points are nondegenerate.

It can be proved using Sard's theorem (Theorem 9.2.1) that Morse functions exist on any smooth manifold. In fact, any smooth function on a smooth manifold can be approximated as closely as desired by a Morse function (Theorem 8.1.1). Nondegenerate critical points are *isolated* (that is, there cannot be a sequence of nondegenerate critical points converging to a nondegenerate critical point); in particular, a Morse function on a compact manifold has only finitely many critical points, and they are isolated.

The fact that nondegenerate critical points are isolated follows from this result, which is proved in [Guillemin 1974, pp. 42], for example:

Lemma 6.1.4 (Morse's Lemma). *If x_0 is a nondegenerate critical point of a function f on a manifold M, there is some open neighborhood of x_0 in M and a set of local coordinates $x^1, \ldots, x^n$ such that, in these coordinates, f has* *the form*

$$f(x) = f(x_0) - (x^1)^2 - \cdots - (x^\lambda)^2 + (x^{\lambda+1})^2 + \cdots + (x^n)^2,$$

where λ is the index of the critical point.

Thus, it is always possible to choose local coordinates in the neighborhood of a nondegenerate critical point so that the function in this neighborhood is a diagonalized quadratic function when expressed in these coordinates. Note that we are dealing here with an exact equality: there are no additional higher-order terms.

6.2. Level Sets Near Nondegenerate Critical Points

We now consider how the level sets $f^{-1}(a)$ of a Morse function f change as a crosses a critical value.

Morse theory of surfaces. We start with the most easily visualizable case: a Morse function on a surface (two-manifold) M. Then $f^{-1}(a)$ is also called the *level curve* of f corresponding to the value a. We also consider M_a, the subset of M "bounded above" by a (See Figure 6.1):

$$M_a = \{x : f(x) \leq a\} = f^{-1}((-\infty, a]).$$

When a is a regular value of f, we know by Theorem 3.4.3 that $f^{-1}(a)$ is a smooth submanifold of M. In addition, M_a is a two-manifold with boundary, the boundary being the level set $f^{-1}(a)$.

If we assume that M is compact, and a is a regular value, $f^{-1}(a)$ is a compact one-manifold, and therefore the union of finitely many smooth circles (Figure 6.1), diffeomorphic to the standard circle. What happens to those circles when we change a?

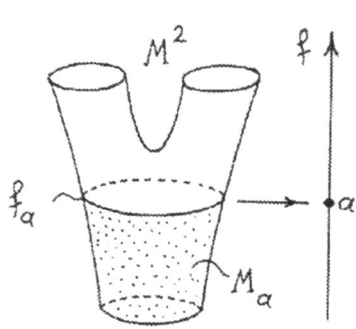

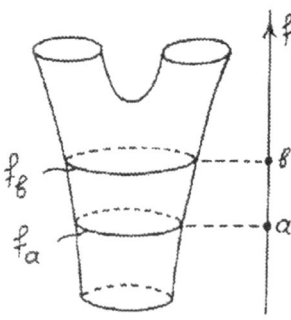

FIGURE 6.1.   FIGURE 6.2.

It is clear that a small change in a does not change the fact that a is regular; that is, if a' is close enough to a, then a' is also regular. Consequently, it is natural to expect that a small change in the regular value a does not change the topological type of the level curve $f^{-1}(a)$. This is indeed so:

Proposition 6.2.1. *Let f be a smooth function on a compact smooth surface M, and assume that the segment $[a, b]$ (where $a < b$) contains no critical value of f, or equivalently that $f^{-1}([a, b]) \subset M$ has no critical point of f (Figure 6.2). Then:*

(a) *The level sets $f^{-1}(a)$ and $f^{-1}(b)$ are diffeomorphic (in particular, they consist of the same number of smooth circles diffeomorphic to a standard circle).*
(b) *M_a and M_b are diffeomorphic as two-manifolds with boundary.*
(c) *Both diffeomorphisms can be realized as shifts along the trajectories of the vector field $-\mathrm{grad}\, f$ on M, for an arbitrary Riemannian metric on M.*

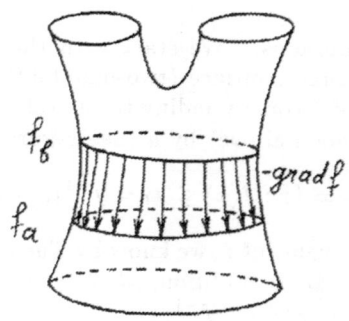

FIGURE 6.3.

Part (c) of Proposition 6.2.1 is illustrated in Figure 6.3. The background to it is the following. Consider an n-dimensional manifold M that has been given a Riemannian metric $\langle\ ,\ \rangle$ (see Definition 2.1.1 and the discussion preceding Lemma 2.1.3). Then, for any smooth function f on M, we can define the *gradient vector field* $\mathrm{grad}\, f$. The value of $\mathrm{grad}\, f$ at x is the unique vector in $T_x M$ with the following property:

$$\langle \mathrm{grad}\, f(x),\ v \rangle = d_x f(v) \quad \text{for all } v \in T_x M.$$

(In other words, $\mathrm{grad}\, f(x)$ is dual to the differential $d_x f$.) Clearly, $\mathrm{grad}\, f(x) = 0$ if and only if x is a critical point. If nonzero, $\mathrm{grad}\, f(x)$ points to a direction of increasing f, and is orthogonal to the local submanifold $f^{-1}(f(x))$, with respect to the Riemannian metric $\langle\ ,\ \rangle$.

Any smooth manifold can be given a Riemannian metric. For example, we can embed the manifold in some $\mathbb{R}^n$, using the Whitney embedding theorem (Theorem 5.1.4), and then choose the Riemannian metric induced from $\mathbb{R}^n$.

Given a smooth vector field $v : U \to \mathbb{R}^n$, where U is a domain in $\mathbb{R}^n$, we can define (locally) the *flow* of v as follows. For each $x_0 \in U$, the differential equation

6.2.2. $\qquad \dfrac{dx}{dt} = v(x)$ with initial condition $x(0) = x_0$

has a unique solution $x(t)$, for t close enough to 0. By definition, the *time-t flow* of v, which we denote φ_t, maps x_0 to $x(t)$:

$$\varphi_t(x_0) = x(t) \quad \text{where } x(t) \text{ is a solution of 6.2.2.}$$

For each x_0, then, $\varphi_t(x_0)$ is defined for t close enough to 0, and morevoer $\varphi_t(x_0)$ varies smoothly with t and x_0. The curves $x(t)$ are called the *trajectories* of φ_t, or of v. A trajectory is tangent to the vector field at all its points. Multiplying the vector field by an everywhere positive scalar function does not change the trajectories, only the speed at which they are traversed.

On a smooth manifold M, the flow of a vector field v is likewise derived from the solutions of the equation $dx/dt = v(x)$, which can be regarded as an equation in $\mathbb{R}^n$ using local coordinates. Moreover, when M is compact, it can be proved that the flow of M is *globally defined*, that is, $\varphi_t(x_0)$ exists for any x_0 and any t, and φ_t is a diffeomorphism of M for each t.

Proof of Proposition 6.2.1. Consider the vector field

$$v(x) = \frac{-\operatorname{grad} f(x)}{\langle \operatorname{grad} f(x), \operatorname{grad} f(x) \rangle}$$

This field is not defined on all of M because $\operatorname{grad} f(x)$ vanishes at the critical points. However, it is defined on $f^{-1}([a, b])$, since by assumption there are no critical points there. In order to obtain a smooth vector field defined on all of M, we multiply $v(x)$ by a smooth function $\rho(x)$ on M that equals 0 in a small neighborhood of each critical point, equals 1 in $f^{-1}([a, b])$, and satisfies $0 \le \rho(x) \le 1$ elsewhere; such a function exists by Lemma 5.2.3.

Now we take the flow φ_t of the smooth vector field ρv. For $t > 0$, this flow *never increases* the value of f, since the vector field points in the direction of $-\operatorname{grad} f$ unless it vanishes. In fact, a simple calculation, which can be performed in $\mathbb{R}^2$ since it refers to an infinitesimal property, shows that the value of f along the trajectories of φ_t changes at a rate equal to $-\rho$:

$$\frac{df(\varphi_t(x_0))}{dt} = -\rho(\varphi_t(x_0)).$$

In particular, in the region $f^{-1}([a, b])$, the value of f decreases at a unit rate. This implies that a point of $f^{-1}(b)$, after being dragged by the flow for a time

$b - a$, ends up on $f^{-1}(a)$—that is, $\varphi_{b-a}(f^{-1}(b)) \subset f^{-1}(a)$. It also implies that $\varphi_{b-a}(M_b) \subset M_a$ (recall that $M_a = f^{-1}((-\infty, a])$, and likewise for M_b).

Now, φ_{b-a} is a diffeomorphism of M because M is compact. Its inverse is $\varphi_{a-b} = \varphi_{b-a}^{-1}$. The reasoning of the preceding paragraph shows that $\varphi_{a-b}(f^{-1}(a)) \subset f^{-1}(b)$, and that $\varphi_{a-b}(M_a) \subset M_b$. Therefore φ_{b-a} maps M_b bijectively to M_a, and $f^{-1}(b)$ bijectively to $f^{-1}(a)$, proving parts (a) and (b) of our result. Part (c) also follows since the vector fields ρv and $-\operatorname{grad} f$ differ by a positive scalar function in $f^{-1}([a, b])$, and so have the same trajectories there. $\square$

We now turn to the evolution of level curves near a nondegenerate critical point of f. Let P_0 be a nondegenerate critical point of f, and assume without loss of generality that $f(P_0) = 0$, so that 0 is a singular value. By Morse's lemma, in a sufficiently small neighborhood of P_0 we can find coordinates x, y such that $f = \pm x^2 \pm y^2$ for different combinations of the signs $\pm$; the index λ is the number of negative terms. We consider the possible cases.

(a) P_0 is a maximum, that is, $f = -x^2 - y^2$, and the index is 2.
(b) P_0 is a minimum, that is, $f = x^2 + y^2$, and the index is 0.
(c) P_0 is a saddle, that is, $f = -x^2 + y^2$, and the index is 1 (the case $f = x^2 - y^2$ can be reduced to this by interchanging x and y).

We assume for simplicity that $f^{-1}(0)$ has only one critical point, and consider what happens to the surface with boundary M_a as a increases past the critical value 0.

Case 1: P_0 is a maximum. If ε is a small positive regular value of f, the level set $f^{-1}(-\varepsilon)$ has a connected component near P_0 that is a circle with center P_0 (in the Morse coordinates). This component tends to P_0 when ε tends to zero, as illustrated in Figure 6.4. As we will prove in Proposition 6.2.4, M_0 is diffeomorphic to the surface obtained from $M_{-\varepsilon}$ by attaching a disk to this component of $f^{-1}(-\varepsilon) = \partial M_{-\varepsilon}$. This disk is foliated by the level curves $f^{-1}(a)$, for $-\varepsilon < a < 0$.

Case 2: P_0 is a minimum. This is case 1 turned upside down. The connected component of M_ε around P_0, for $\varepsilon > 0$ small enough, is a disk foliated by the circles $f^{-1}(a)$, for $0 < a < \varepsilon$, centered at P_0. See Figure 6.5.

Case 3: P_0 is a saddle. Consider, in the neighborhood of P_0, the three level curves $f^{-1}(-\varepsilon)$, $f^{-1}(0)$, and $f^{-1}(\varepsilon)$, where $\varepsilon > 0$ is sufficiently small. These curves are defined by the quadratic equations $-x^2 + y^2 = -\varepsilon, 0, \varepsilon$, and are shown in Figure 6.6. The curve $f^{-1}(0)$ is the union of two lines $x = \pm y$ that intersect at 0. The curves $f^{-1}(\varepsilon)$ and $f^{-1}(-\varepsilon)$ are hyperbolas. Figure 6.7 displays the evolution of the level set in the neighborhood of a critical point of index 1.

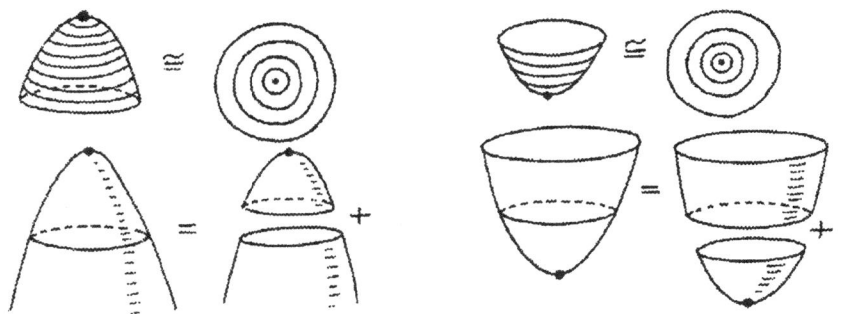

FIGURE 6.4.   FIGURE 6.5.

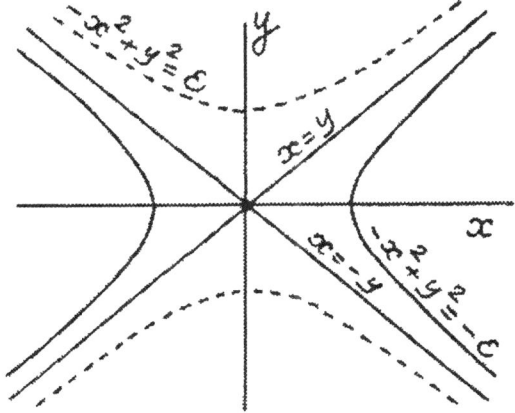

FIGURE 6.6.

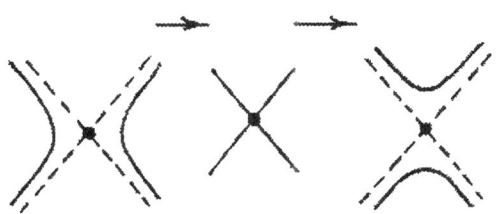

FIGURE 6.7.

Attaching handles. In order to formulate in rigorous terms the change in topology that M_a undergoes as a crosses a critical point, we introduce the notion of a *handle H_λ of index λ*, and the operation of *attaching a handle to M_a*.

Definition 6.2.3. A (two-dimensional) *handle H_λ of index λ* is the direct product $D^\lambda \times D^{2-\lambda}$, where D^p is the closed disk of dimension p.

Thus, H_λ is always homeomorphic to a two-dimensional disk D^2, but we think of H_0 and H_2 as disks, and of H_1 as a rectangle $D^1 \times D^1$ (Figure 6.8).

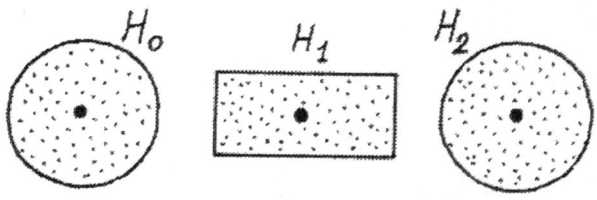

FIGURE 6.8.

Proposition 6.2.4. *Consider a Morse function f on a compact smooth surface M, and let 0 be a critical value. Assume there is only one critical point P_0 in $f^{-1}([-\varepsilon,\,\varepsilon])$, and that it has index λ. Then M_ε is homeomorphic to the surface with boundary obtained by attaching a handle of index λ to $M_{-\varepsilon}$, by which we mean the following:*

(a) *Attaching a handle H_2 (the maximum case) is the operation shown in Figure 6.9(a). We glue the boundary of the disk to the component of the boundary of $M_{-\varepsilon}$ near P_0, and the hole disappears.*

(b) *Attaching a handle H_1 (the saddle case) is the operation shown in Figure 6.9(b). We glue two opposite edges of the rectangle to the two components of the boundary near P_0, without a twist (upper case in the figure).*

(c) *Attaching a handle H_0 (the minimum case) is the simplest operation: we simply form the disjoint union of the surface with the disk D^2, thus creating an additional boundary curve, as shown in Figure 6.9(c).*

Proof. We work as in the proof of Proposition 6.2.1. Outside a small disk U around P_0, we define the vector field ρv as before. Inside U, where f is in Morse form $\pm x^2 \pm y^2$, we define ρ as a smooth function whose value increases from 0 very near P_0 to 1 near the edge of the disk. By choosing ε small enough, we can assume that f takes at least one of the values ε and $-\varepsilon$ inside U.

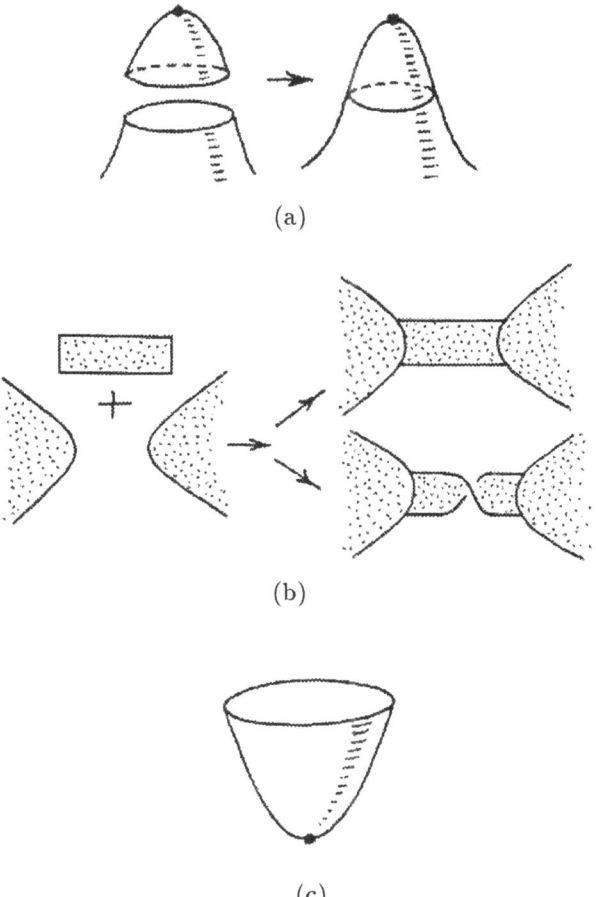

(a)

(b)

(c)

FIGURE 6.9.

Recall that the flow $\varphi_{2\varepsilon}$ decreases the value of f by 2ε if the trajectory lies in a region where $\rho = 1$. Since there are no critical points apart from P_0 in $f^{-1}([-\varepsilon, \varepsilon])$, the argument used in the proof of Proposition 6.2.1 shows that, outside U, $\varphi_{2\varepsilon}$ gives a (partial) diffeomorphism between M_ε and $M_{-\varepsilon}$.

To complete the proof we observe that M_ε is diffeomorphic to $\varphi_{2\varepsilon}(M_{-\varepsilon})$, and we compare this latter set with $M_{-\varepsilon}$ using the explicit form of f inside U:

(a) For $\lambda = 0$ (a minimum), M_ε intersects U but $M_{-\varepsilon}$ does not. We have

$$\varphi_{2\varepsilon}(M_\varepsilon) = M_{-\varepsilon} \cup \varphi_{2\varepsilon}(M_\varepsilon \cap U),$$

the union being disjoint. The set on the right is a round disk if we make the flow radial. The desired result follows.

(b) For $\lambda = 2$ (a maximum), M_ε contains U but $M_{-\varepsilon}$ does not. We have

$$\varphi_{2\varepsilon}(M_\varepsilon) = M_{-\varepsilon} \cup U,$$

and the two sets on the right overlap in an annulus, that is, U is attached to $M_{-\varepsilon}$ around the edge.

(c) For $\lambda = 1$ (a saddle), both $f^{-1}(\varepsilon)$ and $f^{-1}(-\varepsilon)$ intersect U. This time we write

$$\varphi_{2\varepsilon}(M_\varepsilon) = M_{-\varepsilon} \cup R,$$

where $R = \varphi_{2\varepsilon}(M_\varepsilon) - M_{-\varepsilon}$ is the dotted region in Figure 6.10. Because we can control ρ, we can ensure that R really is what it appears to be in the figure: a topological rectangle, with adjacent sides tangent. $\square$

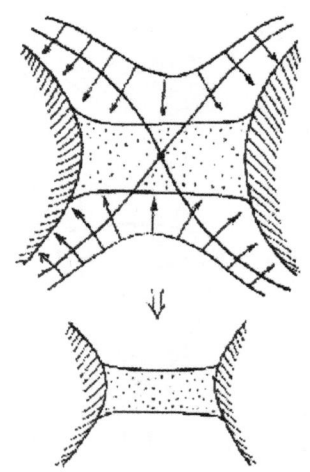

FIGURE 6.10.

Morse theory of three-dimensional manifolds.   Now consider a Morse function on a three-dimensional manifold M. We will follow closely the theory just developed for surfaces. As before, denote by $f^{-1}(a)$ the level set of f with value a, and by M_a the part of M "bounded above" by a:

$$M_a = \{x : f(x) \le a\} = f^{-1}((-\infty, a]).$$

When a is a regular value of f, we know by Theorem 3.4.3 that $f^{-1}(a)$ is a smooth submanifold of M. In addition, M_a is three-manifold with boundary, the boundary being the surface $f^{-1}(a)$. This surface is the union of some number of smooth, compact, closed surfaces. What happens as we change the value of a?

As in the two-dimensional case, the topological type of $f^{-1}(a)$ cannot change unless a goes through a critical value:

Proposition 6.2.5. *Let f be a smooth function on a compact smooth three-manifold M, and assume that the segment $[a, b]$ (where $a < b$) contains no critical value of f. Then:*

(a) *The level sets $f^{-1}(a)$ and $f^{-1}(b)$ are diffeomorphic.*
(b) *M_a and M_b are diffeomorphic as three-manifolds with boundary.*
(c) *Both diffeomorphisms can be realized as shifts along the trajectories of the vector field $-\mathrm{grad}\, f$ on M, for an arbitrary Riemannian metric on M.*

Proof. Literally the same as that of Proposition 6.2.1.   □

We now study the evolution of the level surfaces near a nondegenerate critical point P_0. As before, assume that $f(P_0) = 0$, and use Morse's lemma to find, in a sufficiently small neighborhood of P_0, coordinates x, y, z such that $f = \pm x^2 \pm y^2 \pm z^2$ for different combinations of the signs $\pm$; the index λ is the number of negative terms. We consider the possible cases.

(a) P_0 is a maximum, that is, $f = -x^2 - y^2 - z^2$, and the index is 3.
(b) P_0 is a minimum, that is, $f = x^2 + y^2 + z^2$, and the index is 0.
(c) P_0 is a saddle of index 1, that is, $f = -x^2 + y^2 + z^2$.
(d) P_0 is a saddle of index 2, that is, $f = -x^2 - y^2 + z^2$.

We assume for simplicity that $f^{-1}(0)$ has only one critical point, and consider what happens to the manifold with boundary M_a as a increases past the critical value 0.

Case 1: P_0 is a maximum. If ε is a small positive regular value of f, the level set $f^{-1}(-\varepsilon)$ has a connected component near P_0 that is a sphere S^2 with center P_0. This component tends to P_0 when ε tends to zero, as illustrated in Figure 6.11. As we will prove in Proposition 6.2.7, M_0 is diffeomorphic to the three-manifold obtained from $M_{-\varepsilon}$ by attaching a ball S^3 to this component of $f^{-1}(-\varepsilon) = \partial M_{-\varepsilon}$. This ball is foliated by the level surfaces $f^{-1}(a)$, for $-\varepsilon < a < 0$.

Case 2: P_0 is a minimum. Here the connected component of M_ε around P_0, for $\varepsilon > 0$ small enough, is a ball foliated by the spheres $f^{-1}(a)$, for $0 < a < \varepsilon$, centered at P_0. Figure 6.11, going backwards, illustrates this case.

Case 3: P_0 is a saddle of index 1. Consider, in the neighborhood of P_0, the three level surfaces $f^{-1}(-\varepsilon)$, $f^{-1}(0)$, and $f^{-1}(\varepsilon)$, where $\varepsilon > 0$ is sufficiently small. These surfaces are defined by the quadratic equations $-x^2 + y^2 + z^2 = -\varepsilon, 0, \varepsilon$, and are shown in Figure 6.12. The surface $f^{-1}(0)$ is a double cone with vertex at 0. The surface $f^{-1}(-\varepsilon)$ is a hyperboloid of two sheets, and $f^{-1}(\varepsilon)$ is a hyperboloid of one sheet.

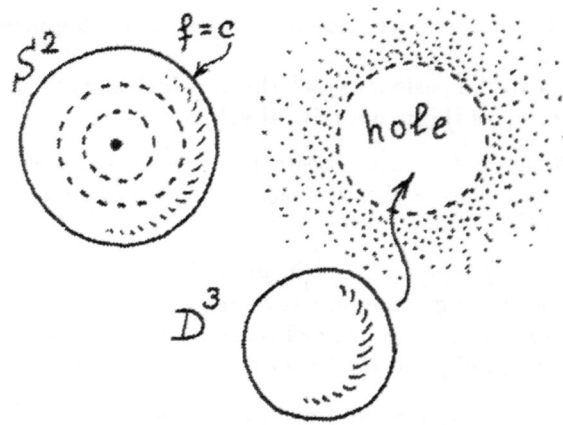

FIGURE 6.11.

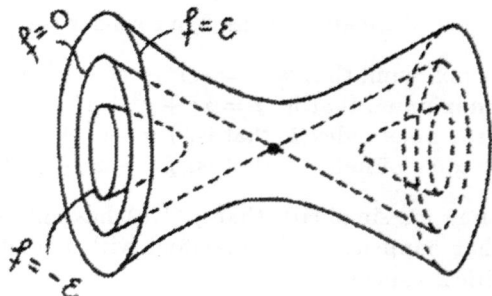

FIGURE 6.12.

Case 4: P_0 is a saddle of index 2. Here the level surfaces $f^{-1}(-\varepsilon)$, $f^{-1}(0)$, and $f^{-1}(\varepsilon)$ are defined by the quadratic equations $-x^2 - y^2 + z^2 = -\varepsilon, 0, \varepsilon$; they, too, are illustrated by Figure 6.12, if we interchange ε and $-\varepsilon$.

Figure 6.13 displays the evolution of the level set in the neighborhood of critical points of different indexes.

Definition 6.2.6. A (three-dimensional) *handle* H_λ *of index* λ is the direct product $D^\lambda \times D^{3-\lambda}$, where D^p is the closed disk (ball) of dimension p.

Thus, H_λ is always homeomorphic to a three-dimensional disk D^3, but we think of H_0 and H_3 as disks, and of H_1 and H_2 as solid cylinders (Figure 6.14). Moreover, H_1 and H_2 will be attached differently.

Proposition 6.2.7. *Consider a Morse function f on a compact smooth three-manifold M, and let 0 be a critical value. Assume there is only one critical*

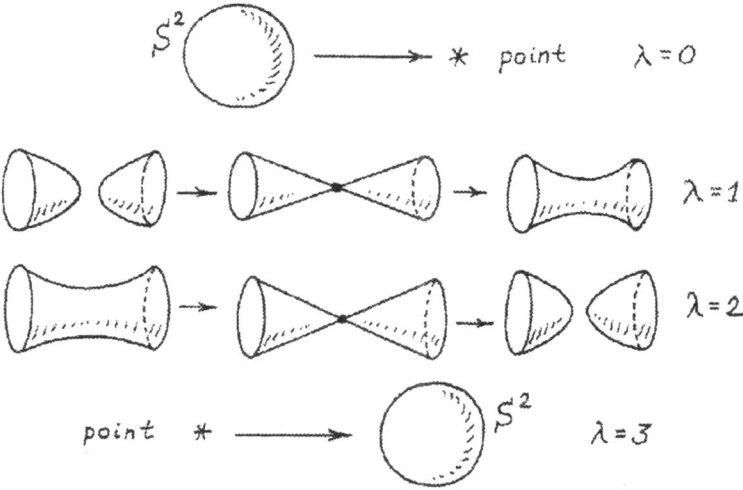

FIGURE 6.13.

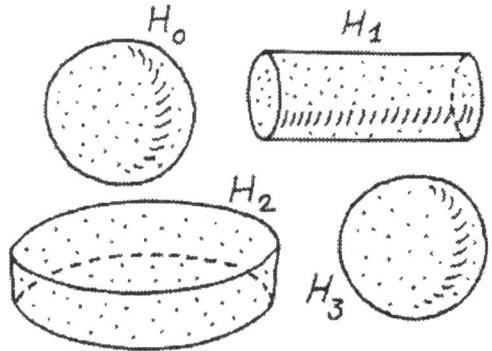

FIGURE 6.14.

point P_0 in $f^{-1}([-\varepsilon, \varepsilon])$, and that it has index λ. Then M_ε is homeomorphic to the three-manifold with boundary obtained by attaching a handle of index λ to $M_{-\varepsilon}$, by which we mean the following:

(a) *Attaching a handle H_3 (the maximum case) is the operation shown in Figure 6.15(a). We glue the boundary of the ball to the component of the boundary of $M_{-\varepsilon}$ near P_0, and the hole disappears.*

(b) *Attaching a handle H_2 (case of a saddle of index 2) is the operation shown in Figure 6.15(b). We glue the lateral wall of the cylinder to the neck of the one-sheeted hyperboloid.*

(c) *Attaching a handle H_1 (case of a saddle of index 1) is the operation shown in Figure 6.15(c). We glue the top and bottom of the cylinder to the extremities of the two-sheeted hyperboloid.*

(d) *Attaching a handle H_0 (the minimum case) consists of forming the disjoint union of the surface with the ball D^3, thus creating an additional boundary surface, as shown in Figure 6.15(d).*

The proof follows the model of the proof of Proposition 6.2.7.

6.3. Handle Decomposition of a Compact Smooth Manifold

We have explained the attachment of handles in the context of the study of a Morse function, in dimensions two and three. We can define the same operation in a more general context, and in arbitrary dimension. Let the dimension n be fixed. We define an *n-handle of index λ* to be

$$H_\lambda = D^\lambda \times D^{n-\lambda},$$

where D^k is the closed k-disk. The handle's *attaching boundary* is $S^{\lambda-1} \times D^{n-\lambda}$; the first factor is the boundary of the first factor of the H_λ. Thus, all n-handles H^λ are homeomorphic, but, for each index, a different part of the boundary is special. The case $\lambda = 0$ relies on the convention that the sphere of dimension -1 is the empty set—that is, the attaching boundary is empty.

Now let M be a smooth n-manifold with boundary, and suppose that in this boundary there exists a submanifold $N \subset \partial M$ diffeomorphic to $S^{\lambda-1} \times D^{n-\lambda}$. *Attaching a handle of index λ* to M consists of taking the union $M \cup H_\lambda$, and identifying N with the attaching boundary of H_λ, by means of some diffeomorphism (of manifolds with boundary).

The result of this process is not obviously a smooth manifold with boundary, but it can be made into one by a process of smoothing, as indicated in Figure 6.16. In any case, what interests us most is the topological type of the result, so after the attachment we can also perform other adjustments, as in Figure 6.17, to make it easier to visualize further attachments. Note that the topological type of the result may depend on the subset N, and also on the identifying diffeomorphism (if $\lambda = 1$).

Note that attaching an n-handle of index n complete closes off one boundary component of M (that is, N is a connected component of ∂N). If, moreover, there are no other boundary components, the result is a manifold with empty boundary (what we original defined as a manifold).

We say that a manifold M (with or without boundary) has a *handle decomposition* if we can build it up by starting with the empty set and attaching a finite number of handles, in sequence. More formally, there should be a sequence of manifolds with boundary

$$\varnothing = M_0 \supset M_1 \supset \cdots \subset M_r = M$$

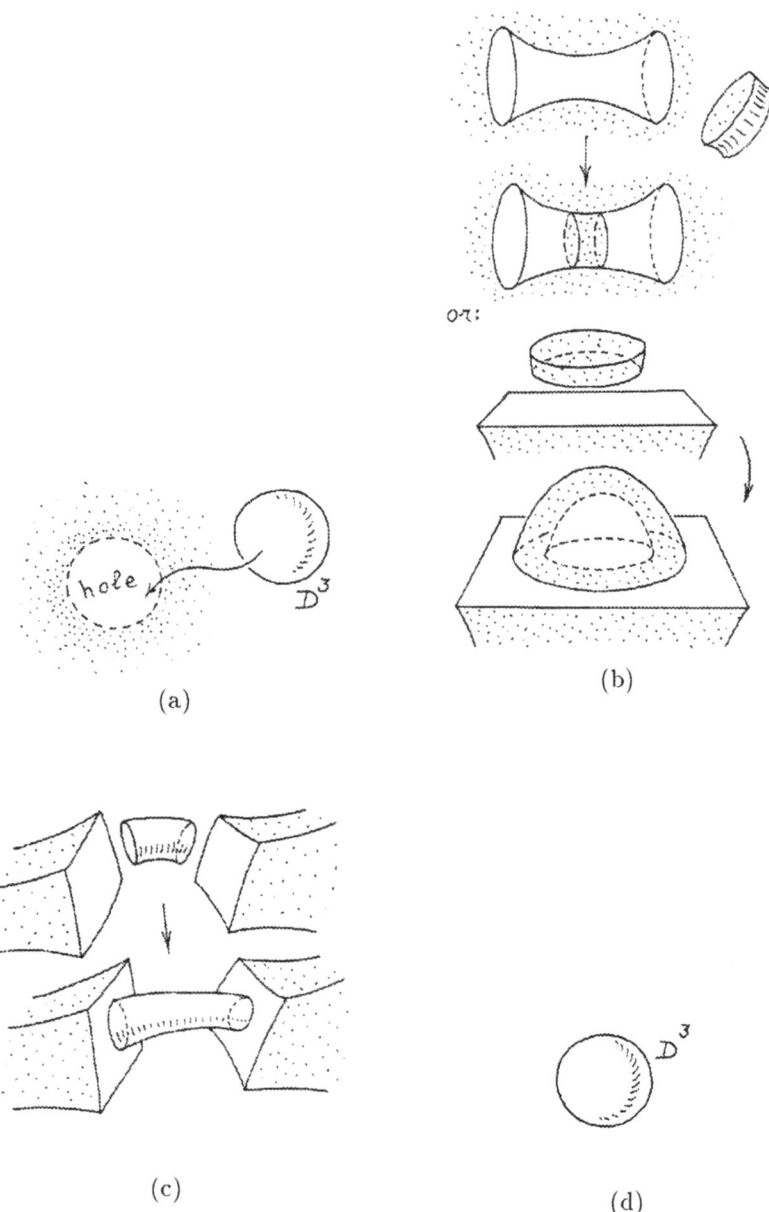

(a)

(b)

(c)

(d)

FIGURE 6.15.

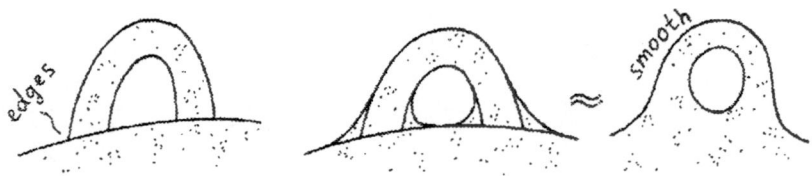

FIGURE 6.16.

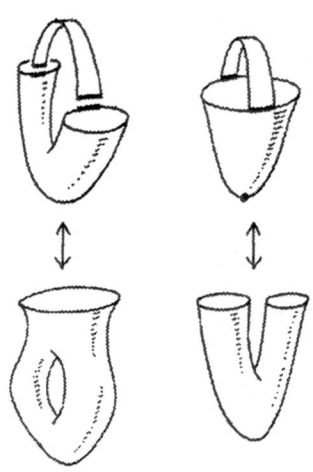

FIGURE 6.17.

such that, for each $i > 0$, we can obtain M_i from M_{i-1} by attaching an n-handle of some index.

Theorem 6.3.1. *Let f be a Morse function on a compact smooth n-manifold M (without boundary), and set $M_a = \{x : f(x) \leq a\}$.*

(a) *Suppose that 0 is a critical value of f, and that there is only one critical point P_0 in $f^{-1}([-\varepsilon, \varepsilon])$, with index λ. Then M_ε can be obtained from $M_{-\varepsilon}$ by attaching an n-dimensional handle of index λ.*

(b) *The manifold M can be built by starting with the empty set, and attaching in succession a finite number of n-dimensional handles H_λ, one H_λ for each critical point of f of index λ. That is, f determines a handle decomposition of M.*

Sketch of proof. Part (a) is proved following the model of the proof of Proposition 6.2.4. Part (b) follows by considering all the critical points of f in ascending order of their values; if several critical points have the same value of f, we can either modify f slightly so they no longer do so, or extend the argument of part (a) so that several handles can be attached simultaneously, which is easy since the critical points are isolated. □

Note that different Morse functions on M generally determine different handle decompositions.

Because every compact smooth manifold admits a Morse function, as mentioned after Definition 6.1.3, the theorem implies the following:

Corollary 6.3.2. *Any compact smooth manifold has a handle decomposition.*

Moreover. given a handle decomposition of a compact smooth manifold, it is possible to construct a Morse function that determines this decomposition. The proof of this is not difficult, but is technical and not very enlightening; therefore we omit it.

Remark. Knowing the order and indices of the critical points of a Morse function—equivalently, knowing the order and indices of the handles of a handle decomposition—is not enough, in general, to allow one to reconstruct the manifold up to homeomorphism. The point is that the topology of the result may depend on what boundary component(s) the handle is glued to, on the homotopy class of the gluing map, and so on. The handle decomposition allows us to reconstruct only the "rough structure" of the manifold; the "fine structure" is based on *how* the handles are glued. The fine structure is important for higher dimensions. We will see in Chapter 8 that in the case of surfaces it is possible to reconstruct the surface just from combinatorial information about critical points.

Theorem 6.3.1 has the following consequence (homotopy equivalences and CW-complexes will be defined in Chapter 10).

Theorem 6.3.3. *Let f be a Morse function on a compact smooth manifold M. Then M is homotopically equivalent to a finite CW-complex containing one cell of dimension λ for each critical point of f of index λ.*

In general. different Morse functions define different CW-decompositions of M.

6.4. Homology and the Morse Inequalities

Consider a compact n-dimensional manifold M, and take the real homology groups $H_k(M)$ of M, for $k = 0, 1, \ldots, n$ (homology groups will be defined in Chapter 10). The dimension of $H_k(M)$ as a real vector space is called the *k-th Betti number* of M, and is denoted by $\beta_k(M)$. It is, of course, a nonnegative integer.

If f is a Morse function on M, the *k-th Morse number* of f is the number of critical points of f having index k, and is denoted by $\mu_k(f)$. For example, $\mu_0(f)$ is the total number of the local minima of f.

Theorem 6.4.1. *Let f be a Morse function on the compact smooth manifold M.*

(a) *For each k, we have $\mu_k(f) \geq \beta_k(M)$.*
(b) *We also have*

$$\sum_{k=0}^{n}(-1)^k \mu_k = \sum_{k=0}^{n}(-1)^k \beta_k.$$

A proof can be found in [Schwarz 1994, Section 6.10], for example.

Theorem 6.4.1(a) says that the Betti numbers β_k of M are lower bounds for the Morse numbers $\mu_k(f)$, for any Morse function f on M. Topologically, this means that *there are no more independent closed cycles* (cycles without boundary) *in dimension k that the total number of k-cells of M.* Recall that each cell corresponds to one critical point of f.

Theorem 6.4.1(b) says that the alternating sum of Morse numbers equals the alternating sum of Betti numbers, also called the *Euler characteristic* of M. The homology groups of a manifold, and therefore the Betti numbers, are topological invariants, that is, they do not change under homeomorphisms (or even under homotopy equivalences; see [Munkres 1984, p. 174], for example). It follows that the alternating sum of Morse numbers does not depend of the function: it is an invariant of the manifold.

6.5. Some Embeddings of Surfaces in $\mathbb{R}^3$ and Their Critical Points

We consider Morse functions on an orientable, compact, connected, smooth surface of genus g, denoted M_g. We know from Lemma 4.4.1 that M_g can be smoothly embedded in $\mathbb{R}^3$. If we take the projection of such an embedding onto a fixed line of $\mathbb{R}^3$ (usually taken as the z-axis) we obtain a *height function* on M_g.

Figure 6.18 shows some embedded surfaces whose height functions are Morse functions. The height function of the standard (round) embedding of sphere $S^2 = M_0$ in $\mathbb{R}^3$ is a Morse function with exactly two critical points: a minimum and a maximum. The embedding of the torus $T^2 = M_1$ as a torus of revolution with a horizontal symmetry axis has a Morse height function with four critical points: one minimum, two saddles, and one maximum. For $g > 1$, a similar Morse height function on M_g can be found, with exactly $2g+2$ nondegenerate critical points: one minimum, $2g$ saddles, and one maximum (Figure 6.18, right).

The Euler characteristic of M_g is $2g - 2$; thus, it follows from Theorem 2 and from the fact that a smooth function on a compact manifold must have a maximum and a minimum that $2g + 2$ is the least number of critical points that a Morse function on M_g can have.

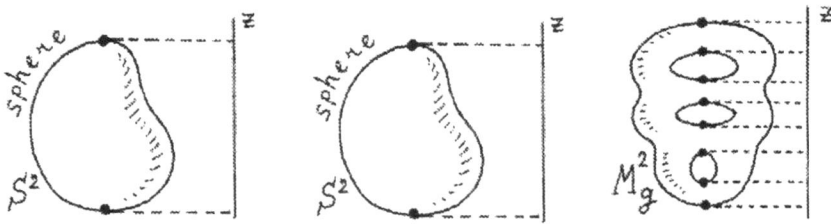

FIGURE 6.18.

It is typically possible to modify a Morse function in such a way that two or more nondegenerate critical points collapse together, forming one degenerate critical point. Thus, if we relax the condition that f be a Morse function, we can get a smooth function on M_g with fewer than $2g + 2$ critical points. Then, evidently, the structure of the degenerate critical points will be more complicated than that of nondegenerate ones. We illustrate this with a simple and visual example.

For $g \geq 1$, there are embeddings of M_g in $\mathbb{R}^3$ whose height function has exactly four critical points: one minimum, one maximum, and two saddle points. The two saddle points will be necessarily *degenerate* if $g > 1$. The case $g = 1$ (the torus) has already been discussed (Figure 6.18, middle; this is a Morse function). The case $g > 1$ is illustrated in Figure 6.19. We simply take the boundary of a small tubular neighborhood of the union of $g + 1$ equally spaced meridians on a sphere, as shown in Figure 6.20.

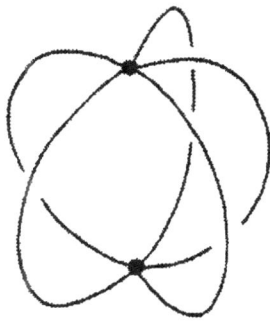

FIGURE 6.19.   FIGURE 6.20.

The height function in a small neighborhood of the degenerate saddles behaves as $\mathrm{Re}\, z^{g+1}$, the real part of the complex function z^{g+1}. It is possible to show that the height function of any embedding of M_g in $\mathbb{R}^3$ (for $g > 0$) must have at least four critical points, so our construction achieves this minimum.

However, if we drop the requirement that f be the height function of some embedding of M_g in $\mathbb{R}^3$, the minimum number of critical points decreases to three (for $g > 0$). In fact, any compact connected surface, orientable or not, admits a smooth function with exactly three critical points: a minimum, a maximum, and a degenerate saddle.

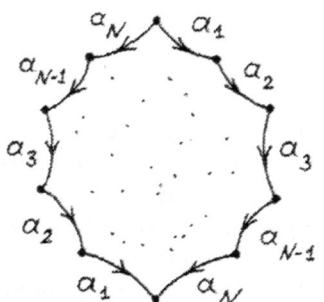

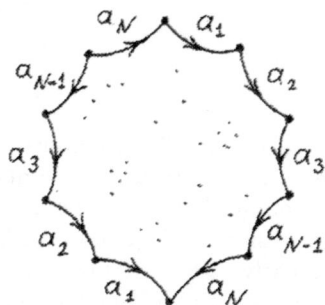

FIGURE 6.21.   FIGURE 6.22.

To construct such a function, we start by using the remark following Theorem 4.2.7: any surface M can be obtained by gluing the edges of a polygon W having the symmetric canonical form

$$W = a_1 \dots a_{N-1} a_N a_1^{-1} \dots a_{N-1}^{-1} a_N^{\pm 1},$$

for some value of N, where the sign -1 (Figure 6.21) gives an orientable surface and the sign $+1$ (Figure 6.22) a nonorientable one. W has $2N$ edges, and N can be taken to be even if the surface is orientable; this ensures that all the vertices of W are identified by the gluing into one point of the surface.

Cut W in half by the diagonal shown in Figure 6.23, then draw level curves of a smooth function f as in Figure 6.23, where in the left half the curves represent increasingly positive values of f, until they reach a maximum at the point marked; and conversely in the right half of W they represent increasingly negative values of f, until they reach a minimum. The third critical point—the degenerate saddle point—is the point of the surface arising from the identification of all the vertices of W.

By choosing f appropriately, we can arrange for it to have the form $\mathrm{Re}\, z^{N+1}$ in a small neighborhood of this degenerate critical point. (You should relate N to the number g of handles in the orientable case or the number k of crosscaps in the nonorientable case.) Notice where we have made use of the symmetric form of W: Since the segment AB separates each edge a_i of W from its match, f has opposite signs on either side of a_i, and we can arrange for the gradient of f to be nonzero at any interior point of a_i. If there had been two edges of W with the same letter on one side of AB, this edge on the

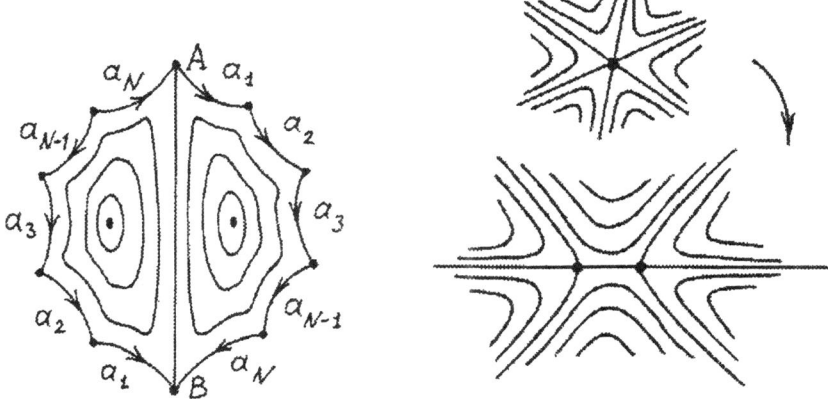

FIGURE 6.23.   FIGURE 6.24.

surface would be entirely made up of degenerate critical points of f, since f would have the same sign on both sides of it.

Under a small perturbation of the constructed smooth function f, the single degenerate critical point splits into several nondegenerate saddles, as shown in Figure 6.24.

Suppose there is a smooth function f with only two, perhaps degenerate, critical points on a compact smooth surface M. Then M is diffeomorphic to the sphere. The corresponding statement in arbitrary dimension, obtained by replacing "surface" with "n-manifold", is also true. This was proved by Reeb.

7. Application: Analyzing Human Body Motions Using Manifolds and Critical Points

This chapter discusses how we can use manifolds and critical points to analyze *human body motions* on the computer. A robot or a human body can be modeled as a *multibody*, that is, a system of rigid bodies connected by joints. The analysis of human body motions is, however, far more complex than that of a robot. A robot consists of several segments, each being a rigid body. A human body, by contrast, has many more parts, and they are not really rigid, although we model them as if they were. We first modeled a human body as a multibody consisting of fifty segments, as shown in Figure 7.1. A recent extension has added more segments, actually of hands, and now the multibody has 82 segments.

FIGURE 7.1. A modeled human body.

From a sequence of human body images processed through computer vision, we computed the *changing power* (that is, dynamic relationships) between the segments, based on *inverse dynamics*. Six Lagrangian equations specify the dynamics of each segment. Computing the power vector change of a multibody with 82 segments from one frame of an animation to the next requires solving 6×82 Lagrangian equations; thus, for real-time animation at 30 frames per second, we need to solve close to 15,000 Lagrangian equations per second. We also compute the changing power exerted from one person on the other by the same method [Kunii and Sun 1990]. The computed powers are displayed as vectors on computer graphics equipment (Figure 7.2).

Sports coaches are finding the results very useful in studying athletes' movements and in designing improved training programs. Medical doctors are showing strong interest in applying the method to rehabilitation programs.

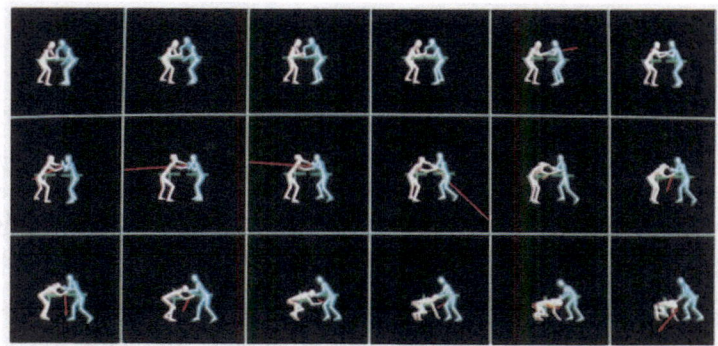

FIGURE 7.2. Analytical results indicating the forces exerted on wrists.

7.1. A Detailed Example: The Hand and Arm

We study in detail hands and arms, which are extremely important tools for humans. They can perform various motions, such as pinching, seizing, grasping, pressing, scooping, wrenching, and slapping. This dexterity primarily comes from their having a large number of joints.

Hands and arms can be modeled as a multibody. Once the model is fixed, a procedure to determine whether a particular motion is possible can be found [Kunii et al. 1993]. Mathematical concepts such as critical points and maps between manifolds are essential tools in the procedure. The details of our model and procedure are described later, in Sections 7.2 and 7.3.

Many researchers [Gourret et al. 1989; Lee et al. 1990; Monheit and Badler 1991; Rijpkema and Girard 1990] have investigated the various motions of human body, including the hand and the arm. However, they have typically used statistical average data or intuitive imaginary geometric data. Although the animations they produce look realistic enough, their models are powerless to analyze the real motion of a person.

The data required for our model are:

(1) the relative position and orientation of the adjoining joint axes,
(2) the joint rotation limits, and

(3) the dependencies among the joint angles [Landsmeer 1963].

Since these data are different from person to person, we have to acquire the real data of individual persons. In order to analyze real motions of a person, we develop a system to acquire the real geometric data of the person. The system is based on a stereo method with five video cameras. Using the system, we obtain the positions of hands and arms by fitting our model to the images.

We have applied our method to a technique of martial arts, *Shorinji Kempo*, and found that experts take advantage of the joint rotation limits and the dependencies among the joints of hands and arms. This will be discussed in Section 7.4.

7.2. A Model of the Hand and Arm

Multibody modeling of hands and arms. As mentioned earlier, we model the hand and arm as a multibody system [Huston 1990]. The system is assumed to be connected, rigid, and open. A multibody system of this type is often called an *open-chain system* or an *open-tree system*. We define the upper arm as the root body of the system; the position and orientation of the root body represent those of the whole system.

When a body comes into contact with another body belonging either to the same or to a different multibody, it receives force through the contact surface, and its motion is restricted. Conversely, a body applies force to another body through the contact surface. For simplicity, we approximate the contact surface by a finite set of representative points of contact through which the representative forces are applied. We call the representative points of contact *end-effectors*. The position of the end-effector is defined to be the position of the representative point of contact, and its orientation is defined as the normal vector of the tangent plane at the point.

Configuration spaces and workspaces as manifolds. The ordered set of variables q, consisting of the position and orientation of the upper arm and the joint angles, uniquely determines the posture of the hand and arm. This set q is called a *configuration*. The set of all possible configurations is called the *configuration space*.

Figure 7.3 shows a simple example of a configuration space. The configuration of the wrist can be determined by the two angles θ and φ. As we know, there are physical limits on these angles. As indicated in the figure, the set of all possible pairs (θ, φ) is a subset of the direct product $[-\pi, \pi] \times [-\pi, \pi]$. The interior and boundary of the set of possible pairs is the configuration space, which we denote by $\mathcal{D}$.

The constraints on one angle depend on the other (see Figure 7.4). We can rotate our wrist widely (that is, the interval of allowable values of θ is wide) when the hand is parallel to the body, as shown in part (a) of the figure. We

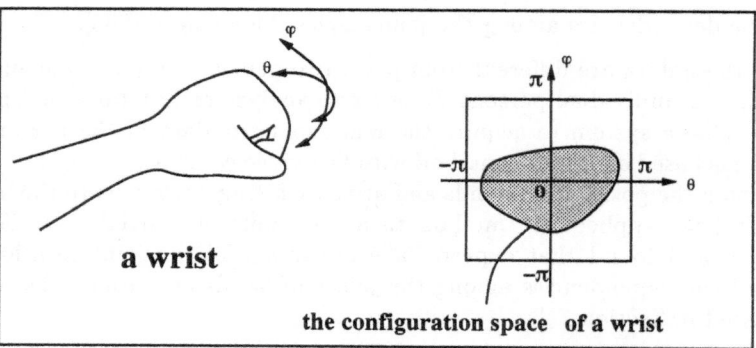

FIGURE 7.3. A wrist and its configuration space.

cannot rotate our wrist widely to left and right (that is, the interval of θ is narrow) if the wrist is at a right angle with the arm, as shown in part (b). Thus the configuration space can represent the allowable range of motion of a joint, and the dependency of one angle on another.

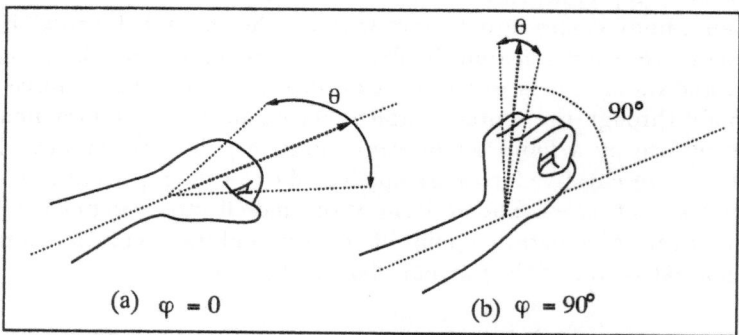

FIGURE 7.4. Angle dependency in the wrist joint.

A configuration determines the positions and orientations of end-effectors. The set of all possible positions and orientations of an end-effector is called a *workspace*. Configuration spaces and workspaces are described with several local coordinates systems (*charts*), and these spaces can be interpreted as manifolds with boundaries (see Definition 5.1.5).

If there is no dependency between joints, the configuration space is the manifold with boundary

$$\mathcal{C} = \underbrace{I_1 \times I_2 \times \cdots \times I_n}_{\subset \mathbb{R}^n} \times SO(3) \times \mathbb{R}^3 \subset \mathbb{R}^n \times SO(3) \times \mathbb{R}^3,$$

where each I_j is a closed interval, and SO(3) is the parameter space for rotation of the end-effector (the group of rotations in three-space).

Now consider the manifold representation of the workspace. Suppose a multibody has m representative points of contact. Attach a Euclidean frame to the i-th representative point of contact of the multibody. The i-th workspace manifold $\mathcal{W}_i$ is

$$\{(x, y, z, \varphi_1, \varphi_2, \varphi_3)\} \subset \mathcal{O} = \mathbb{R}^3 \times \text{SO}(3).$$

$\mathcal{O}$ is usually called an output space in robotics.

Assume that we have m end-effectors. Let $\boldsymbol{x}_i$ be the state (position and orientation) of the i-th end-effector. The state $\boldsymbol{x}_i$ is related to the configuration $\boldsymbol{q}$ by the forward kinematic function $\boldsymbol{f}_i$:

$$\boldsymbol{x}_i = \boldsymbol{f}_i(\boldsymbol{q}).$$

The forward kinematic function of the i-th end-effector is considered as a map $\mathcal{D} \rightarrow \mathcal{W}_i$ between manifolds:

$$\boldsymbol{f}_i : \mathcal{D} \subset \mathcal{C} \rightarrow \mathcal{W}_i \subset \mathcal{O}, \qquad \text{for } i = 1, \ldots, m.$$

Decomposition of the configuration space manifold based on joint dependencies. The joint limits and the dependencies should be considered in the configuration space manifold. The configuration space manifold can be decomposed into the direct product of manifolds where the variables of a manifold are independent of those of other manifolds.

We subdivide the set of all joint axes into p sets of joint axes $\mathcal{L}_1, \mathcal{L}_2, \ldots,$ $\mathcal{L}_p$, in such a way that any two joints j_1 and j_2, for $j_1, j_2 \in \mathcal{L}_k$ are dependent on each other, and any two joints j_1' and j_2', for $j_1' \in \mathcal{L}_k$ and $j_2' \in \mathcal{L}_i$, $k \neq i$, are independent. The configuration space manifold $\mathcal{D}$ can be regarded as the direct product of manifolds:

$$\mathcal{D} = \mathcal{D}_{\mathcal{L}_1} \times \mathcal{D}_{\mathcal{L}_2} \times \cdots \times \mathcal{D}_{\mathcal{L}_p} \times \mathbb{R}^n \times \text{SO}(3),$$

where $\mathcal{D}_{\mathcal{L}_k}$, for $1 \leq k \leq p$, is the domain of variables for the joint axes in the set $\mathcal{L}_k$. Note that $\mathcal{D}$ is a subset of $\mathcal{C}$, the configuration space of a multibody without joint dependencies.

Example decomposition of the configuration space manifold. As an example, assume that $\mathcal{L}_1 = \{1\}$, $\mathcal{L}_2 = \{2, 3\}$, $\mathcal{L}_3 = \{4, 5\}$, $\mathcal{L}_4 = \{6, 7, 8\}$, and that joints 4 and 5 are completely dependent on each other. The configuration space manifold $\mathcal{D}$ is expressed as follows:

$$\mathcal{D} = \mathcal{D}_{\mathcal{L}_1} \times \mathcal{D}_{\mathcal{L}_2} \times \mathcal{D}_{\mathcal{L}_3} \times \mathcal{D}_{\mathcal{L}_4} \times \mathbb{R}^3 \times \text{SO}(3).$$

Note that

$$\begin{aligned}
\mathcal{D}_{\mathcal{L}_1} & \quad \text{is the domain of } \theta_1, \\
\mathcal{D}_{\mathcal{L}_2} & \quad \text{is the domain of } (\theta_2, \theta_3), \\
\mathcal{D}_{\mathcal{L}_3} & \quad \text{is the domain of } (\theta_4, \theta_5), \quad \text{and} \\
\mathcal{D}_{\mathcal{L}_4} & \quad \text{is the domain of } (\theta_6, \theta_7, \theta_8),
\end{aligned}$$

where $\theta_1, \theta_2, \ldots, \theta_8$ are the joint angles.

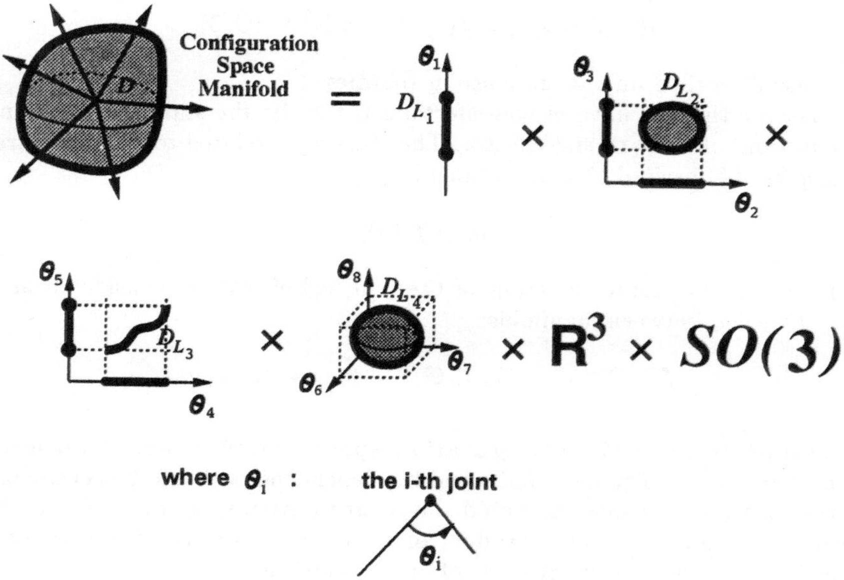

FIGURE 7.5. Decomposition of the configuration space manifold.

7.3. A Procedure for Checking Task Feasibility

A *task* is specified in terms of the velocities, the angular velocities, or both the velocities and the angular velocities of the end-effectors. In this section, we introduce the concept of *task feasibility* in the framework of the manifold maps. We consider the Jacobian matrix of the forward kinematic function in order to check task feasibility. We define regular and singular configurations using submatrices extracted from the Jacobian matrix. We then define task feasibility to mean whether the system can perform the given task or not. Finally, we present a method to check task feasibility, using the concepts of configuration space manifold and singular configurations.

Task and task feasibility.   A task is conceptually an instantaneous motion given to an end-effector. Tasks are classified into *positional*, *spherical*, or *spatial*, depending on whether they require only a translation, only a rotation, or both a translation and a rotation of an end-effector.

For example, a push, a twist, or a pinch are simple tasks for the hand and arm. More complicated tasks include, for example, writing with a brush in Shodo (the art of calligraphy), and throwing down the opponent in *Shorinji Kempo*.

Task feasibility is defined to be true if the given task can be performed, and false if not.

Partial translational velocities and partial angular velocities.   We describe a task by the velocities, the angular velocities, or both the velocities and the angular velocities of the end-effectors. This section is devoted to an explanation of these concepts.

Let q_j, for $0 \le j \le N$, denote the variables describing the joint angles and the position and orientation of the whole hand-and-arm system. Let the velocity (relative to the global coordinate system) of the origin of the i-th end-effector frame and the angular velocity of the i-th end-effector frame be denoted by the 3×1 vectors $\boldsymbol{V}^i$ and $\boldsymbol{w}^i$, respectively. The end-effector frame velocities are expressed in terms of the individual joint velocities and root body velocities as

$$\boldsymbol{V}^i = \sum_{j=0}^{N} \tilde{\boldsymbol{V}}^i_j \dot{q}_j \quad \text{and} \quad \boldsymbol{w}^i = \sum_{j=0}^{N} \tilde{\boldsymbol{w}}^i_j \dot{q}_j,$$

where $\tilde{\boldsymbol{V}}^i_j$ and $\tilde{\boldsymbol{w}}^i_j$, for $j = 1, \ldots, N$, are 3×1 vectors called the *partial translational velocity* and *partial angular velocity*. They are given by the equations

$$\tilde{\boldsymbol{V}}^i_j = \frac{\partial \boldsymbol{V}^i}{\partial \dot{q}_j}, \qquad \tilde{\boldsymbol{w}}^i_j = \frac{\partial \boldsymbol{w}^i}{\partial \dot{q}_j}.$$

Jacobian matrices.   When we are interested only in the translational motion of an end-effector, we call the end-effector *positional*. When the rotation of an end-effector is of concern, we call the end-effector *spherical*. When we are interested in both the translation and rotation of an end-effector, we call the end-effector *spatial*. In our application of the motion of the arm, the end-effectors are positional, so we limit our subsequent discussion to the positional case.

Using the partial translational velocity and the partial angular velocity vectors, the Jacobian matrix for the i-th end-effector can be written as

$$\boldsymbol{J}_i = (\tilde{\boldsymbol{V}}^i_1 \ \tilde{\boldsymbol{V}}^i_2 \ \ldots \ \tilde{\boldsymbol{V}}^i_N).$$

The Jacobian matrix is $3 \times N$ if the end-effector is positional. The columns of the matrix for a spatial end-effector can also be interpreted as *partial screw vectors*, since they express the contribution of each joint rotation and the position and orientation of the whole system to the screw motion of the output frame.

Singular configurations.   In order to consider singularity, a square matrix is extracted from the Jacobian matrix. In the case of a positional end-effector, we extract three arbitrary column vectors from the Jacobian matrix. We call the joint axes corresponding to the extracted column vectors the *extracted joint axes*. We call the extracted 3×3 square matrix the *extracted submatrix* for the extracted joint axes.

If the extracted submatrix is regular, that is, if it has nonzero determinant, we say that the extracted joint axes are in a *regular configuration*. If the extracted submatrix is singular, that is, if its determinant vanishes, we say that the extracted joint axes are in a *singular configuration*.

Singular configurations of the extracted joints can be studied in a more geometric fashion by considering the dimension of the space spanned by the column vectors in the extracted submatrices. In fact, the rank of the extracted submatrix is equivalent to the number of linearly independent column vectors of the matrix. If the dimension of the space spanned by the column vectors is less than the number of column vectors, the system is in a singular configuration. In other words, in a singular configuration, the column vectors become dependent and form a space with dimension less than the number of the column vectors.

Infeasible tasks.   A task is infeasible in the following cases.

Singular configuration. If the extracted joint axes are in a *singular configuration*, the extracted column vectors span a space whose dimension is less than the number of extracted column vectors. If the vector representing the task is not contained in this space, the task is infeasible in terms of the extracted joint axes.

Boundary transgression. Even if the vector representing the task is in the space spanned by all the column vectors in Jacobian matrix, when the system configuration is at the boundary of the configuration space, motions directed toward the outside of the configuration space manifold cannot be performed. We call this *boundary transgression*.

Disharmony. When the given task requires the (cooperative) simultaneous motions of multiple end-effectors, there are situations where the simultaneous motions are impossible even if the tasks given to the end-effectors could be performed separately. We call this *disharmony*.

Task feasibility checking. Now we introduce a procedure for checking task feasibility. Let A^i be a task given to the i-th end-effector. Then A^i is a vector of quantities V^i_j and w^i_j.

First step: check for singular configurations. Check the task feasibility of individual end-effectors one by one. Consider the i-th end-effector. Solve the following system of linear equations in order to find out the possible $\dot{q}$:

$$A^i = J_i \dot{q}.$$

If this system has no solution, the task is unfeasible at the i-th end-effector, so the overall task feasibility of the given task is false.

If this system of linear equations has a single solution or a set of solutions, record the constraints that determine the single solution or the set of solutions as $\mathrm{Co}_i(t)$.

Second step: check for disharmony. Check if all the end-effectors can perform the given tasks concurrently. There exists the possibility that the end-effectors can perform the tasks only sequentially, not concurrently.

For each i we know the constraint $\mathrm{Co}_i(t)$. Together the constraints form a system of linear equations. If this system has no solution, the end-effectors cannot perform the required motions simultaneously, and task feasibility becomes false.

If the system has a single solution or a set of solutions, record the constraints that determine the single solution or the set of solutions as $\mathrm{Co}(t)$.

Third step: check for boundary transgression. If the configuration $q(t)$ is in the interior of the configuration space manifold $\mathcal{D}$, the task is feasible.

If the configuration $q(t)$ is at the boundary of $\mathcal{D}$, check if there exists a solution that satisfies the constraint $\mathrm{Co}(t)$ obtained in the second step, and is directed to the interior of the configuration space manifold. This is done by checking one by one the projections $\mathcal{D}_{\mathcal{L}_k}$, for $1 \leq k \leq p$, of the vector representing the solution on the manifolds, representing the domains of variables for mutually independent sets of the joint axes. If such a solution exists, the task is feasible. Otherwise, it is infeasible.

7.4. Analysis of a *Shorinji Kempo* Technique

In this section, our model of hand and arm is applied to a type of martial arts, the *Shorinji Kempo*. One of its techniques, *Kirigote*, is analyzed using the concept of the task feasibility.

Suppose that two representative points of contact with the opponent are chosen, and that the forces applied at the points are given as shown in Figure 7.6. Green dots indicate the representative points of contact and the lines originating from the dots indicate the task given to the end-effectors. We consider that the tasks are positional. The lines indicate the velocities required

FIGURE 7.6. Tasks and the singular configuration in *Kirigote*.

for the end-effectors to perform the given tasks. Because the opponent loses balance and cannot resist the forces applied at the representative points, the task given to him is to perform the motion that results from the applied forces.

Using the task feasibility check procedure presented in Section 7.3, it is shown that this task is infeasible without shoulder translation. Let us analyze the reasons why this task is infeasible.

Singular configurations. First of all, consider the end-effector farther from the body. The three extracted joint axes, namely the elbow axis, the principal axis of the forearm, and the back and forth rotation axis of the wrist, are in a *singular configuration*. The absolute value of the determinant of the normalized extracted submatrix in terms of the extracted joint axes is computed and turns out to equal 0.007895. This value very near zero means that the matrix is almost singular. The column vectors are the partial velocity vectors. They are shown as the red lines originating from the end-effector in Figure 7.6. This figure shows that the partial velocity vectors in terms of the three axes are essentially on a plane. The task velocity vector is not contained in this plane. Accordingly, rotation about the shoulder joint is needed in order in order to perform the task.

Disharmony. Suppose that end-effector 1 is the end-effector nearer the body and end-effector 2 is the farther one.

Solving the equation $A^1 = J_1\dot{q}$ yields the solution: angular velocity 4.599256 about the shoulder joint axis $(0.466886, 0.260193, -0.845173)$. Solving $A^2 = J_2\dot{q}$ yields a possible solution: angular velocity 1.294783 about the shoulder joint axis $(0.097477, 0.247768, -0.963903)$. *Disharmony* occurs because the shoulder rotation is insufficient to satisfy these tasks concurrently.

Boundary transgression. Solving simultaneously the above two systems of equations yields a single possible solution, including the rotation of the wrist, which, however, cannot be performed due to *boundary transgression*. The configuration of the opponent at this time is at the boundary of the configuration space manifold. Consider $\mathcal{D}_{\mathcal{L}_k}$, which represents the manifold for the rotation angles of two joint axes of the wrist. The situation is depicted in Figure 7.7. The shape of $\mathcal{D}_{\mathcal{L}_k}$ was measured in the experiment. The black dot represents the projection of the configuration on $\mathcal{D}_{\mathcal{L}_k}$. The arrow originating from the dot represents the required joint angle velocities of the wrist. By calculation, it is $(-25.879189, -14.296033)$. The picture shows that the velocity is directed to the outside of $\mathcal{D}_{\mathcal{L}_k}$. The task is thus shown to be *infeasible* without shoulder translation. This means that the opponent must be thrown down since the shoulder position must move down in order to perform the task.

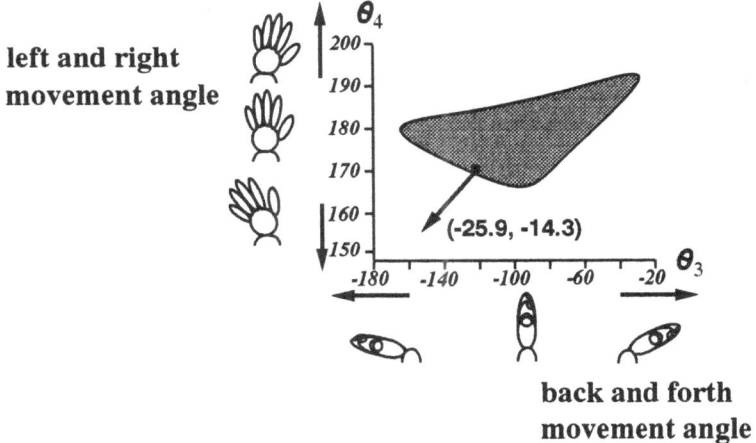

FIGURE 7.7. Boundary transgression in *Kirigote*.

8. Computer Examination of Surfaces and Morse Functions

This chapter considers the problem of encoding essential information about a compact surface in terms of a finite amount of data that can be stored and manipulated by computer. If all we want is the topological type of a surface, the genus and the orientability are sufficient. But in many applications, we would like to have a rough version of the geometry, and yet we don't have, or don't care about, an exact geometric description of the surface. Using Morse theory, it is possible to assign a *code*, or schematic representation, to a surface. We consider three problems:

(a) How to code a Morse function on a smooth manifold, and in particular on a surface.
(b) How to code a surface using a Morse function defined on it.
(c) How to reconstruct a surface if we know a coding of it.

8.1. The Reeb Graph and Molecules

Simple and complicated Morse functions. Recall that a smooth function on a smooth manifold is called a *Morse function* if all its critical points are nondegenerate.

Theorem 8.1.1 (Morse). *Let X be a smooth manifold. Morse functions are everywhere dense in the space of all smooth functions on X. Equivalently, any smooth function on X can be converted into a Morse function as a result of a perturbation as slight as desired.*

The perturbation splits degenerate critical points into a certain number of nondegenerate singularities, as illustrated in Figure 6.24.

In the set of all Morse functions, there exists a dense subset consisting of Morse functions such that different critical points have different values. In other words, critical points falling on the same level can be moved to closely spaced levels (see Figure 8.1). Morse functions with this property are called *simple*. Morse functions that have several critical points at the same level will be called *complicated*.

The fact that a complicated Morse function can be changed by small perturbations into a simple Morse function is useful in many applications

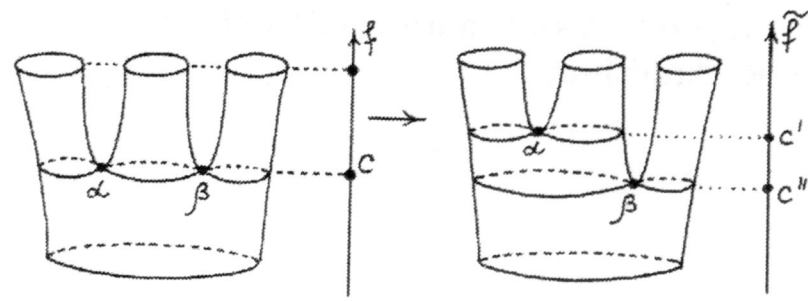

FIGURE 8.1.

and theoretical questions. There are many cases, however, when one cannot perturb the complicated Morse function to transform it into a simple one: for example, when the function has symmetries, and you want to preserve the symmetries. We will discuss this problem later.

The Reeb graph.   Now we restrict our attention to Morse functions f on a *compact surface M*. Recall that, for any regular value a, the level set $f^{-1}(a)$ is a union of circles. For any singular value, the level set has the possible local structures discussed in the preceding chapter, and dictated by Morse's Lemma: an isolated point (at a maximum or minimum), a transversal crossing of two curves (at a saddle), or a smooth curve (at other points of the level set). The global topological types for the connected components of the level set can then be easily determined if f is *simple*: a point, a circle, or a figure eight, that is, two circles joined at a transversal crossing. As we go through a critical level, components can merge, split, appear, or disappear, in ways that we will detail shortly.

Reeb [1946] proposed a way to condense the information about this evolution and rearrangement of level curves. His idea was to consider the quotient topological space obtained by collapsing to a point each connected component of each level set of f. More precisely:

Definition 8.1.2. Let $f : M \to \mathbb{R}$ be a real-valued function on a compact manifold M. The *Reeb graph* of f is the quotient space of M by the equivalence relation $\sim$ defined by "$x_1 \sim x_2$ holds if and only if $f(x_1) = f(x_2)$ and x_1, x_2 are in the same connected component of $f^{-1}(f(x_1))$."

For example, the Reeb graph of the height function of a surface embedded in $\mathbb{R}^3$ is the quotient space of the surface under the equivalence relation that identifies (x_1, y_1, z) and (x_2, y_2, z) if these two points lie in the same connected component of the horizontal cross section of the surface at height z. Each connected contour on each horizontal plane is represented by a point in the Reeb graph. Thus, the Reeb graph of the height function for the torus of Figure 8.2(a) is as shown in Figure 8.2(c); this is easy to see from Figure 8.2(b), which shows the cross sections.

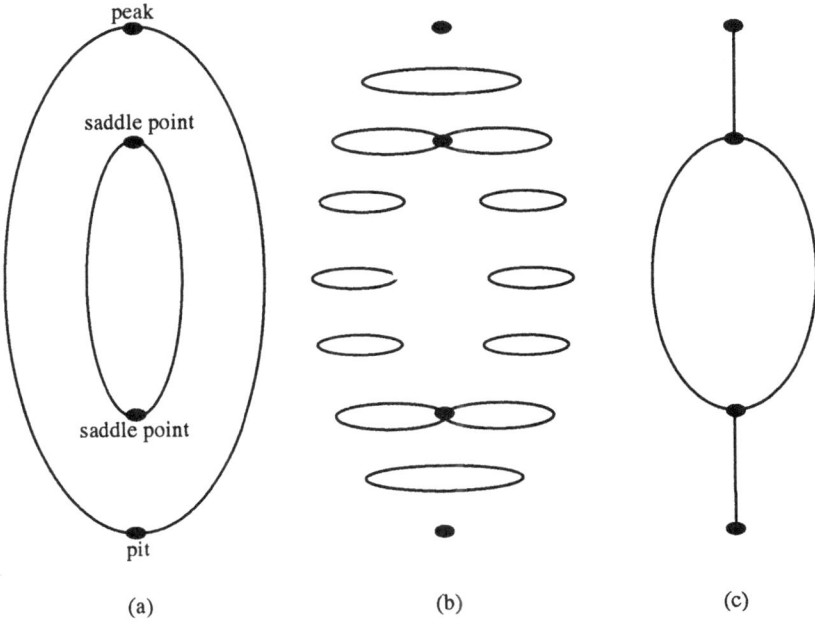

FIGURE 8.2. (a) A torus and the critical points of its height function. (b) Cross sections of that same torus. (c) The Reeb graph of the height function on the torus.

The name *Reeb graph* was introduced by René Thom. Reeb graphs can be quite helpful in the visualization of certain aspects of surfaces or higher-dimensional objects. Modern applications of Reeb graphs to computer geometry were developed by T. L. Kunii and Y. Shinagawa [1992], and will be discussed later in this chapter. Right now we discuss a refinement of Reeb graphs of surfaces, introduced by A. Fomenko, initially for the purposes of topological classification of integrable Hamiltonian systems; the idea was then extended to other topological problems and applications.

Labeled Reeb graphs, or molecules.   Definition 8.1.2 defines the Reeb graph as an abstract topological space. For ease of visualization, we can think of each connected component of a regular level set $f^{-1}(a)$ as a point at height a. As a changes, this point moves and sweeps a segment or arc—it cannot do anything else until we reach a critical level. At a critical level interesting things can happen: the arc may disappear, or branch off, etc.

Thus, the Reeb graph consists of two types of elements: arcs or edges, coming from the connected components of the part of the surface that is situated strictly between two critical levels; and nodes or vertices, coming from critical points. (In fact even at a critical level an arc of the graph may continue undisturbed, if the connected component that it represents does not contain a critical value at that level; in this case we don't consider the corresponding point of the graph a node, but an interior edge point.)

To distinguish different types of critical points, we label the nodes of the graph. In the simplest case, that of a *simple Morse function on an orientable manifold*, we need just two labels. Nodes that represent a maximum or a minimum are labeled A; they are leaves of the graph, that is, only one edge goes out of them, pointing up in the case of a minimum, or down in the case of a maximum. This is illustrated in Figure 8.3.

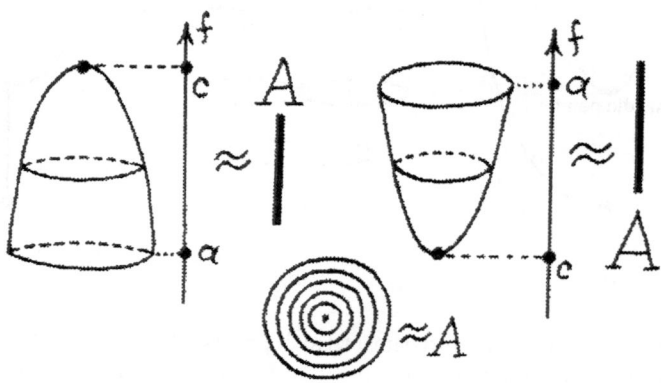

FIGURE 8.3.

What about nodes that represent the connected component of a saddle? Because the function is simple, the component is topologically a figure eight, as already mentioned. To understand the evolution of the level curves, we must then determine what the surface looks like in the neighborhood of this figure eight. Morse's Lemma says that on opposite sides of the critical point the function has the same sign. The assumption that the surface is orientable then requires that the adjacent level curves be as in Figure 8.4. On one side the level set evolves into a single circle, and on the other into two circles.

In the Reeb graph, then, we have a vertex with three edges coming out: two pointing up and one down, or vice versa. We label such vertices B; they represent an *orientable saddle transformation*. See Figure 8.5.

We call the labeled nodes of the Reeb graph *atoms*, and the labeled graph itself a *molecule*. Two examples of molecules are given in Figure 8.6.

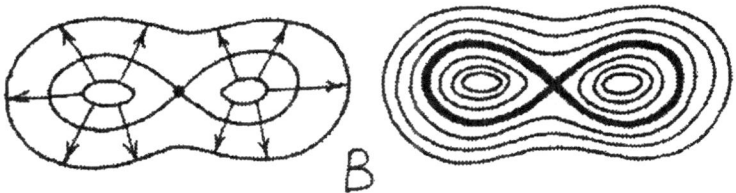

FIGURE 8.4.

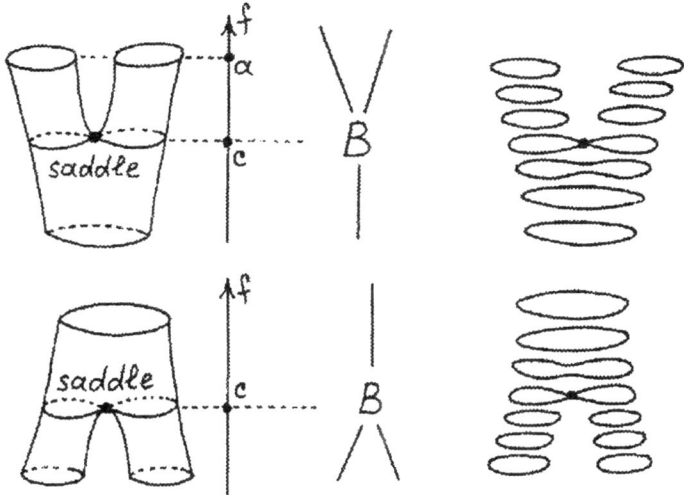

FIGURE 8.5.

Let us note that the edges incident to an atom of the molecule can be non-equivalent and therefore we should distinguish them. To show what we mean, consider, for example, the atom B. This atom denotes the bifurcation under which one circle (say S_1) transforms into two new circles (say S_2 and S_3). see Figure 8.4. These two circles S_2 and S_3 are naturally equivalent in the sence that there exists a diffeomorphism of a neighborhood of the singular level line (i.e., the figure eight) into itself that preserves the function f but permutates these circles (on Figure 8.4 this diffeomorphism is just the central symmetry with respect to the singular point). On the other hand there is no such a diffeomorphism which maps S_2 or S_3 onto S_1. Thus, three edges incident to B can be naturally separated into two classes $\{S_1\}$ and $\{S_2, S_3\}$. In other words, we do not need to distinguish S_2 and S_3, but should distinguish S_1 from them.

To be able to extract this information from the molecule we can make all its edges oriented according to increasing of f (by putting arrows on them showing the direction in which f increases). As a result, the edges incident to a certain atom will be separated into two classes(with respect to the atom), namely, incoming and outgoing ones.

However, in the case of simple Morse functions we can use the following agreement: we draw the molecule in such a way that the direction of increasing of f is "upwards"(see Figure 8.6).

Remark. In the case of complicated atoms in order to distinguish the edges of the molecule it is not sufficient to use the orientation only, because even the edges of the same type can be non-equivalent.

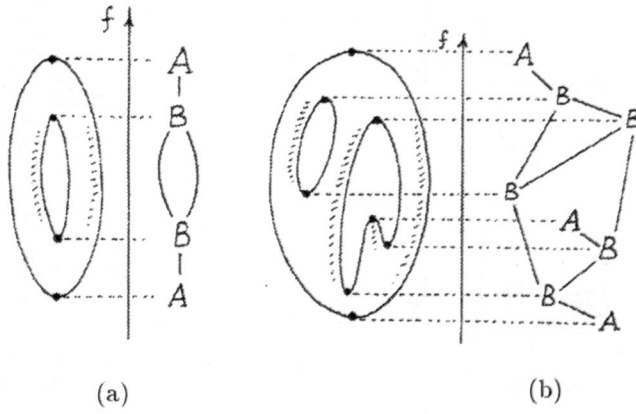

(a)   (b)

FIGURE 8.6.

The molecule of a Morse function is not necessarily a planar graph, that is, it may not be possible to arrange it in the plane without edges crossing. It is of course possible to arrange it in $\mathbb{R}^3$, and there are many nonequivalent ways of doing this.

The molecule as a code.   Having assigned a molecule to any simple Morse function on an orientable compact surface, we can now ask whether we can reconstruct the surface and the function from the molecule. Let us introduce a natural equivalent relation on the set of pairs (M, f), where M is a closed oriented surface and f is a Morse function on it. We will say that two pairs (M, f) and (M', f') are fiber equivalent if there exists a diffeomorphism $\xi :$ $M \to M'$ which maps level lines of f to those of f'.

In the case of a simple Morse function f its molecule can be considered as a code that allows to define the equivalence class of (M, f) uniquely. Before formulating this result let us note that the pairs (M, f) and $(M, -f)$ are obviously equivalent because the functions f and $-f$ have the same level lines. The only difference between the corresponding molecules is the orientations of their edges. For f and $-f$ these orientations are opposite and we come to the following natural definition. We will say that two molecules W and W' with oriented edges are *isomorphic* if there is a homeomorphism between then which

(a) preserves its atom labels and
(b) either preserves the orientation on the edges or change it on all the edges simultaneously.

Proposition 8.1.3. *Let (M, f) and (M', f') be two compact orientable surfaces provided with simple Morse functions f and f'. Then (M, f) and (M', f') are fiber equivalent if and only if the corresponding molecules W and W' are isomophic. In particular, if $W \approx W'$ then the surfaces M and M' are diffeomorphic.*

Note that if we have two diffeomorphic surfaces embedded in $\mathbb{R}^3$, with Morse height functions, the corresponding molecules may be different even if the embeddings are equivalent—think of a round sphere and a cactus. (The right notion of equivalence for embeddings is *isotopy*; informally, two embeddings $M \to \mathbb{R}^3$ are isotopic if one can be smoothly transformed into the other without self-intersections arising.)

Conversely, the molecules can be isomorphic even if the embeddings are not equivalent. As an example, consider Figure 8.7. Here we see two non-isotopic embeddings of the torus T^2 in $\mathbb{R}^3$, and the corresponding height functions. Their molecules are isomorphic. The first embedding is isotopic to the standard embedding, but the second is knotted, and so cannot be isotopically deformed into the standard embedding.

We have assumed so far that the Morse function under   consideration is simple. For a complicated Morse function, some vertices of the Reeb graph may correspond to (components of) level sets containing several critical points, so the topology of these level sets may be complicated. We will return to this point later.

8.2. Surface Codings for Computer-Aided Design

Solid modeling involves the representation of the shapes of three-dimensional objects. Many types of representation are possible. In the boundary representation method, for example, an object is represented by means of the vertices, edges, and faces of its boundary; faces need not be planar, but can be taken from a family of functions large enough to approximate the surface of the object being modeled to within the desired precision—splines, Bézier surfaces, and so on. In constructive solid geometry (CSG), by contrast, an object is represented as the result of set operations (union, intersection, set

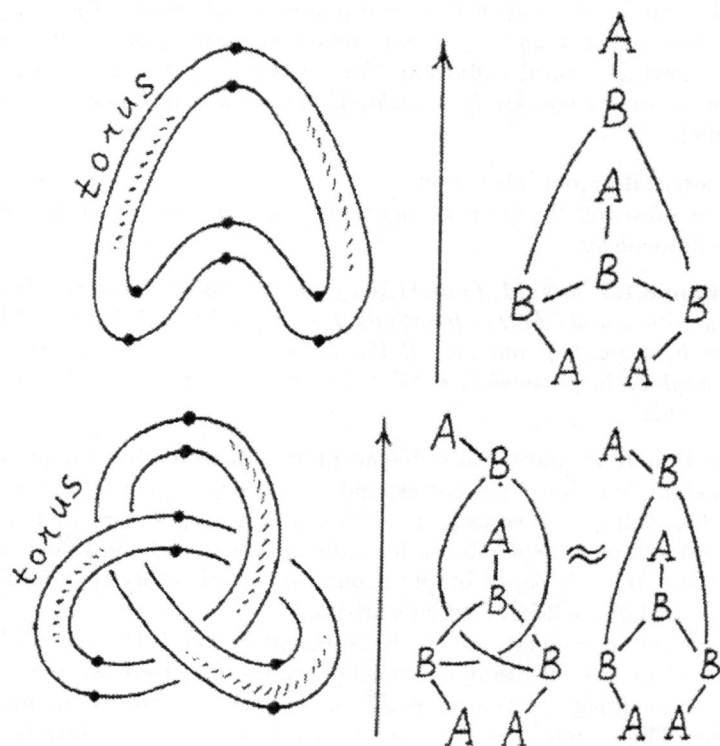

FIGURE 8.7.

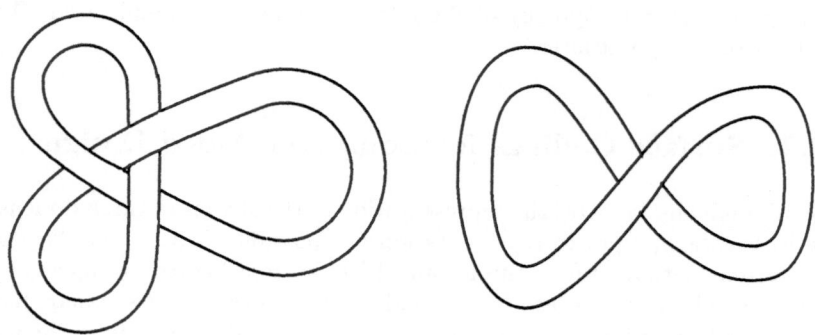

FIGURE 8.8. A pair of surfaces with the same molecule.

difference) on primitives such as cubes, spheres and cylinders. Each type of representation has certain advantages, but the traditional methods are often inadequate for the modeling of complex shapes of natural objects such as human organs.

We will describe here a method of surface representation, introduced in [Shinagawa et al. 1991], that has certain advantages over traditional methods, in that it makes it easy to capture the topology of a surface or set of surfaces, and is powerful enough to handle even very complex shapes. The method codes a surface by describing the topological evolution of the level sets of a Morse height function defined on it.

Extending the Morse coding. Consider a smooth (or C^2) compact surface embedded in $\mathbb{R}^3$, and assume that its height function is a Morse function. We are interested in representing the surface by means of the evolution of its horizontal cross sections (level sets) as the height changes. A nonsingular cross section is made of (topological) circles embedded in a plane, and by the Jordan curve theorem, which says that a closed curve in the plane has an inside and an outside, there is a natural hierarchy among the circles in a given level set, defined by what circles are inside what other circles. This will be discussed in detail shortly. Moreover, as long as we don't cross a level set, this relationship cannot change. When we do cross a level set, we have a change in the topology of the cross section, as explained in the preceding section. Such changes are recorded by the Reeb graph, but as we have seen the Reeb graph is too coarse to distinguish between different embeddings of the same surface (Figure 8.8).

What an extended coding has to add to the Reeb graph is a record of the way the cylinders (which correspond to edges of the Reeb graph) are exchanged and turn around each other between two successive critical values. One way of doing this is to record the configurations of enough intermediary levels such that an equivalent of the surface can be reconstructed using interpolation [Shinagawa and Kunii 1991]. Such a section-oriented approximation is always possible under the hypotheses of a C^2 embedded compact surface with Morse height function.

An example of a coding system. We now present an example of a coding system that describes a surface by the changes in the hierarchical structure of contours on cross sectional planes caused by attaching k-handles (see page 112). Operators are introduced to describe the attachment of handles, and the surface is coded using them. An iconic representation of handles attached by the operators helps visualize the structure of the surface being coded. We still assume that the surface is compact, smooth (or C^2), embedded in $\mathbb{R}^3$, and that its height function is a Morse function.

The major advantage of this coding system is that the topological integrity of the resulting code is guaranteed under these assumptions. Not all changes in the hierarchical structure of contours between cross sectional planes define a valid physical surface, as shown in Figure 8.9. As can be seen from this example, the topological integrity is not guaranteed by merely coding the hierarchical structure of contours on the cross sectional planes. It is necessary to code the operations that cause the changes in the hierarchical structure of contours. This situation is similar to that in [Baumgart 1975; Mantyla and Sulonen 1982], where topological integrity is maintained by coding changes in the structure of vertices, edges, and faces using so-called *Euler operators*.

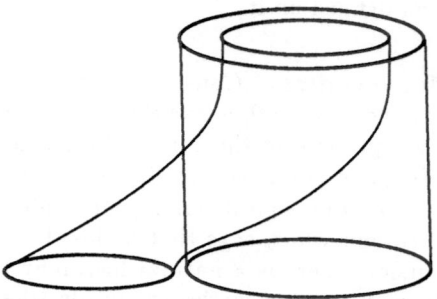

FIGURE 8.9. Unacceptable change in the contour structure.

The parent-child relation.   Recall that we introduced a hierarchy among the circles (contours) that make up a noncritical level set of our surface. A contour that is enclosed in another, with no other intervening contours, is said to be a *child* of it; the enclosing contour is the enclosed contour's *parent*. The parent-child relation can be nested, as shown in Figure 8.10(a), where we have numbered the contours for convenience; #1 represents contour 1, and so on. Thus #1 is the parent of #2, which in turn is the parent of #4. For convenience, contours that have no parents, such as #1 and #7, are considered children of a *virtual contour* #0. Then every contour apart from #0 has exactly one parent, and we can describe the hierarchy in the form of a tree, as shown in Figure 8.10(b). Contours having the same parent, like #2 and #3 in Figure 8.10, are called *siblings*.

One last bit of terminology will be important. Because the contours describe a surface that divides space into an inside and an outside (and which, in practical applications, is the boundary of a solid object), they can be divided into two types. Suppose we shade the areas of the plane that correspond to the region inside the surface, as in Figure 8.10(a). If the area immediately inside a contour is shaded, the contour is *solid*; if it is the area immediately outside that is shaded, the contour is *hollow*. Thus, contours #1, #4, #5, #6 and #7 are solid, while #2 and #3 are hollow. It is easy to see that solid and hollow contours alternate as we go down any particular branch of the

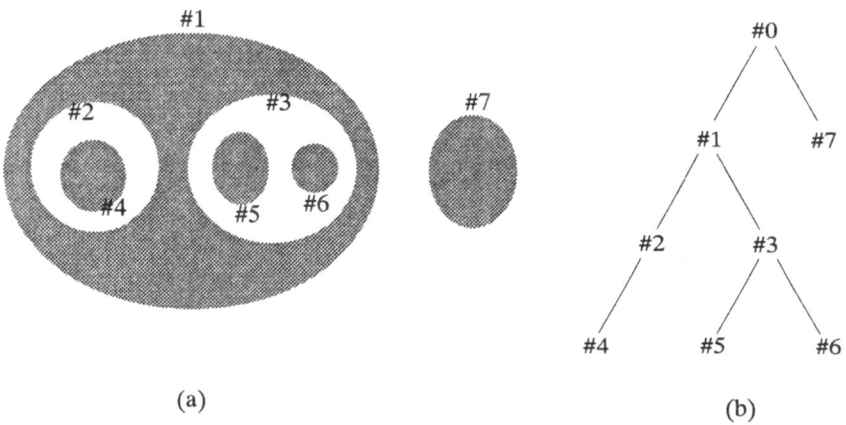

FIGURE 8.10. The parent-child relation among contours (a) and its tree representation (b).

tree; in particular, all outer contours, children of the virtual contour #0, are solid. We call contour #0 hollow by convention.

Building up the surface. To proceed with the description of our technique, we will represent the parent of countour n by $parent[n]$, and the children of countour n by an array $children[n]$, where the indexes of the children are listed in some order. For example, in Figure 8.10 we have $parent[1] = 0$ and $parent[2] = 1$; we also might have $children[3] = (5, 6)$, in which case we write

$$children[3] \uparrow [1] = 5 \quad \text{and} \quad children[3] \uparrow [2] = 6.$$

This abstract notation does not commit us to a particular type of storage structure for the information; in an actual implementation one might use simple arrays or more sophisticated structures for tree handling, so long as the storage method chosen makes it possible to perform efficiently the operations we now describe.

The building of a surface is done by the application of four operators, whose geometric function is the attaching of k-dimensional handles H_k, as described on pages 112ff. The operators are PUT_H0, PUT_H1_MERGE, PUT_H1_DIVIDE, and PUT_H2. The attaching of a one-handle can either split a contour or merge two contours, so there are different operators to deal with the two cases. The construction proceeds from the top to the bottom of the object, and ends when it is no longer possible to attach handles. The status of the surface being constructed by the operators is represented by the contour structure on the cross sectional planes.

As an example of how the operators work, we take the construction of a torus, shown in Figure 8.11. The three columns of the figure represent the evolution in three schematic ways. Column (a) shows icons that evoke a lateral view of the surface; column (b) shows a perspective view; and column (c) shows the approximate cross-section at each stage.

(1) First of all, $PUT\_H2(0)$ is called to add a two-handle H_2, as shown in row (1) of Figure 8.11. This creates a new contour, as shown in column (c). The single parameter of PUT_H2 says within what existing contour the new contour is to be created, and the value returned is the index of the new contour; this is a number not used before. We may as well number contours in the order in which they are created. Here, then, $PUT\_H2(0)$ returns the value 0, which means that contour #1 is created inside #0, the virtual contour.

One nonobvious bit of information that we have to attach to each contour is whether it is present at the current level, or, equivalently, whether handles can be attached to it. This will be referred to by saying that the status of the contour is $ENABLED$. The status of a newly created contour is always $ENABLED$ at first.

(2) Next, a one-handle H_1 is attached to the existing handle. The operation that does this is $PUT\_H1\_DIVIDE(1, nil, INSIDE)$; see row (2) of Figure 8.11. This splits the contour into two; one of them keeps the index of the original contour (the first argument to PUT_H1_DIVIDE), while the other gets a new index (the return value of the operator). Two more arguments are necessary for PUT_H1_DIVIDE. When the third parameter is $INSIDE$, the resulting contours are siblings. (The alternative will be explained later.) The second parameter is the list of children of the old contour that should be transferred to the new one. Here this list is empty.

(3) We then attach another H_1, to reconnect #1 and #2. The command for this is $PUT\_H1\_MERGE(1, 2)$; see row (3) of the figure. The arguments 1 and 2 represent the two contours to be merged, and the index of the result is the first argument. The second contour is removed from the list of children of its parent, and its status is changed from $ENABLED$ to $DISABLED$, indicating that no handle can be attached to the contour anymore.

(4) Finally, $PUT\_H0(1)$ eliminates contour #1 by capping it off with a zero-handle H_0. The single argument is the index of the contour that's being capped off; its status is changed from $ENABLED$ to $DISABLED$. Note that by this time all the children of the contour that's being eliminated must be $DISABLED$. This concludes our example; there are no more $ENABLED$ contours, so no more attachments are possible.

We now go systematically through all the cases, describing the effect of the handle-attaching operators and the meaning of the arguments with which they are called.

An H_2 has to be created inside another contour, and again the contour number is the only information needed; hence $PUT\_H2(n)$ is necessary and sufficient for this case.

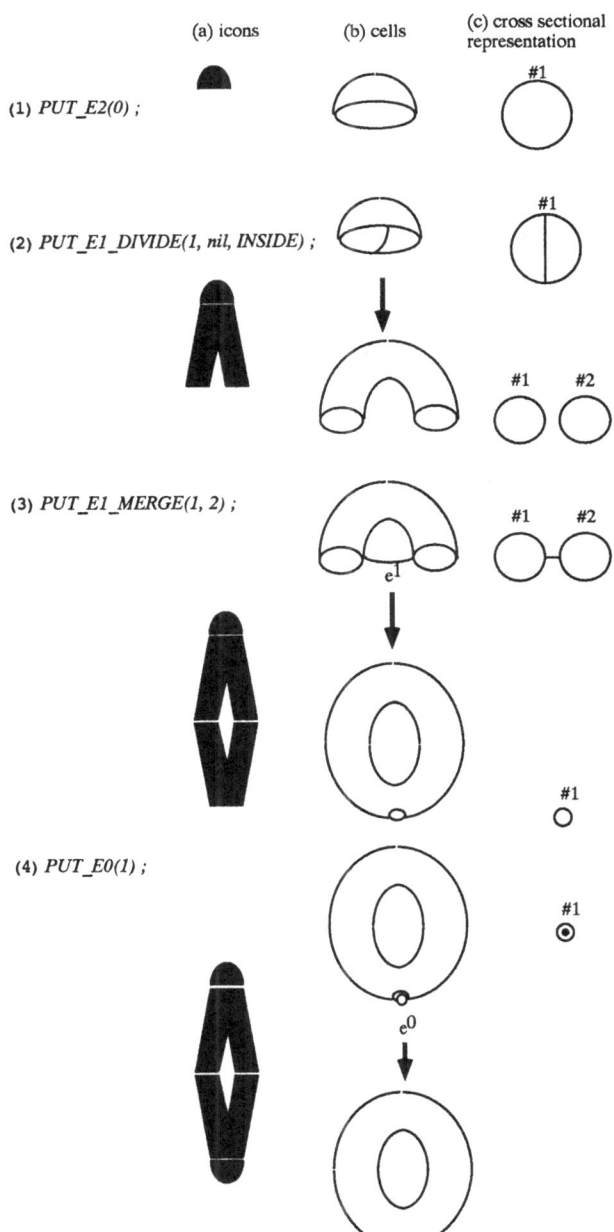

(a) icons   (b) cells   (c) cross sectional representation

(1) *PUT_E2(0)* ;

(2) *PUT_E1_DIVIDE(1, nil, INSIDE)* ;

(3) *PUT_E1_MERGE(1, 2)* ;

(4) *PUT_E0(1)* ;

FIGURE 8.11. Constructing a torus.

An H_0 has to be attached to a contour other than the virtual contour, and again that is the only information needed. Therefore $PUT\_H0(n)$ is necessary and sufficient for this case.

An H_1 has to be attached to two contours (not necessarily distinct) with the condition that no other contour separates them. That is, it can connect a contour with its parent, a sibling, a child, or itself. We take each case in turn; refer to Figure 8.12, top.

(A,C) By reversing roles, we see that the case of a contour connecting with its parent is equivalent to that of a contour connecting with its child. See segments (A) and (C) in the figure, which represent where the bridges (H_1's) are to be attached. When contour $\#n$ is merged with its parent, it vanishes (becomes *DISABLED*), and the children of $\#n$ become children of $parent[parent[n]]$. This is done by $PUT\_H1\_MERGE(parent[n], n)$.

(B) When an H_1 is attached between two sibling contours m and n, we arbitrarily choose one index, say n, to be that of the resulting contour, and the other disappears. All the children of $\#m$ become children of $\#n$. This is done by $PUT\_H1\_MERGE(n, m)$.

When an H_1 is attached to the same contour at both ends, it can be either outside or inside the contour. These two possibilities are indicated by segments (D) and (E) in Figure 8.12, top.

(D) An H_1 looping from a contour to itself on the outside transforms it into two nested contours. The parent contour gets the index of the original contour, say $\#n$, and the child gets a new number, say $\#k$. Thus $\#n$ gains exactly one child in the process. The contours lying outside of $\#n$, enclosed by $\#n$ and the handle, become the children of $\#k$, and have to be removed from the list of children of $parent[n]$. If these contours are listed in *clist*, the whole operation is performed by calling $PUT\_H1\_DIVIDE(n, clist, OUTSIDE)$. This operation is the inverse of Case A.

(E) An H_1 looping from a contour to itself on the inside transforms it into two sibling contours; one is given a new number, say $\#k$, and the other inherits the number $\#n$ of the original contour. The children of $\#n$ must be split between $\#n$ and $\#k$. If the children to be passed to $\#k$ are listed in *clist*, the operation is performed by $PUT\_H1\_DIVIDE(n, clist, INSIDE)$. This is the inverse of Case B.

This takes care of all possible cases.

To help visualize the structure of the surface being coded by the operators, we use a graphical representation of the handles that make up the Reeb graph of the surface, by means of the icons depicted in Figure 8.13. Gluing two handles is done by letting the flat top of one and the flat bottom of the other coincide. A dummy icon can be used to connect handles that do not align, as the example in Figure 8.14 will make clear. Icons can also be adapted to fit with each other; for example, reflecting an icon in a vertical line is permissible, and does not change its meaning. Handles for hollow contours

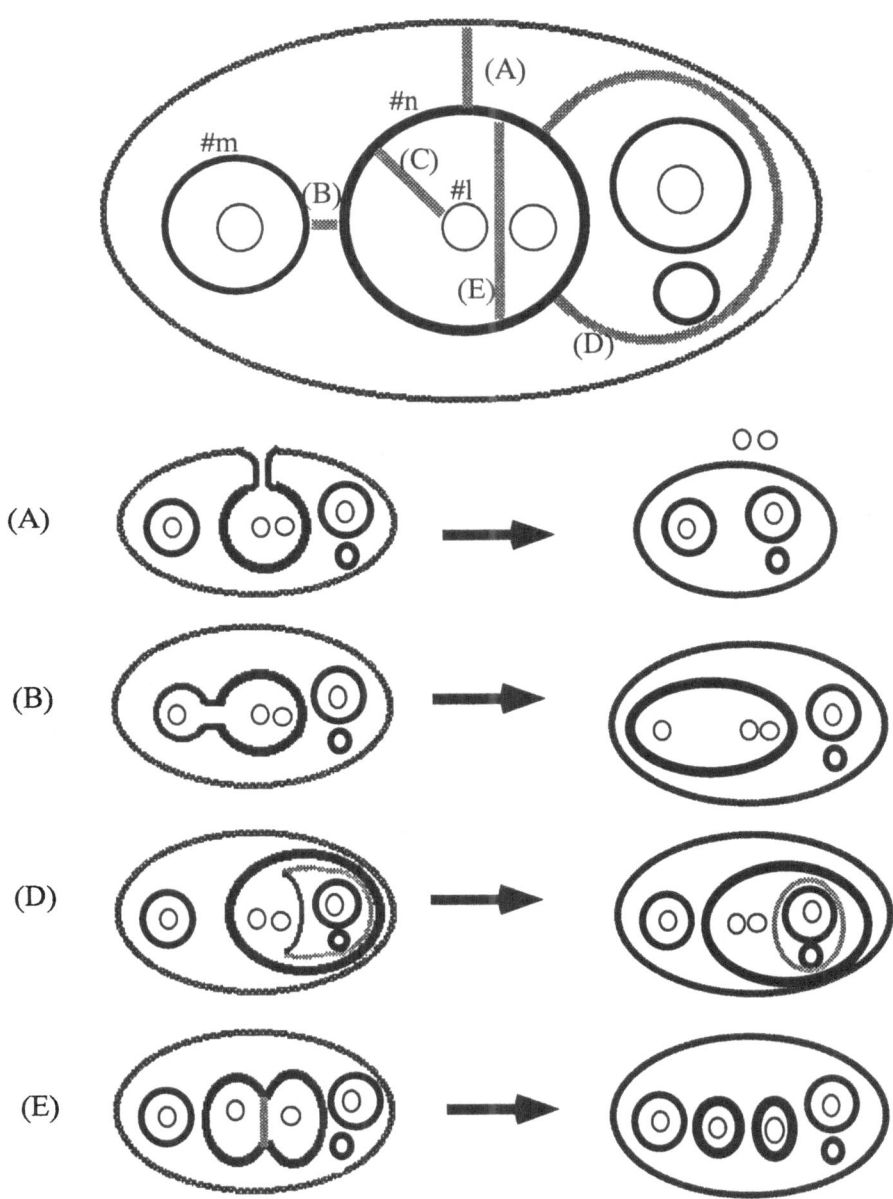

FIGURE 8.12. Classification of the effects of a one-dimensional handle.

are depicted by white icons, while those for solid contours are depicted by black icons. For H_1, there are four types of icons, corresponding to the cases stated above; note that the icons for the mutually inverse Cases D and A are symmetric by reversal of the vertical coordinate, and likewise for Cases E and B.

The icons of children are drawn inside the icons of their parents, as shown in Figure 8.14; thus the figure represents the Reeb graph as well as the parent-child relationship. The icons preserve the contour structure to which they are attached. For example, when an H_1 is attached to the handles as in Figure 8.14, it preserves the parent-child relationships of each of the contours involved. The attachment of the icons should obey the rules of the operators; H_1 for Case B, for instance, can be attached only to sibling contours.

Dummy icons are inserted where necessary to adjust the height of handles (Figure 8.14). Because the Reeb graph may not be a planar graph, dummy icons can also be used to interchange handles (together with their inner structure) so that H_1 can be attached to adjacent handles (Figure 8.15).

To preserve the hierarchical contour structure, a dummy icon cannot exceed the bounds of the parent of the contour to which it is attached, or intrude into other contours. For this reason, only sibling contours can be interchanged using the dummy icons. Note also that when the icon for *PUT_H1_MERGE* of Case A shown in Figure 8.13 is attached to the diagram at a contour, say #n, the triangle at the top must match the rightmost child of #n, say #m, and the children of #m are transfered to the outside of #n. When the mirror image icon is used, the triangle must match the leftmost child of #n.

8.3. Modeling the Contact of Two Complex Objects Using the Reeb Graph

This section presents another application of surface coding based on Morse theory and the Reeb graph, this time to the field of dental diagnosis. The interference of the motion paths of objects has been an important research subject. Much research has been done on collision detection in the fields of CAD/CAM (computer-aided design and manufacturing) and robotics; see, for example, [Boyse 1979; Cameron 1989; Uchiki et al. 1983; Noborio et al. 1987; Moore and Wilhelms 1988]. Voxel-based and octree-based methods were presented in [Jackins and Tanimoto 1980; Meager 1982]. *Motion planning* [Lozano-Pérez and Wesley 1979; Lozano-Pérez 1983; Schwartz and Sharir 1983; Leven and Sharir 1987; Level and Sharir 1987; Inamoto 1993] is another field where collision detection has been studied. Inamoto [1993] characterized this problem as four-dimensional interference detection, and solved it using hexadecimal trees.

There has been, however, no work on characterizing the mutual interference of objects that are in close contact in irregular ways. An example of a

FIGURE 8.13. Icons used in the coding method.

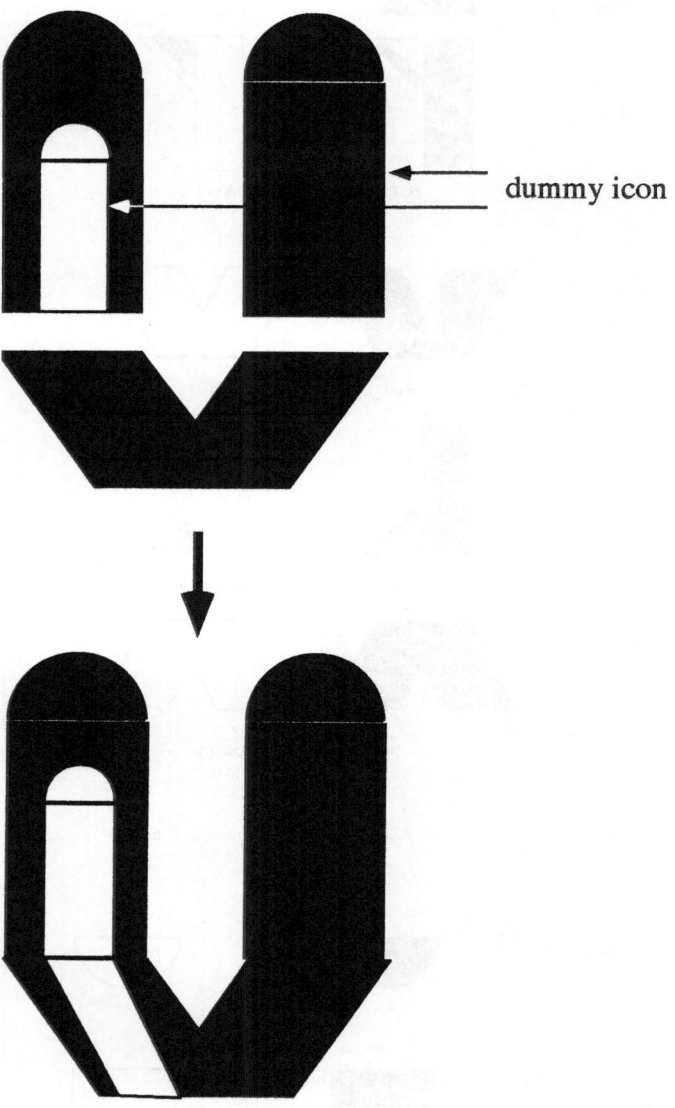

dummy icon

FIGURE 8.14. When the icon for an H_1 handle is attached to two sibling contours, it is modified so that the structure inside the contours is preserved.

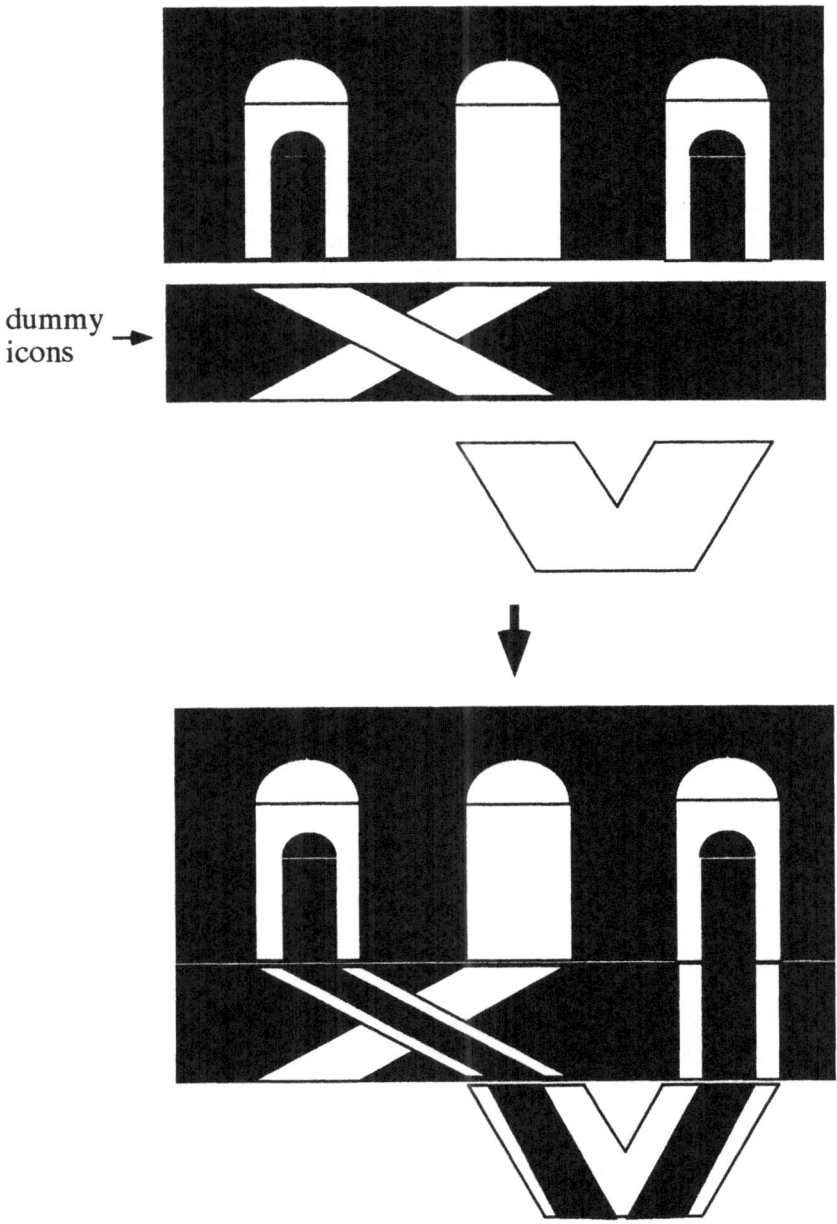

dummy
icons

FIGURE 8.15. To make contours that should be connected by an H_1 adjacent in the iconic representation, one can interchange them using dummy icons.

problem of this type is the process of chewing: analyzing the spaces between teeth, and the exact way in which they enter into contact, is important in understanding how the food gets cut and ground down. In other words, we are interested in the structure of the spaces between solid objects, and of their regions of contact. Previous work on detecting and locating collisions does not help understand or model such processes.

This section proposes a novel method to characterize the contact of objects. The method is applied to the actual problem of *dental articulations*, to demonstrate its effectiveness. Our characterization is based on the topology of the complement of the three-dimensional objects, as it changes over time; this temporal evolution of a three-dimensional region of space can be regarded as a set in four-dimensional space $\mathbb{R}^4$. Changes in the *grinding area*—the portion of the complement whose thickness is less than a given threshold δ—are examined. The topological structures are analyzed using a coding method based on Morse theory and the Reeb graph [Shinagawa et al. 1991].

Topological characterization of the interference of objects. Consider two three-dimensional objects O_1 and O_2, whose position varies with time. Let $T_1(t) \subset \mathbb{R}^3$ be the set occupied by at time t by O_1, and likewise for $T_2(t)$; we assume that $T_1(t)$ and $T_2(t)$ are closed sets. In our example, O_1 and O_2 are an upper and a lower tooth that come into contact. The complementary space $C(t)$ at time t is defined by

$$C(t) = \overline{T_1(t)^\circ \cup T_2(t)^\circ},$$

where T_i° denotes the interior of T_i, for $i = 1, 2$, and the bar denotes complement.

The grinding area G. Next, we analyze the topology of the *grinding area* $G(t)$ at time t. This set is defined by

$$G(t) = \{p \in C(t) : d(p) \leq \delta\},$$

where δ is a given threshold value and $d : \mathbb{R}^3 \to \mathbb{R}$, called the *thickness function*, is defined as follows. Let l be the line passing through p and parallel to the z-axis; then $d(p)$ gives the length of the connected component of $G(t) \cap l$ that contains p. We call $d(p)$ the *thickness* of $C(t)$ at p in the z-direction.

When the boundaries of T_1 and T_2 can be locally represented by functions $z = f_1(x, y)$ and $z = f_2(x, y)$, the thickness at $p = (a, b, c)$ is given by

$$d(p) = \left| f_1(a, b) - f_2(a, b) \right|,$$

assuming that $f_2(a, b) < c < f_1(a, b)$ (see Figure 8.16).

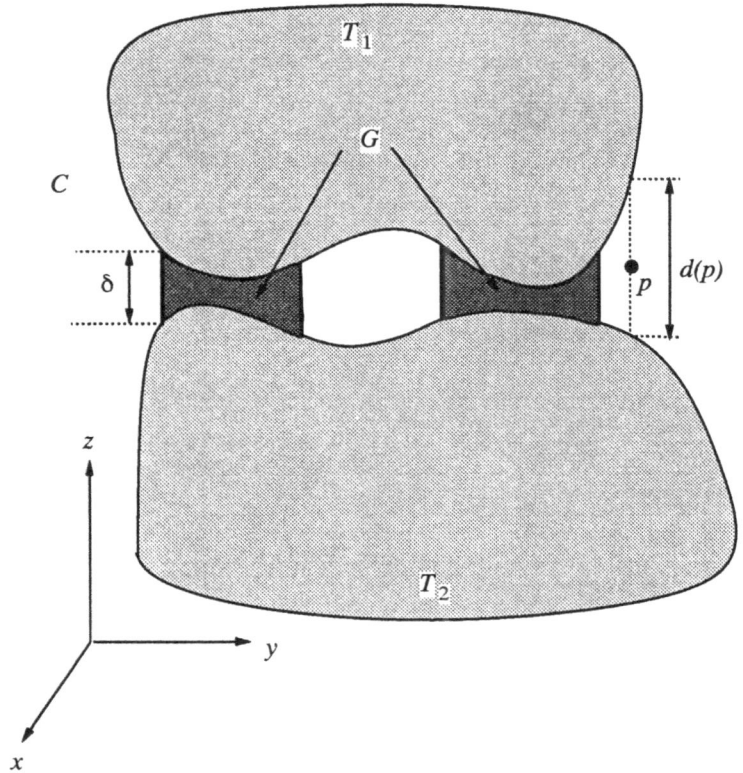

FIGURE 8.16. Thickness at the point $p \in C(t)$.

Representing structural changes to the grinding area. Our model assumes the existence of a coordinate system such that l intersects $C(t)$ at most once, in the sense that the intersection is either empty or connected (a line segment). In the case of teeth, the xy-plane can be taken as the the *occlusal plane* (a plane approximating the grinding or biting surface of the tooth; see Figure 8.17).

When such a coordinate system exists, the topological structure of $G(t)$ is determined by the projection of this set onto the xy-plane, namely

$$\pi(G(t)) = \{(x, y) : (x, y, z) \in G(t)\}.$$

This projection sweeps, as t varies, a subset

$$H(G) := \{(x, y, t) : t \in \mathbb{R}, (x, y) \in \pi(G(t))\} \subset \mathbb{R}^3,$$

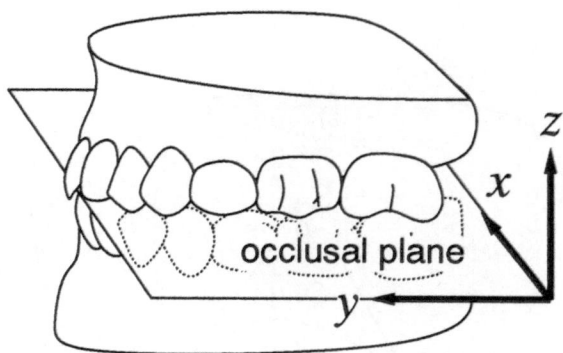

FIGURE 8.17. The coordinate system used in the dental articulation problem.

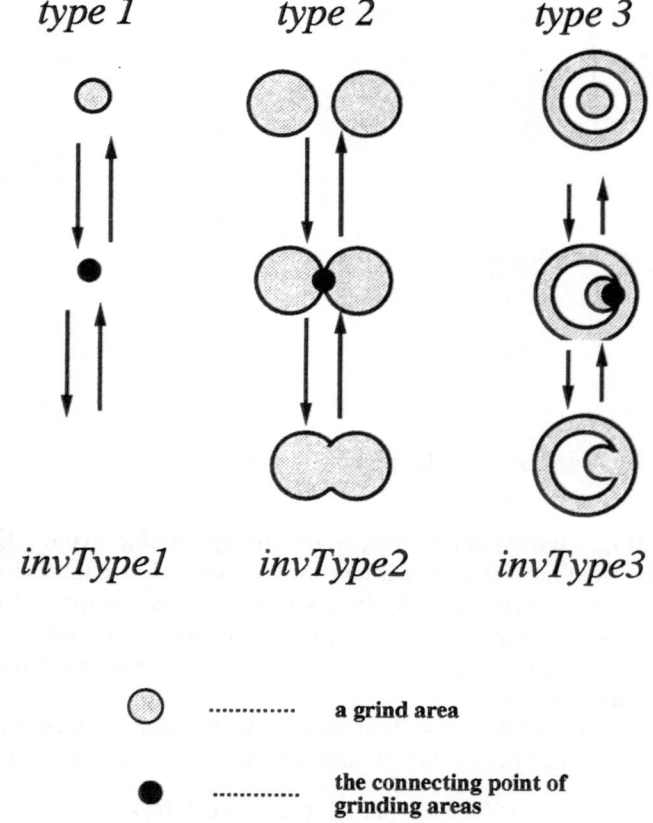

FIGURE 8.18. Possible types of change in the projected image of G and its complement.

and the structural features of this time evolution are encoded in the topology of $H(G)$. We will use the Reeb graph of $H(G)$ and of its complement to represent the structural change in the regions.

We regard the boundary of $\pi(G(t))$ (for a fixed t) as a collection of circles (closed contours) in the plane, at least in the generic case. As in Section 8.2, we can then talk about parent, child, and sibling contours.

The three types of changes in the topology of the grinding area. The changes in topological structure of regions in $\pi(G(t))$ can be classified into three types and their inverses, as shown in Figure 8.18.

In a type 1 change, a new contour appears, enclosing a new region. In the inverse change, a contour and the enclosed region disappear. This corresponds to the operators *PUT_H2* and *PUT_H0* of the preceding section.

In a type 2 change, a contour merges with a sibling, and the regions they enclose are combined. In the inverse change, a region splits into sibling regions. This corresponds to the case B of *PUT_H1_MERGE* and case E of *PUT_H1_DIVIDE*.

In a type 3 change, a parent and a child contour merge; the region enclosed by the child and the one outside the parent are combined. In the inverse change, a region splits into two, one inside the other.

The corresponding changes in the topology of $G(t)$ are shown in Figure 8.19. The relevant portions of the Reeb graphs of $H(G)$ and of its complement $\overline{H(G)}$ are shown in Figure 8.20. Note that here we are representing the Reeb graphs of *solid regions*, rather than that of surfaces. Again, the Reeb graph cannot distinguish between types 2 and 3; in order to distinguish between the two types, one must use the parent-child relationships in the projected image [Shinagawa et al. 1991].

Examples. Consider the structural changes of the projected images of the grinding area on the xy-plane shown in Figure 8.21. Shaded regions represent G. The Reeb graphs of G and $\overline{G}$ are also shown.

Between time t_1 and t_2, an inverse type 1 change occurs in G, leading to the appearance of region #2, a child of #0. (Here we are numbering regions rather than contours.) The number of holes of #0 increases by one.

Between time t_2 and t_3, a type 2 change occurs in G, and #2 becomes connected to its sibling region #1. The number of holes of #0 decreases by one. At the same time, an inverse type 1 change occurs in $\overline{G}$, increasing by one the number of holes of #1.

Between time t_3 and t_4, a type 3 change occurs in G, connecting #2 to its grandparent region #0 (that is, connecting the parent-child contours). The number of holes of #1 decreases by one.

Application to chewing. We are now ready to discuss how chewing can be described by our model. Foods are ground as the topological structure of

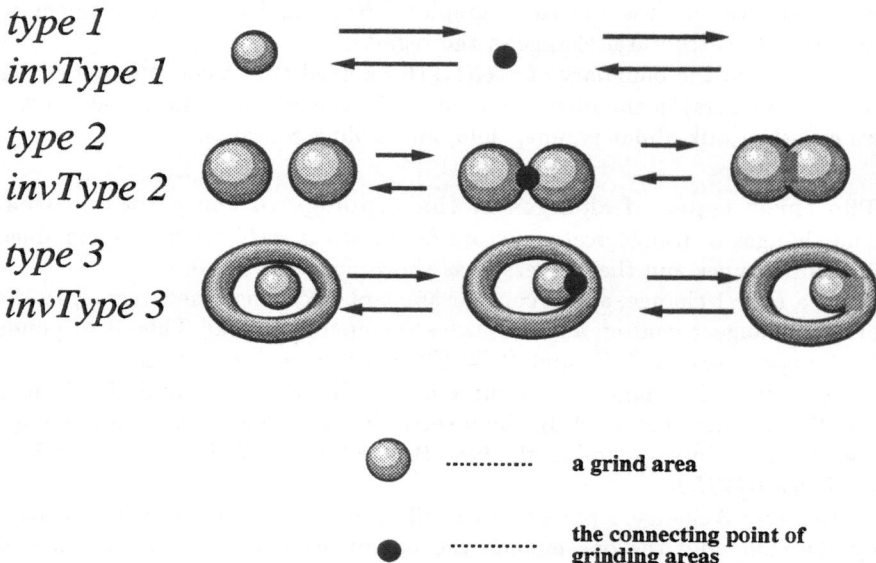

FIGURE 8.19. Possible types of change in G and in its complement $\overline{G}$.

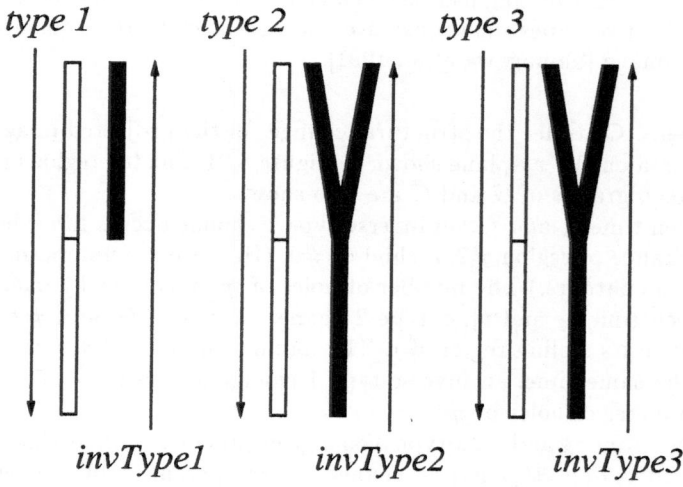

FIGURE 8.20. Portions of the Reeb graph of G (black) and of its complement (white) corresponding to the three types of changes in topology.

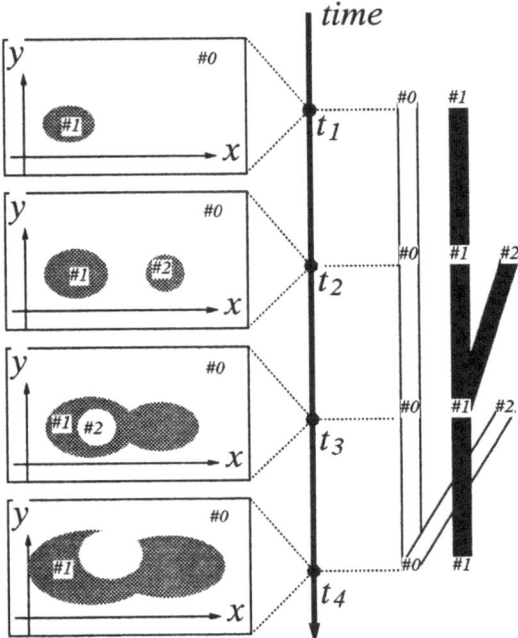

FIGURE 8.21. Changes in the projection $\pi(G)$ of the grinding area (left), and the corresponding Reeb graphs (right).

the grinding area changes with time. For example, the growth of the area of G narrows the spaces where the food can exist. Okubo [1992] used the change in the total area of G to describe chewing. The process, however, cannot be modeled simply by the total area: we cannot chew well with flat teeth, although the total area is large, because neither the upper nor the lower tooth surfaces can accommodate the food, nor they can grind it. The model proposed in this section is based instead on the topological changes undergone by the grinding area.

As seen in Figure 8.22, a change of inverse type 1 in G, meaning the birth of a new grinding area in $\bar{G}$, corresponds to holding the food and making a hole in it. Changes of types 2 and 3 in G mean the merging of two regions in G; this corresponds to cutting the food. The same is true for changes of inverse types 2 and 3 in $\bar{G}$ (Figure 8.23). A changes of type 1 in $\bar{G}$ grinds the food.

The opposite changes to the ones listed above make room for food, namely: type 1 and inverse types 2 and 3 in G, and types 2, 3, and inverse type 1 in $\bar{G}$. See Figures 8.22 and 8.23, reversing all the arrows.

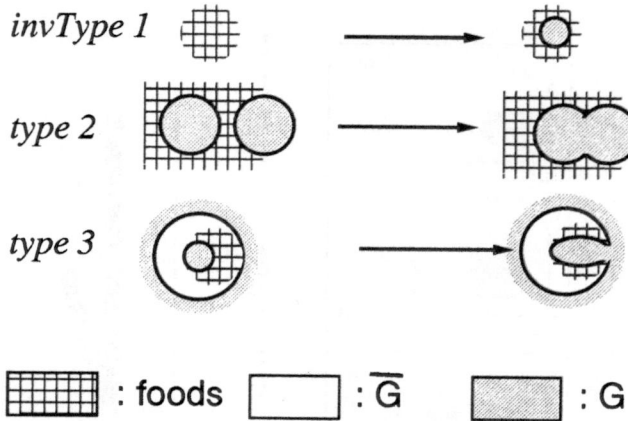

FIGURE 8.22. Physical meanings of the types of changes in G.

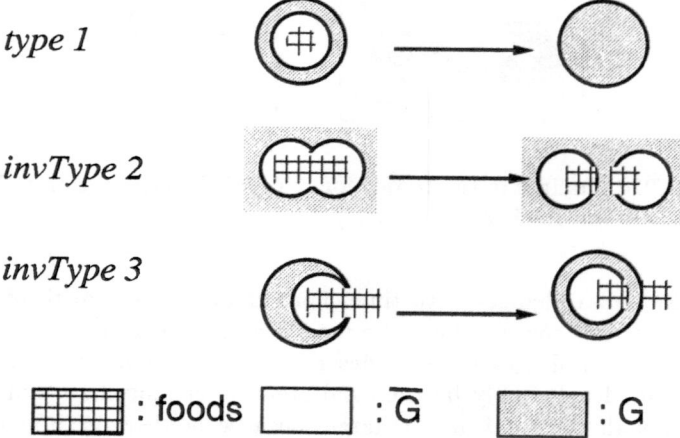

FIGURE 8.23. Physical meanings of the types of changes in $\overline{G}$.

A simulation run. We describe how our model can be used to study the movements of molars during chewing. First, the shape data of human teeth and artificial teeth are obtained by measuring plaster models with a numerically-controlled three-dimensional digitizer. The coordinate system is defined so that the xy-plane coincides with the occlusal plane (Figure 8.17). Each tooth surface is measured at sampling intervals of 0.1 mm. The surface is then interpolated using the sampled points, linearly for simplicity.

The particular movement we study consists of a lower tooth moving toward its *central occlusal position*, that is, the position where it rests stably against the corresponding upper tooth. The movement of the lower jaw is horizontal, and is specified by the relative coordinates of the starting position, as shown in Figure 8.24: The angle θ specifies the direction of movement of the lower tooth, and r specifies how far it is from the central occlusal position. At each moment in the motion, the z-values are calculated so that the upper and lower teeth touch.

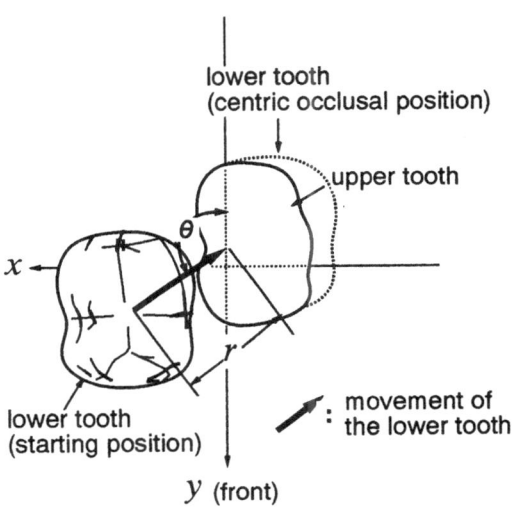

FIGURE 8.24. Specification of the starting position of the lower tooth (view from above).

Next we examine the topological structure of $H(G)$. The starting position and direction of movement are selected consistently with values that occur in actual chewing, as determined by clinical dentistry. Figures 8.25 and 8.26 show examples of Reeb graphs obtained by simulating the movement. We see in each example patterns of increase and decrease in the number of connected components of G and $\overline{G}$ as the lower tooth approaches its resting position. For example, in Figure 8.25 (left), when the number of $\overline{G}$ regions (white edges) increases, the number of G regions (black edges) decreases, and vice versa. Figure 8.26 shows a smaller number of changes; the teeth in this case had been smoothed by fillings.

As r approaches zero, the number of edges in the Reeb graph decreases. This is expected, because it is known in clinical dentistry that the upper and lower teeth have the maximum contact area if they are in the central occlusal position. Considering only the area, however, we cannot express the role of the irregularities in the teeth surfaces. If contact area were the only thing that mattered, flat teeth would be best, whereas, as explained before, they in

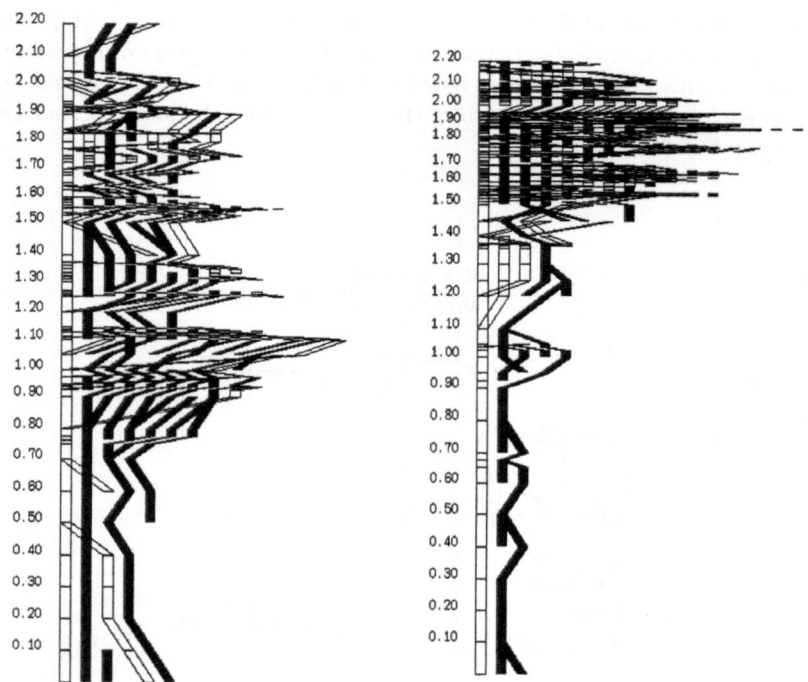

FIGURE 8.25. The Reeb graph of $H(G)$ and $H(\overline{G})$ of a sample of human teeth (Sample 1). The threshold δ is 1.275 mm. The scale to the left of each graph gives the distance r from the central occlusal position, for (left) the first right molar and (right) the second right molar. In each case, $\theta = 75°$.

fact cannot work well. Our model enables us to express the role of the teeth's irregularities by considering the number of changes of each type.

We conclude this section with an analysis of how the number of structural changes in G and $\overline{G}$ depends on the direction θ in the simulation. For each direction, sampled at intervals of 15°, two categories of changes were counted: *stage 1* changes (comprising changes of inverse type 1, 2, and 3 in G), and *stage 2* changes (comprising changes of types 2 and 3 in G, and changes of type 1 and inverse types 2 and 3 in $\overline{G}$. The results are graphed in Figure 8.27.

In clinical dentistry, it is known that food is mainly ground by a person's dominant-side teeth, which is either the left or the right side of the jaw. The dominant-side teeth are called the working teeth in clinical dentistry. If the dominant side is the right side, the lower teeth move from the right side to the central occlusal position in the chewing processes. This movement corresponds to the *lateral movement* of the working teeth in clinical dentistry. The lateral movement is regarded as the main movement in the chewing process, and corresponds to values of θ around 90°; this is considered to be

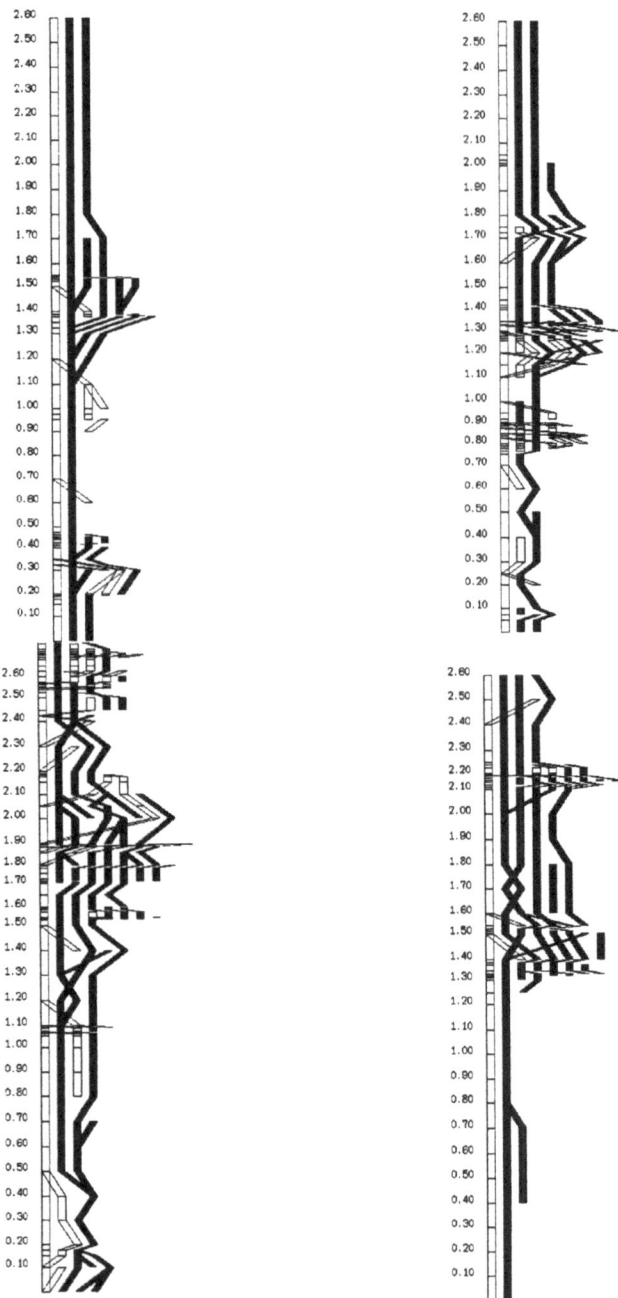

FIGURE 8.26. Same as Figure 8.25, but with teeth that have been treated by cavity filling (Sample 2). As before, $\delta = 1.275\,\text{mm}$; left and right graphs are for the first and second right molars, respectively; $\theta = 75°$ (top) and $\theta = 285°$ (bottom).

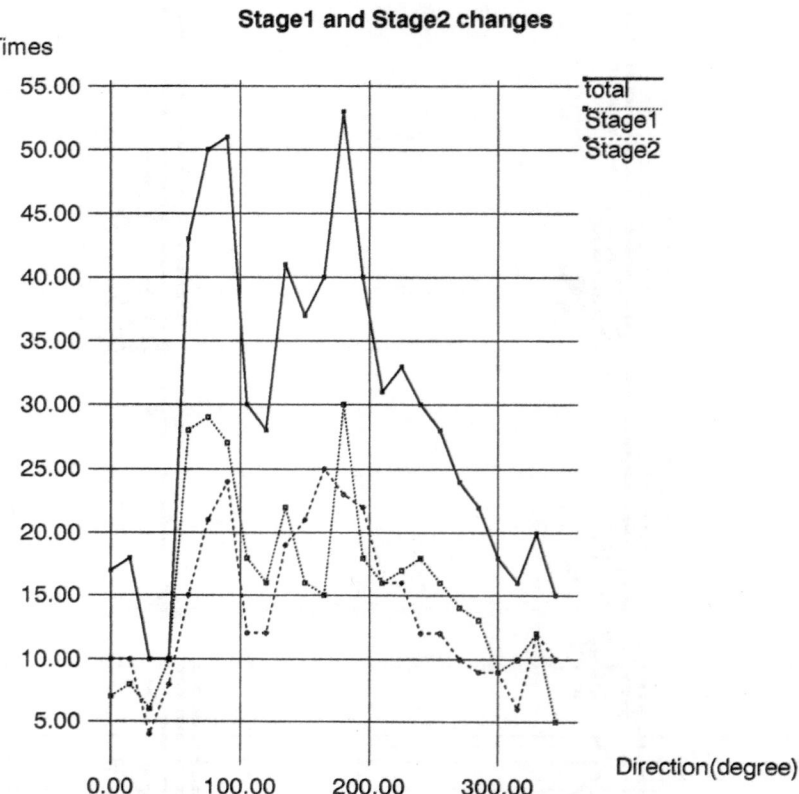

FIGURE 8.27. The number of stage 1 and stage 2 changes during a chewing simulation involving the first right molar of Sample 1, as a function of θ.

the direction in which the lower teeth should move in order to grind food effectively. The number of stage 1 and stage 2 changes should, therefore, be large when $\theta \approx 90°$ in the simulation.

This is confirmed by Figure 8.27, where there is a peak around $\theta = 90°$. Perhaps surprisingly, the graph also shows a peak around $\theta = 180°$, corresponding to what is called in clinical dentistry the *anterior movement*; the simulation from this sample shows that food should be ground well by movement in this direction, though this result has not been reported in clinical dentistry.

8.4. Coding of Nonorientable Surfaces

We now return to molecules, picking up where we left off in Section 8.1. The goal now is to understand how the molecule code can describe a nonorientable surface, still assuming the Morse function to be simple. Section 8.5 will deal with complicated Morse functions.

The A^*-atom.   The classification of possible contours in Section 8.1 did not depend on the orientability of the surface: a connected component is either a point, a circle, or a figure eight. The local picture near a minimum or a maximum also does not depend on whether the surface is orientable; the A-atoms of Figure 8.3 remain the only possibility. However, the behavior of the surface in a neighborhood of a figure eight (a contour containing a saddle) is no longer restricted to the B-atom shown in Figures 8.4 and 8.5. That picture was based on the assumption that the loops of the figure eight at the critical level *preserve orientation* (orientation-preserving and orientation-reversing loops were discussed in Section 4.3).

What happens instead if at least one of the loops of the figure eight *reverses orientation*? The local picture very near the saddle point must still look like Figure 6.6, by the Morse Lemma. If one of the loops reverses orientation, the only way for consistent signs to be assigned to the function in the four quadrants determined by saddle is the one depicted in Figure 8.28: *Both* loops must reverse orientation, and the ends of each loop come into the saddle from opposite directions.

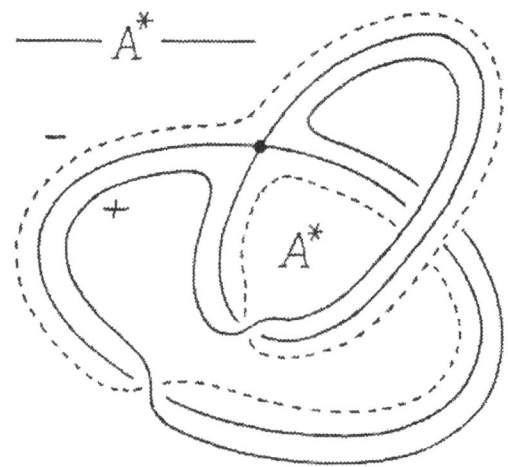

FIGURE 8.28.

We see in Figure 8.28 that a neighborhood of the critical level can be thought of as the union of two Möbius strips, whose central curves intersect at a point. On either side of the central figure eight, the level set is a single circle; this contrasts with the case of a B-atom, where there are two components on one side and one on the other. (The $+$ in the figure marks the circle lying on the level $c + \varepsilon$, and the $-$ marks the circle on the level $c - \varepsilon$, if c is the critical level.)

We will call the type of critical level just described an A^*-*atom*. In the labeled Reeb graph, two edges come out of an A^*-atom, as for example in Figure 8.29.

To summarize, a simple Morse function on a compact closed surface, orientable or not, can be represented by a molecule (labeled Reeb graph) whose vertices have one the types A, B, or A^* whose connectivities are shown in Figure 8.29. An A^*-atom requires a nonorientable surface.

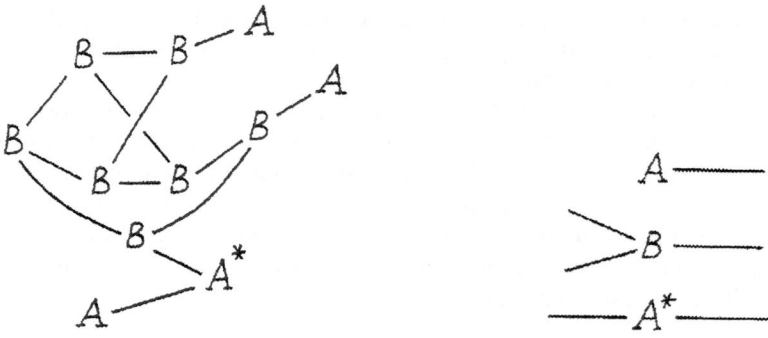

FIGURE 8.29.   FIGURE 8.30.

Some simple Morse functions on the projective plane and the Klein bottle.   Consider the simple Morse function f determined on the projective plane by the set of level curves shown in Figure 8.31, left. (Of course there are many smooth functions with this same set of level curves, but among them simple Morse functions are essentially all equivalent, if we disregard the precise value of the contours and consider only the qualitative behavior.) The function f has exactly one minimum, marked m_- in the figure, and exactly one maximum, marked m_+. It also has exactly one saddle, at S (in the center of the square). The level curve containing the saddle is obtained from the two diagonals of the square by identification of their endpoints P and Q (notice that P and Q remain distinct on $\mathbb{RP}^2$ after the gluing). Thus, the critical saddle level is a figure eight. This is even more obvious if we deform the function slightly, obtaining the representation in Figure 8.31, right.

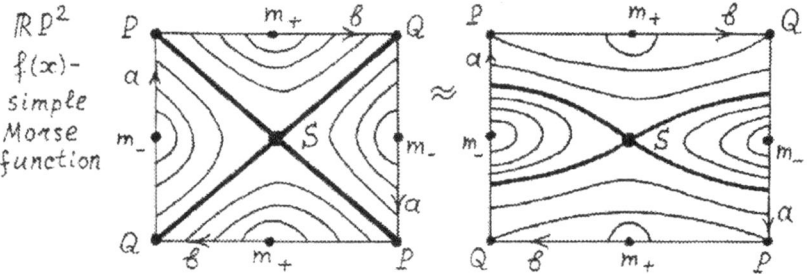

FIGURE 8.31.

Because the neighborhood of the figure eight is nonorientable, the molecule for this function is

$$
\begin{array}{ccc}
m_+ & & A \\
| & & | \\
A^* & = & A^* \\
| & & | \\
m_- & & A.
\end{array}
$$

We thus get a decomposition of $\mathbb{RP}^2$ as $D^2 \cup A^* \cup D^2$, where the first D^2 is a disk around m_-, the second is a disk around m_+, and A^* is an ε-neighborhood of the saddle level curve. This A^* is of course the same surface of Figure 8.28, and is obtained from the X-shaped neighborhood of the diagonals of the square by identification of its ends, as seen in Figure 8.32.

Figure 8.33, based on the deformed version of f shown on the right in Figure 8.31, shows that A^* is also homeomorphic to a Möbius strip with a hole (a fact that could also be derived from the arguments used in our classification of surfaces).

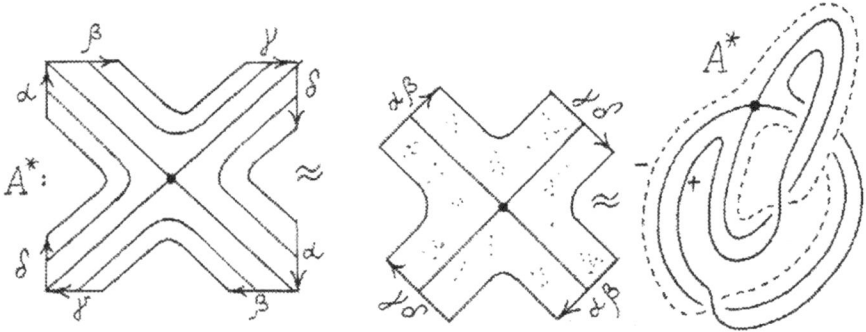

FIGURE 8.32.

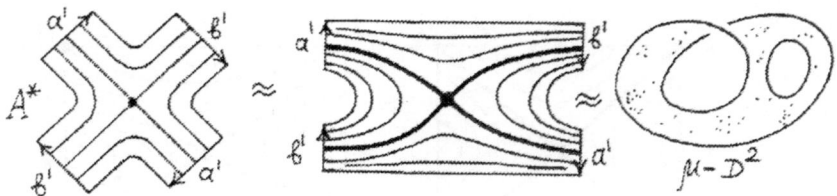

FIGURE 8.33.

Turning now to the Klein bottle, we consider its representation as the gluing of two Möbius strips along their boundary circles (Figures 4.34 and 4.37). We can use this decomposition to construct a simple Morse function on the Klein bottle, using for example the set of level curves in Figure 8.34. Here we have one minimum m_-, one maximum m_+, and two nondegenerate saddles R and S. Each saddle level set is a figure eight, and has a nonorientable neighborhood. that is, corresponds to an A^*-atom. Consequently, the whole Morse function is described by the molecule

$$
\begin{array}{c}
A \\
| \\
A^* \\
| \\
A^* \\
| \\
A.
\end{array}
$$

The functions just studied are the simple Morse functions on the projective plane and the Klein bottle having the smallest number of critical points.

8.5. Coding of Complicated Morse Functions

Morse functions with symmetries.   Consider a smooth manifold X and a Morse function f on X. Assume that this function is invariant under the action of some discrete group G acting on X. We recall what this means: Saying that G acts on X means that we have, for each element $g \in G$. a diffeomorphism of X, also denoted by the same letter $g : X \to X$, and these diffeomorphisms satisfy $g'(g(x)) = (g'g)(x)$, for all $g, g' \in G$ and all $x \in X$. Saying that f is *invariant* under the action means that $f(g(x)) = f(x)$ for all $g \in G$ and all $x \in X$. In other words, f is constant on every orbit of the group G in X, where the *orbit* $G(x)$ of x is the set $\{g(x) : g \in G\}$. Such functions f are usually called *functions with symmetries*; G is called the group of symmetries of the function.

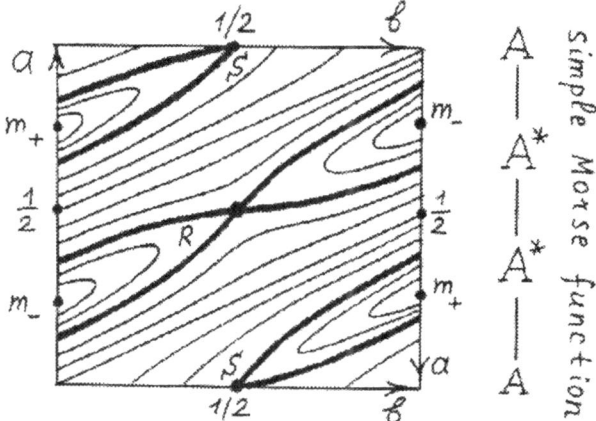

FIGURE 8.34.

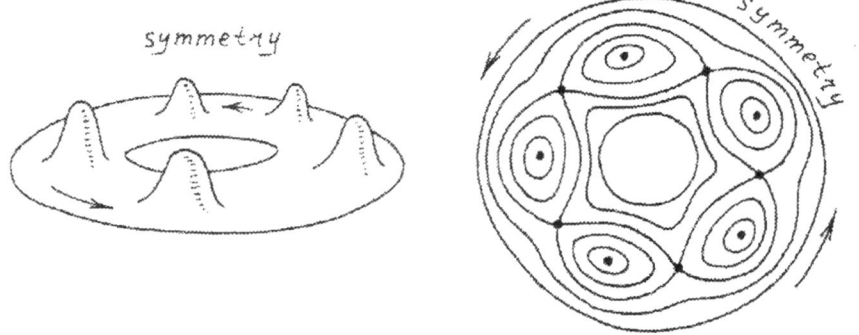

FIGURE 8.35.

Consider the simple example in Figure 8.35: a height function defined on the annulus X, having five maxima and five saddles, for a total of ten critical points. Here $G = \mathbb{Z}_5$, the cyclic abelian group of order 5, acting on X by rotations about the vertical z-axis in $\mathbb{R}^3$. This group is generated by a rotation through $2\pi/5$. Similar functions occur in many applied problems of modern geometry and topology.

Returning to the general situation of the first paragraph, let x_0 be a critical point of the G-invariant Morse function f. Then it easy to see that all points in the orbit of x_0 also are critical points of f, having the same index as x_0; and of course f has the same value on all these points. Therefore, if the orbit is nontrivial (that is, has more than one point), the critical level $f^{-1}(f(x_0))$ has several critical points, and f cannot be a simple Morse function—it is a *complicated Morse function*.

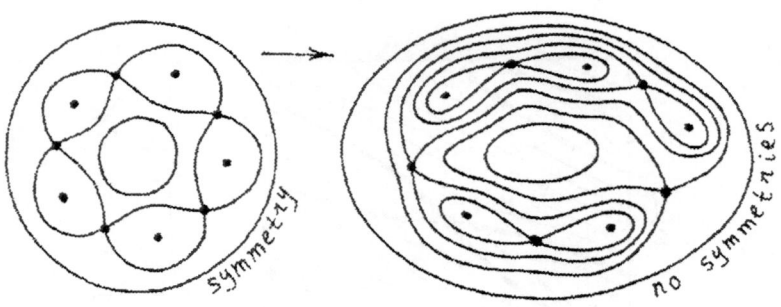

FIGURE 8.36.

In the example in Figure 8.35 the critical level set of the saddle contains five critical points. Topologically, it is obtained by gluing five circles, each two neighboring circles being glued at one point, as shown in Figure 8.35, right. The maximal critical level set consists of five isolated maxima.

As we mentioned at the beginning of this chapter, any complicated Morse function on a compact manifold can be disturbed slightly to yield a simple Morse function; this process is shown in Figure 8.36, where a perturbation turns the function of Figure 8.35 into a simple Morse function, having five saddles at different levels, and thus five figure eight curves instead of a complicated critical set.

But if we start from a G-invariant Morse function having nontrivial critical orbits, it is clear that we cannot approximate it by a simple Morse function and at the same time preserve G-invariance, since, as we have just seen, the resulting G-invariant function would still have nontrivial critical orbits. (In a small perturbation, the critical points move very little, so a nontrivial critical orbit cannot coalesce.)

Thus, if we want to investigate the properties of *symmetric* invariant Morse functions, we are forced to consider complicated Morse functions. Our theory of coding of Morse functions must be expanded with new types of atoms to describe complicated critical level curves having several critical points. This description is a particular case of general theory for classification of bifurcations in Liouville foliations corresponding to the integrable Hamiltonian dynamical systems. This was, to our knowledge, first done in [Fomenko 1989]; we follow the treatment there.

Suppose that X is a compact surface and that f is an arbitary Morse function on X. Let c be a critical value of f and K_c be the corresponding singular level curve, that is, the connected component of $f^{-1}(c)$ which contains one or several critical points. It is easily seen that K_c is either an isolated point – in the case of a maximum or a minimun – or a number of closed curves that meet at the saddles of f. Choose ε so small that no other critical value occurs in the interval $[c - \varepsilon, c + \varepsilon]$ and consider the connected component P_c

of the subset $f^{-1}([c-\varepsilon, c+\varepsilon])$ such that $K_c \subset P_c$. It is clear that P_c can be considered as a regular neighborhood of the critical level curve K_c.

Definition 8.5.1. The pair (P_c, K_c) is calld an *atom*.

Note that the atom (P_c, K_c) completely describes the type of the singularity up to fiber equivalence. It means that two Morse functions f and f' are fiber equivalent it some neighborhoods of their critical level lines K_c and K'_c if and only if the corresponding atoms (P_c, K_c) and (P'_c, K'_c) are homeomorphic. i.e., there exists a homeomorphism $\xi : P_c \to P'_c$ such that $\xi(K_c) = K'_c$. Sometimes, for the sake of simplicity, we will say that the surface P_c itself is an atom taking into account that there is K_c embedded into it.

Now, P_c has a boundary consisting of a certain number of circles, some at the level $c - \varepsilon$ (*negative circles*), and some at the level $c + \varepsilon$ (*positive circles*). The number of positive and negative circles can be different. The critical points of f that lie on K_c are called the *vertices* of the atom (and of K_c). We can regard K_c as a graph; the vertices have just been defined, and the edges are the connected components of the complement in K_c of the set of vertices. Some edges may be loops, that is, starting and ending at the same vertex. The vertices of K_c may only have a valence of zero (maximum or minimum) or four (saddle).

We call K_c the *spine* of the atom P_c. The spine is a *deformation retract* of P_c; this means. roughly, that P_c can be continuously retracted, or shrunk, into K_c (formally, there is a homotopy of P_c into P_c whose initial stage is the identity map. whose final stage has image K_c, and where at each stage each point of K_c is fixed). This retraction can be carried out along the gradient lines of the function f. If we remove the spine from the atom, it disintegrates into a union of cylinders (annuli). Associated with each edge of the spine there is exactly one positive and exactly one negative boundary circle.

We will give an outline of a partial classification of atoms. It is tempting to classify atoms by their spine, but nonhomeomorphic atoms may have homeomorphic spines. Therefore, the spine itself, considered as an abstract graph without some immersion in the plane, does not define the atom uniquely.

As observed above, if P_c contains a maximum or a minimum, this is its only critical point, and P_c is necessarily a disk. We continue to represent such an atom by the letter A. To study the case of saddles, we first assume that the surface is orientable.

Complicated atoms for orientable surfaces. If the surface X is *orientable*. each atom P_c must also be orientable.

Figure 8.37 shows an example of an atom that we denote by D_1 (the justification for this and related symbols is not important here; the notation was introduced in [Fomenko 1989], in the context of Hamiltonian physics). D_1 is a planar surface, that is, it can be realized as a subset of the plane. However, not all atoms are planar. An example of a nonplanar atom is shown

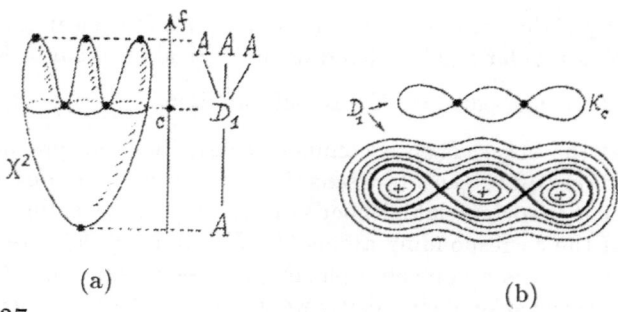

(a)

(b)

FIGURE 8.37.

in Figure 8.38; it is an easy exercise to construct a Morse function realizing this atom. Although this atom cannot be embedded in the plane, it can be *immersed* there; in fact, Figure 8.38 shows exactly such an immersion.

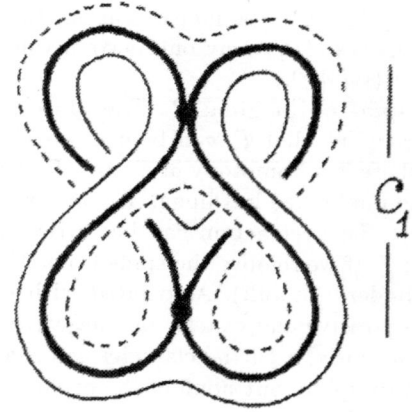

FIGURE 8.38.

Clearly, the number of types of atoms is infinitely large; but they can easily be ordered and classified as their complexity increases. We describe a useful and graphical method of depicting atoms. It is known from topology that any orientable surface with a boundary can be immersed in a two-dimensional sphere. Therefore, any atom can be visualized as immersed in a sphere. Naturally, we shall not distinguish between the immersions of an atom that can be obtained from one another by a smooth deformation (isotopy) within the sphere. Besides, we agree not to distinguish between immersions differing from one another only by loops such as the one shown in Figure 8.39, top left. By removing a point (outside the atom) from the sphere, we can depict the atom as immersed in the plane.

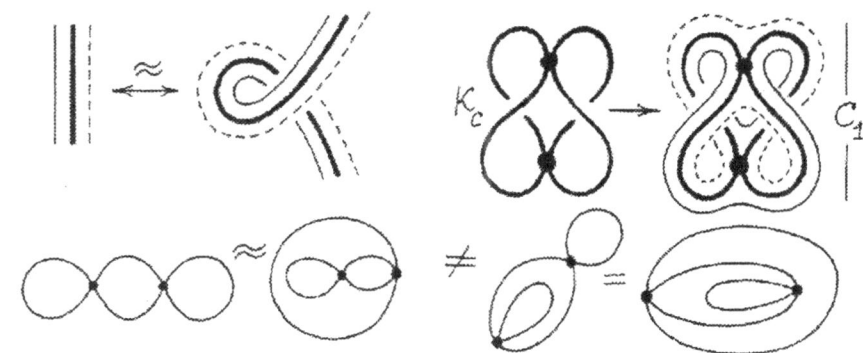

FIGURE 8.39.

The picture of an atom can be simplified still further. Indeed, we shall define the atom completely (and uniquely) by specifying only the *immersion* in the sphere of its graph K_c. If the immersion of the graph is specified, the atom can be reconstructed by taking a small tubular neighborhood of this immersion. This idea is illustrated in Figure 8.39, top right, for the atom called C_1'.

Thus, *an atom can be characterized by an immersion of the graph K_c in S^2.* Two immersions isotopic in S^2 are considered equivalent, as are two immersions that differ only by a loop.

For convenience, we will depict our immersions in S^2 as immersions in $\mathbb{R}^2$, by removing a point from the sphere that does not lie on the graph; but we should keep in mind that they really are immersions in S^2, because isotopy in $\mathbb{R}^2$ and isotopy in S^2 give rise to different equivalence relations.

Henceforth, when depicting atoms P_c, we will show only their spines K_c immersed in the plane. The bottom row of Figure 8.39 also shows examples of equivalent and nonequivalent immersions of the same graph K_c.

Thus, the set of inequivalent atoms forms an infinitely long *discrete* list. We shall define the *weight* of an atom as the number of vertices it has. There exists an algorithm (realized on a computer) that enumerates consecutively all atoms, in order of increasing weight. Figure 8.40 shows the beginning of this list.

To get the *molecule* of a complicated Morse function, we label the vertices of the Reeb graph with the symbols for the function's atoms. Figure 8.41 shows an example of a possible molecule.

As an interesting example, consider the complicated Morse function defined on the torus by the level curves shown in Figure 8.42, left. This function

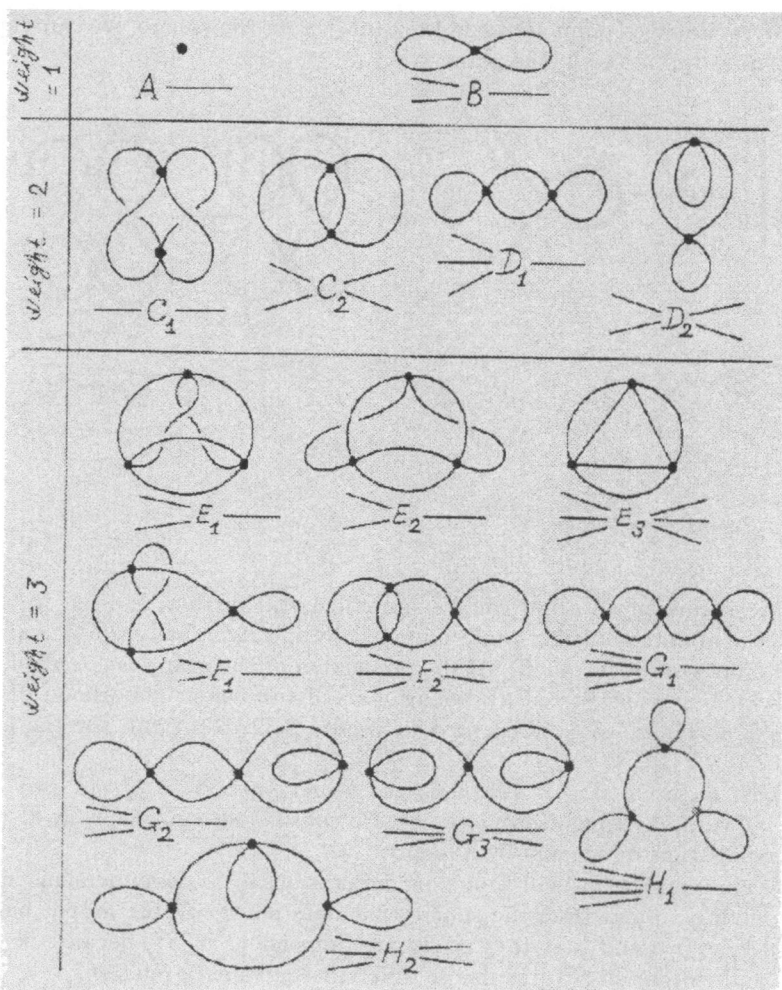

FIGURE 8.40. Orientable atoms of weight 1, 2, and 3.

has two nondegenerate saddles at the same level, one minimum, and one maximum. The saddle atom is the atom C_1 of in Figure 8.38. Figure 8.42 shows the decomposition.

Complicated atoms for nonorientable surfaces. When the surface X is allowed to be nonorientable, the theory of atoms of general Morse functions is more complicated. We will not go into its details because in the applications

$$W = \begin{array}{c} A - C_1 \\ \qquad\searrow \\ \qquad\quad B - A^* \\ A \searrow C_2' \nearrow \\ A \nearrow \end{array}$$

FIGURE 8.41.

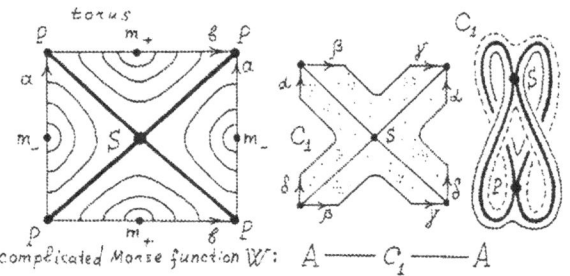

complicated Morse function W: $A \underline{\qquad} C_1 \underline{\qquad} A$

FIGURE 8.42.

of computer geometry nonorientable surfaces occur but rarely. We will just give one example.

Take the Morse function determined on the Klein bottle by the set of level curves of Figure 8.43. This complicated Morse function has two saddles P and S, one minimum, and one maximum. The complicated saddle atom, which we call M^{**}, can be obtained from Figure 8.44 by performing the indicated identifications. It is easy to see that the result of this gluing has the form shown in Figure 8.45, or the more symmetric form in Figure 8.46. The corresponding molecule is therefore

$$\begin{array}{c} A \\ | \\ M^{**} \\ | \\ A. \end{array}$$

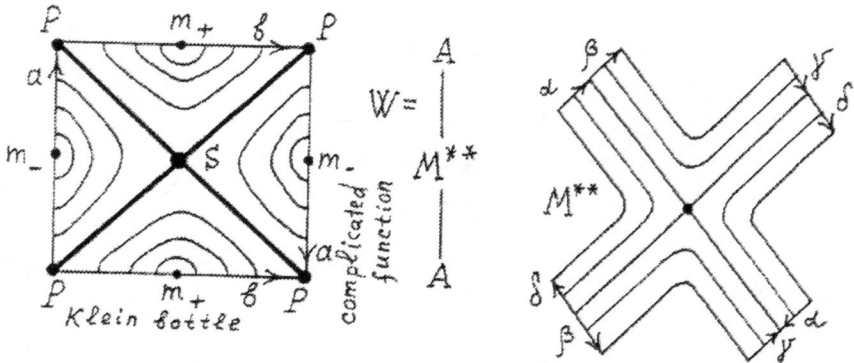

FIGURE 8.43.   FIGURE 8.44.

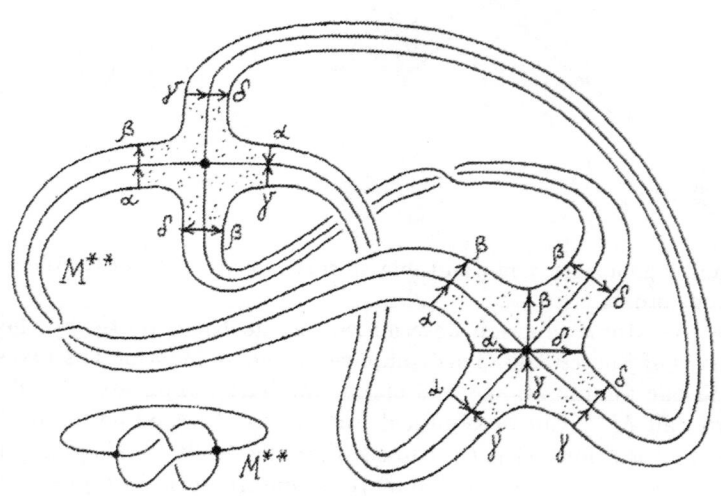

FIGURE 8.45.

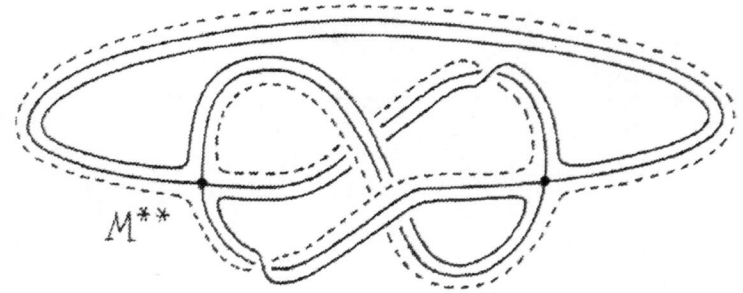

FIGURE 8.46.

9. Height Functions and Distance Functions

9.1. Almost All Height Functions Are Morse Functions

We now restrict our attention to two classes of functions on surfaces, which have particular importance in computer geometry: *height functions* and *distance functions*. In particular, we investigate when such functions are Morse functions. We consider a solid, bounded, connected object in $\mathbb{R}^3$, as in Figure 9.1, and assume that it has a smooth boundary. The boundary, therefore, is an orientable compact surface embedded in $\mathbb{R}^3$.

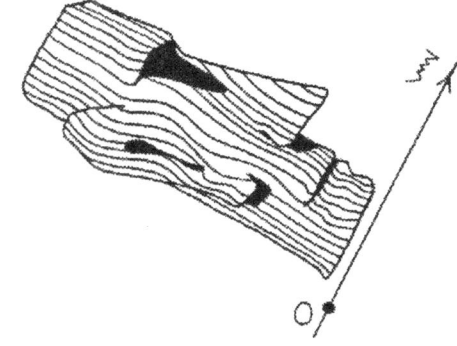

FIGURE 9.1.   FIGURE 9.2.

We recall the following definition. If ξ is a line in $\mathbb{R}^3$, with a fixed orientation and a fixed origin, we can identify ξ with $\mathbb{R}$ (we are considering $\mathbb{R}^3$ with its usual Euclidean metric). The orthogonal projection $f_\xi : \mathbb{R}^3 \rightarrow \xi = \mathbb{R}$ then defines the *height function* in $\mathbb{R}^3$ with respect to ξ. This function can be restricted to any subset of $\mathbb{R}^3$, and in particular to any smooth surface $M \in \mathbb{R}^3$; the restriction is the height function of M (with respect to ξ), which we still denote by f_ξ.

The level sets of this function are the intersections of M with planes orthogonal to ξ (Figure 9.2). Let's denote the plane at height a by Q_a. Clearly a level set can be arbitrarily complicated: it might have isolated points, or curves, and might even contain an open subset of the plane, like the top level in Figure 9.3. It may be connected or disconnected. The curves may contain arbitrarily complicated singularities. For example, consider the real part of the complex function $(x + iy)^n$ on the xy-plane. The level set at height zero for the graph of this function is a union of n lines through the origin; see Figure 9.4 for the case $n = 3$, which displays also some nearby level sets.

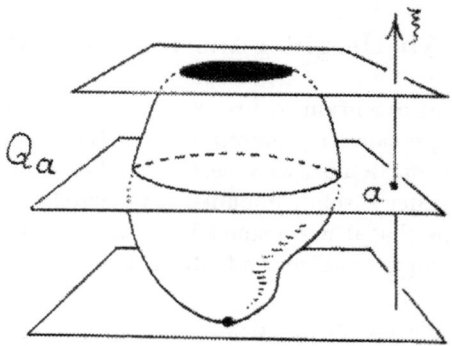

FIGURE 9.3.

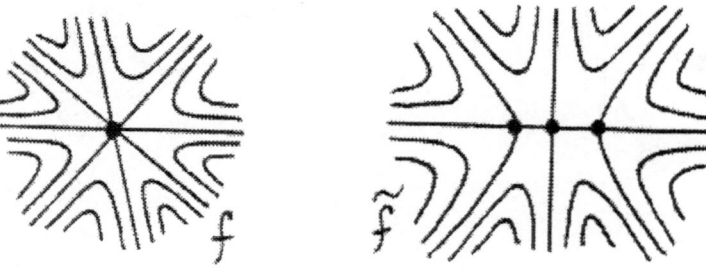

FIGURE 9.4.   FIGURE 9.5.

As we know, for Morse functions the situation is much simpler. It is natural to ask, then, can we choose the line ξ so as to make f_ξ a Morse function?

As geometric motivation, we take a few concrete examples that show that we really can simplify the level sets by rotating the object in space (or, equivalently, the line ξ) by an arbitrarily small amount. For the function

$\mathrm{Re}(x+iy)^n$ discussed above, we get the behavior shown in Figure 9.5 if we tilt the line ξ slightly: One complicated critical point disintegrates into several simple (nondegenerate) saddles.

The same idea makes it easy to get rid of two-dimensional pieces in a level set. Figure 9.6 shows how a small change in ξ eliminates the flat top of the function, so that all the level sets become one-dimensional (piecewise smooth curves).

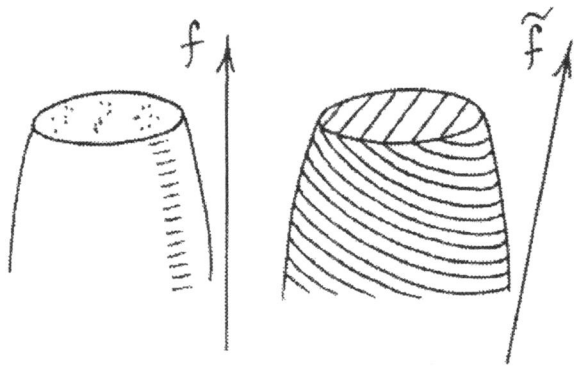

FIGURE 9.6.

Experimentation shows that for a smooth surface it is always possible to change ξ slightly so that complicated level sets are resolved into those expected for a Morse function. There is a very simple formulation for this "experimental result". Recall that, unless we say otherwise, by a "manifold" or "surface" we always mean one without boundary.

Theorem 9.1.1. *Let a smooth surface immersed in $\mathbb{R}^3$ be given. Then, for almost all directions of lines ξ, the height function f_ξ is a Morse function.*

The notion of "almost all" directions is to be understood as follows. Let P be the space of possible directions; P is the projective plane $\mathbb{RP}^2$. The set of directions that *do not* satisfy the desired property has *measure zero*, that is, it can be covered by open sets whose total area (with respect to a fixed Riemannian metric on P) is as small as one wishes. This implies, in particular, that the set of directions that *do* satisfy the property is *dense*, that is, intersects every open set in P. In particular, for any line ξ one can, by rotating ξ through an angle as small as desired, ensure that f_ξ is a Morse function.

We will prove Theorem 9.1.1, because the proof demonstrates important notions of geometry and topology that occur in many other problems.

9.2. Sard's theorem

Consider a smooth map $f : M \to N$, where M and N are smooth compact manifolds, of arbitrary dimensions m and n, respectively. For our purposes we can assume that both manifolds are endowed with a Riemannian metric.

Recall from Definition 3.4.2 that a *regular point* of f is any point $x \in M$ for which the differential $df_x : T_x M \to T_{f(x)} N$ is surjective, that is, has rank n; a *singular* or *critical point* is one for which this is not so. A *critical value* is the image of a critical point; a *regular value* is a point of N that is not the image of a critical point. In particular, a point of N not in the image of f is automatically a regular value; this is very useful in the formulation of some theorems. We denote the set of critical points of f by C.

We denote by vol_n the n-dimensional measure (volume) on the n-manifold N. For our purposes we need to consider only "reasonable" subsets of N, so we won't provide a formal definition of n-dimensional volume—the intuitive concept is sufficient, and the reader is probably familiar with the notion of volume for an open set. It is worth remarking that a set of *measure zero* can be covered by open sets the sum of whose volumes is as small as one wishes; we can take this to be the definition of a set of measure zero.

Theorem 9.2.1 (Sard). *If $f : M \to N$ is a smooth map between smooth manifolds, the set $f(C)$ of critical values of f has measure zero in N.*

For a proof, see [Guillemin 1974], for example.

Remark. It is not always true that the m-dimensional measure of the set C of critical points is zero: far from it, sometimes. If $m < n$, for example, all points are critical, since the differential map can have rank at most m. Likewise, if f is the constant map, all points are critical.

Corollary 9.2.2. *The set $N - f(C)$ of regular values of f is dense in N.*

Corollary 9.2.3. *If $\dim M < \dim N$, the image $f(M)$ has measure zero in N; in particular, it does not cover all of N.*

9.3. Proof That Almost All Height Functions Are Morse Functions

We return to the consideration of height functions f_ξ on a fixed surface in $\mathbb{R}^3$. We assume for simplicity that the surface $M \subset \mathbb{R}^3$ is *embedded* and *orientable*, but the observant reader will notice that the proof works essentially without change for the case of an immersion $M \to \mathbb{R}^3$; this also takes care of the nonorientable case.

Apart from the addition of a constant, which does not affect the question of whether something is a Morse function, the function f_ξ remains unchanged when ξ is moved parallel to itself, or when its origin is moved. Thus, we may restrict our attention to oriented lines ξ going through the origin of $\mathbb{R}^3$ and whose origin coincides with the origin of $\mathbb{R}^3$. The space of such lines is the sphere S^2—in fact, we may consider each ξ as a unit vector, that is, an element of S^2. Thus the space of Morse functions that interests us is in one-to-one correspondence with S^2.

Next, $P \in M$ is a critical point for the height function f_ξ if and only if ξ is orthogonal to the tangent plane to the surface at P. This is obvious, since the differential of f_ξ at P, applied to a vector $v \in T_P M$, is exactly the scalar product $\langle v, \xi \rangle$, where ξ is regarded as a unit vector.

Now we need to find out when a critical point for the height function is *nondegenerate*, since we are interested in functions all of whose critical points are nondegenerate. For this we use the important concept of the *Gauss map*. This is the map $g : M \to S^2$ that takes each point $P \in M$ to the unit normal vector to M at P, considered as a point in the unit sphere S^2. See Figure 9.7. (Since M is orientable, a unit normal vector can be chosen consistently over the whole surface.) This map is clearly smooth, since M is smooth.

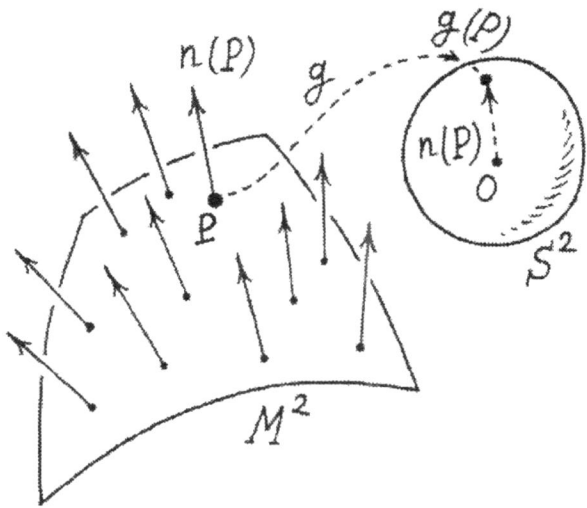

FIGURE 9.7.

Proposition 9.3.1. *A critical point of the height function f_ξ on M is nondegenerate if and only if it is a regular point for the Gauss map $g : M \to S^2$.*

Proof. By an orthogonal change of coordinates, we can arrange for the point in question to be the origin $(x, y, z) = (0, 0, 0)$, and for the z-axis to be parallel to ξ (Figure 9.8). In these coordinates, M is locally the graph of its height function, $z = h(x, y)$, where we have set $h = f_\xi$ for convenience.

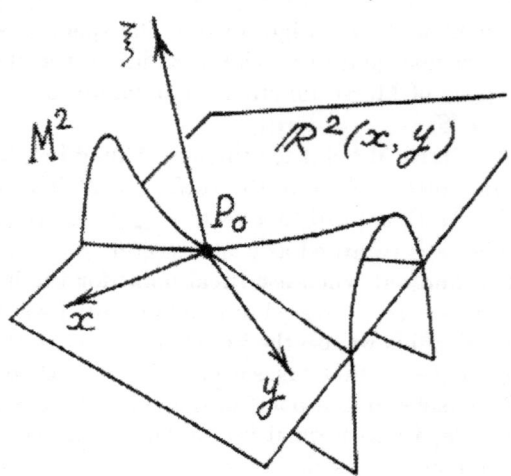

FIGURE 9.8.

The unit normal at the point of M parametrized by (x, y) can be taken to be

9.3.2.
$$g(x, y) = \left(-\frac{h_x}{r}, -\frac{h_y}{r}, \frac{1}{r}\right),$$

where $h_x = \partial h/\partial x$, $h_y = \partial h/\partial y$, and $r = (1 + h_x^2 + h_y^2)^{1/2}$. This can be easily verified by, for example, taking the cross product of the tangent vectors $(1, 0, h_x)$ and $(0, 1, h_y)$, and then normalizing. By differentiating we see that, *at the point $x = y = 0$*, we have

$$\frac{\partial}{\partial x}\frac{1}{r} = \frac{\partial}{\partial y}\frac{1}{r} = 0, \quad \frac{\partial}{\partial x}\frac{h_x}{r} = h_{xx}, \quad \frac{\partial}{\partial x}\frac{h_y}{r} = h_{yx}, \quad \frac{\partial}{\partial y}\frac{h_x}{r} = h_{xy}, \quad \frac{\partial}{\partial y}\frac{h_y}{r} = h_{yy},$$

since $h_x = h_y = 0$ at that point. Therefore the matrix of partial derivatives of g at $(x, y) = (0, 0)$ is

$$\begin{pmatrix} -h_{xx} & -h_{xy} \\ -h_{yx} & -h_{xx} \\ 0 & 0 \end{pmatrix},$$

and so has rank two if and only if $h_{xx}h_{yy} - h_{xy}h_{yx} = 0$, that is, if and only if the Hessian of the height function h is nondegenerate at the same point.   □

We can compose the Gauss map g with the quotient map $S^2 \to \mathbb{RP}^2$ (identification of antipodal points), obtaining a map $\bar{g} : M \to \mathbb{RP}^2$. The critical points of $\bar{g}$ are the same as those of g, since the quotient map $S^2 \to \mathbb{RP}^2$ is a local diffeomorphism (its differential has rank two everywhere). Now, we have seen that a point $P \in M$ is critical for f_ξ if and only if $\bar{g}(P) = \pm\xi$, where $\pm\xi$ is the point in $\mathbb{RP}^2$ coresponding to $\xi \in S^2$—that is, the inverse image $\bar{g}^{-1}(\pm\xi)$ consists exactly of all points on the surface M whose tangent planes are orthogonal to ξ (Figure 9.9). We have also seen that such a critical point is nondegenerate if and only if it is a *regular point* for $\bar{g}$. Thus $\pm\xi$ is a regular value of $\bar{g}$ if and only if all the critical points of f_ξ are nondegenerate! We have proved:

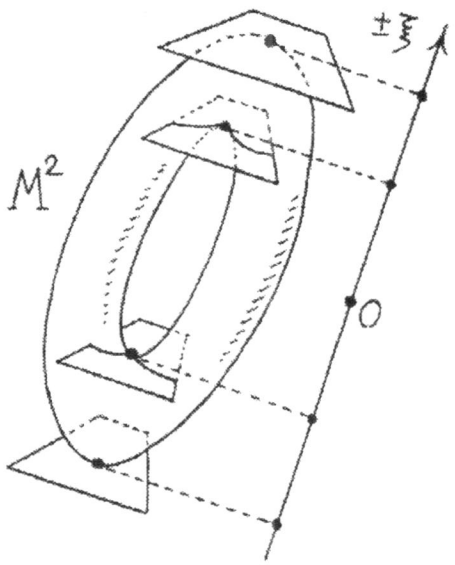

FIGURE 9.9.

Proposition 9.3.3. *The height function f_ξ on M is a Morse function if and only if the corresponding point $\pm\xi \in \mathbb{RP}^2$ is a regular value for the Gauss map $\bar{g} : M^2 \to \mathbb{RP}^2$.*

Now we can apply Sard's theorem to the map $\bar{g} : M \to \mathbb{RP}^2$, to conclude that the set of $\pm\xi$ for which the height function f_ξ is not a Morse function has measure zero in $\mathbb{RP}^2$. This proves Theorem 9.1.1.

9.4. Almost All Distance Functions are Morse Functions

We turn to our second important class of functions, and prove that in this class, too, Morse functions are dense. In a sense this idea is similar to the previous one, but the technical tools used are different. In the case of a height

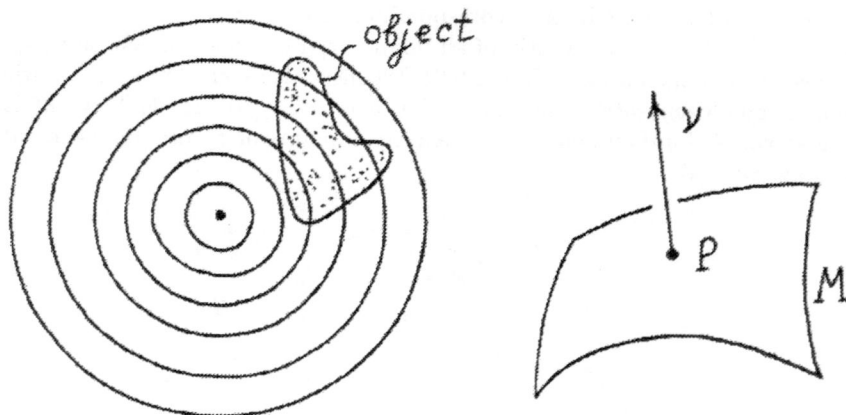

FIGURE 9.10.

function we considered a surface in three-space and foliated space into the union of parallel planes in such a way that they all intersect our surface in a regular way. We can reconstruct the original object if we know all its sections by these parallel planes. (Such cross sections can be determined in physical experiments for real objects, for example, in medicine, geology, etc.)

There are another reasonable ways to foliate $\mathbb{R}^3$: for instance, as the union of concentric spheres (plus their center), as shown in Figure 9.10. We can reconstruct an object if we know its intersections with a family of concentric spheres. Concentric spheres are the level sets of what is called, naturally enough, a *distance function*; these are the functions we turn to now.

Let $q \in \mathbb{R}^3$ be a fixed point; then the function $l_q : \mathbb{R}^3 \to \mathbb{R}$ given by

$$l_q(P) = |q - P|^2,$$

where $|\cdot|$ is Euclidean length, is called the *distance function* to q. We use the square of the length rather than the length itself so that l_q is everywhere smooth (otherwise it would not be smooth at the point q.) We can restrict l_q to any subset of $\mathbb{R}^3$, and in particular to any embedded smooth surface $M \subset \mathbb{R}^3$; the restriction is called the distance function of M to q, and we still denote it by l_q.

Theorem 9.4.1. *Let a smooth surface immersed in $\mathbb{R}^3$ be given. Then, for almost all points $q \in \mathbb{R}^3$, the distance function l_q is a Morse function.*

As in the proof of Theorem 9.1.1, we assume here that the surface $M \subset \mathbb{R}^3$ is embedded and orientable, but these assumptions are not essential.

It turns out that we can effectively describe all points q in $\mathbb{R}^3$ such that $l_q(P)$ is not a Morse function.

We start by considering the three-manifold N formed by all pairs (P, ν), where P belongs to M and ν is a normal vector to M at P (Figure 9.10). It is clear that the set of these pairs really does form a smooth three-manifold; in fact, since M is orientable and embedded in $\mathbb{R}^3$, the manifold N is topologically the direct product $N = M \times R^1$. (You should imagine this manifold as existing outside of $\mathbb{R}^3$—as a subset of $\mathbb{R}^3 \times \mathbb{R}^2$, or as an abstract manifold.)

Next we take the smooth map $F : N \to \mathbb{R}^3$ that associates to each pair $(P.\nu) \in N$ the point $P + \nu$, the endpoint of the vector ν of Figure 9.10.

Definition 9.4.2. A point $q \in \mathbb{R}^3$ is called a *focal point* of the surface M if there exists $(P, \nu) \in N$ such that $q = F(P, \nu)$ and the differential dF of F at (P, ν) has rank < 3. In other words, q is a focal point if it is the image of a critical point of the map $F : N \to \mathbb{R}^3$. (The difference $3 - \operatorname{rank} dF$ is called the *multiplicity* of the focal point.)

It follows from Sard's theorem that the measure of the set of all focal points in $\mathbb{R}^3$ is equal to zero. In other words, almost all points in $\mathbb{R}^3$ are nonfocal. In fact, we have the following result, which is not necessary for the proof of Theorem 9.4.1 but is interesting in its own right:

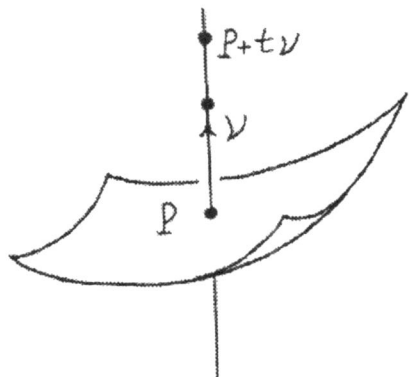

FIGURE 9.11.

Proposition 9.4.3. *Let ν be a unit normal to M at P. A point on the line $\{(P, t\nu) \in N : t \in \mathbb{R}\}$ is critical for the map F if and only if t equals λ_1^{-1} or λ_2^{-1}, where λ_1 and λ_2 are the principal curvatures of M at P (their signs having been chosen in accordance with the choice of ν).*

Proof. We use the same coordinate system as in the proof of Proposition 9.3.1, where P is at the origin and ν points in the direction of the positive z-axis; the surface is locally the graph of the function $h(x, y)$. Using Equation 9.3.2, we can write the map F in these coordinates as

$$F(x, y, t) = (x, y, h(x, y)) + t\left(-\frac{h_x}{r}, -\frac{h_y}{r}, \frac{1}{r}\right) = \left(x - \frac{h_x}{r}t, \ y - \frac{h_y}{r}t, \ h + \frac{1}{r}t\right).$$

The matrix of partial derivatives of this map at $(x, y) = (0, 0)$ is

$$\begin{pmatrix} 1 - h_{xx}t & -h_{xy}t & 0 \\ -h_{yx}t & 1 - h_{yy}t & 0 \\ 0 & 0 & 1 \end{pmatrix},$$

since $h_x = h_y = 0$ at that point. This matrix is singular if and only if its 2×2 minor on the top left is singular. The minor equals $I - tH$, where I is the identity matrix and H is the Hessian of h, so the desired values of t are the inverses of the eigenvalues of H. But the eigenvalues of the Hessian are the principal curvatures, by Definition 3.6.1. $\qquad\square$

Returning to the proof of Theorem 9.4.1, we will show that l_q is a Morse function if and only if q is not a focal point. First, observe that P is a critical point of l_q if and only if the vector $P - q$ is normal to M at P; this is because the gradient of l_q at P lies in the direction of $P - q$. To find out when a critical point P is degenerate, we again use the coordinate system from the proof of Proposition 9.3.1, where $q = (0, 0, t) = F(P, t\nu)$ for some $t \in \mathbb{R}$, by the preceding observation. (As before, ν is the unit normal in the positive z-direction.)

We can write
$$l_q(x, y) = x^2 + y^2 + (t - h)^2,$$

if M is locally the graph of the function $h(x, y)$. Differentiating we get

$$(l_q)_x = 2x - 2h_x(t - h), \qquad (l_q)_y = 2y - 2h_y(t - h),$$

so the Hessian of l_q at $x = y = 0$ is

$$2\begin{pmatrix} 1 - h_{xx}t & -h_{xy}t \\ -h_{yx}t & 1 - h_{yy}t \end{pmatrix}.$$

The degeneracy of this matrix is, as we saw in the proof of Proposition 9.4.3, exactly the condition for the point $(P, t\nu) \in N$ to be a critical point of F.

It follows that l_q has a degenerate critical point exactly when q is the image of *some* critical point of F—that is, when q is a focal point. Equivalently, l_q is a Morse function if q is not a focal point. This, together with the fact that the set of focal points has measure zero, concludes the proof of Theorem 9.4.1.

10. Homotopies and Surface Generation

10.1. Homotopies and Homotopy Equivalence

Let $f : X \to Y$ and $g : X \to Y$ be two continuous maps between topological spaces X and Y. These maps are called *homotopic* if there exists a family φ_t, for $0 \le t \le 1$, of continuous maps

$$\varphi_t : X \to Y,$$

continuous with respect to t and $x \in X$ simultaneously, and satisfying $\varphi_0 = f$, $\varphi_1 = g$. In words, two maps are homotopic if we can go from one to the other by means of a continuous deformation with parameter $t \in [0, 1]$. The family of maps φ_t is called a *homotopy* between X and Y; it can also be regarded as a continuous map $\Phi : X \times [0, 1] \to Y$.

Definition 10.1.1. Two spaces X and Y are called *homotopically equivalent* if there are continuous maps $f : X \to Y$ and $g : Y \to X$ such that the composition $fg : Y \to Y$ is homotopic to the identity map of Y, and the composition $gf : X \to X$ is homotopic to the identity map of X.

Some examples of homotopically equivalent spaces are depicted in Figure 10.1:

a) $\mathbb{R}^n$ and a point;
b) a circle and an annulus;
c) a Möbius strip and a circle:
d) a sphere with three holes and bouquet of two circles;
e) a torus with hole and a bouquet of two circles.

As we see from these examples, homotopy equivalence is a weaker relation than homeomorphism. For example, a circle, a Möbius strip, and an annulus are all nonhomeomorphic, but all homotopically equivalent. The other parts of Figure 10.1 are examples of the same phenomenon.

10.2. A Homotopy-Based Model for Smooth Surface Generation from Cross-Sectional Data

As an application of homotopy, this section presents a *homotopy model* for reconstructing surfaces from cross sections of objects. The idea is to use ho-

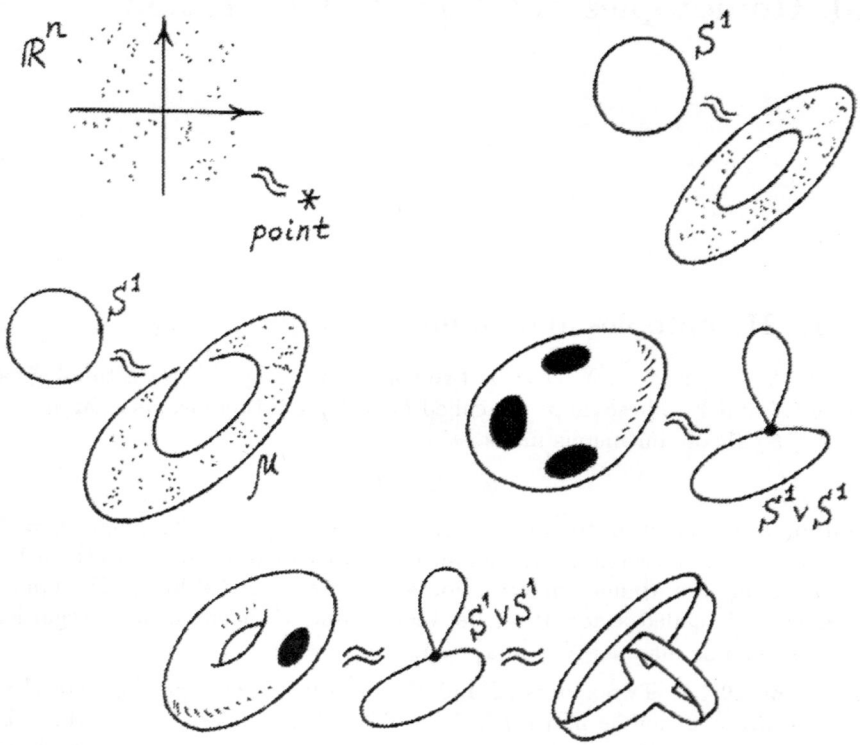

FIGURE 10.1.

motopy to generate surfaces connecting consecutive contours. At the critical levels, branch handling is necessary.

The homotopy model consists of two parts: a *continuous toroidal graph representation*, and homotopy-based generation of surfaces from the representation. A continuous toroidal graph shows the correspondence of parameters of the shape functions of two consecutive contours. In the case of natural data such as CT images, the correspondence of parameters is nontrivial, and automatic generation of paths in toroidal graphs is necessary. We present a heuristic method to find an optimal or nearly optimal path on the toroidal graph.

The homotopy model includes triangulations as a special case, and generates smooth parametric surfaces from contour definitions. The model can be applied to contours represented by parametric curves as well as polygonal curves. Two examples are given: a generalization of the notion of suspensions involving a straight-line homotopy, and interpolation by a cardinal spline surface. At the end of the chapter, example results (surface reconstruction of human ear ossicles) illustrate the advantages of the homotopy model over others.

Overview of existing techniques for surface reconstruction. There is often a need for reconstructing a three-dimensional object from its cross-sectional data. Particularly in the medical field, to construct the entire shape of an organ from a set of sections is of great significance, because it is difficult to envisage the three-dimensional structure of the organ by viewing individual slices. Tam and Davis [1988] categorize the approaches to three-dimensional medical dataset generation into four main classes, the first two classes being based on surface models and the last two on solid models:

(a) *triangular tile techniques*,
(b) *spline approximation*,
(c) *the cuberille method*, and
(d) *octree-encoding*.

Surface models are advantageous for a walk-through animation [Shinagawa et al. 1990], and also when modification of contour data is necessary. Our new model includes the two surface models (1) and (2) as special cases, overcoming their drawbacks. The resulting surface is a generalization of classical parametric surfaces, and includes spline approximations. Two examples are shown. one being a generalization of the notion of suspensions, using a straight-line homotopy, and the other using cardinal spline surfaces.

Regarding the triangulation technique, Fuchs, Kedem, and Uselton [Fuchs et al. 1977] used the toroidal graph representation and, based on graph theory, gave an example that minimizes surface area. This approach requires a great deal of computation to get the optimal solution. Christiansen and Sederberg [1978] provided a simpler triangulation scheme based on the shortest diagonal algorithm, which can be regarded as a kind of greedy algorithm. Their approach also included branch handling, which is adopted in this section. While this scheme works well when adjacent loops are similar in shape, it produces defective triangles if the consecutive contours vary widely. Ganapathy and Dennehy [1982] improved the coherence between contour pairs by transforming the contours in such a way that the perimeter of each contour is equal to one. Their method was based on local constraints only. Kaneda, Harada, Nakamae. Yasuda, and Sato [Kaneda et al. 1987a] proposed the addition of another condition to the shortest diagonal algorithm to remedy this problem. This condition, however, cannot always be satisfied and the algorithm is still greedy. Ekoule et al. [1991] improved the coherence by projecting every point of a contour to its convex hull. The result, however, is affected by the distribution of the points of the contour.

Essentially, triangulation algorithms generate defective triangles if the consecutive contours differ widely in shape. "Defective" means that the normal vectors of the triangles are perturbed and the surface generated appears to have artificial wrinkles or folds. Figure 10.2(a), for example, shows two contours p_0p_1 and $q_0q_1 \ldots q_5$, at different levels; the interpolating surface between them is chosen as a union of triangles. The normal at the point A is very different from the normal at the point B (which is near A), yet is the same as the normal at C (which is far from A). Thus, the surface appears to have a wrinkle around A. The wire-frame representation of this example is in Figure 10.2(b), and a shaded version in Figure 10.2(c), where the defective triangle is painted in red. When we have another contour that has the same shape as the contour p_0p_1 above the contour $q_0q_1 \ldots q_5$, the surface generated is as in Figure 10.2(d), where the defect becomes even more visible.

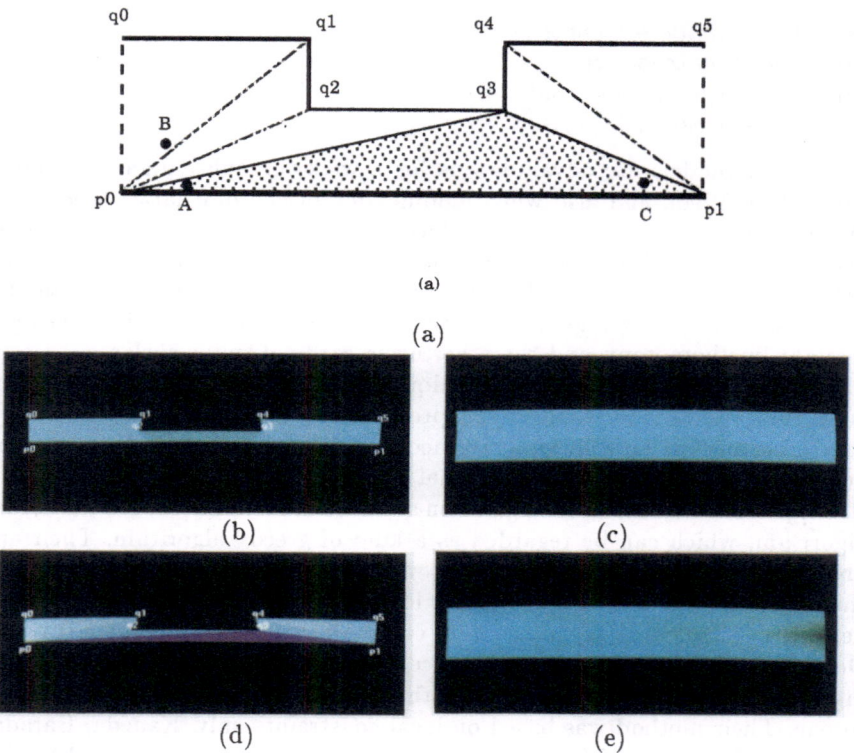

(a)

(a)

(b)   (c)

(d)   (e)

FIGURE 10.2. Defective triangles are inevitable when the triangulation method is used: (a) a union of triangles interpolating between two very different contours; (b) the wire-frame representation of the example; (c) the shaded image; (d) the corresponding surface with the position of the contours interchanged; (e) an image obtained by smooth shading.

The artificial wrinkle thus generated cannot be eliminated even when we use smooth shading, as shown in Figure 10.2(e). Nor can the problem be solved by modifying the connections between nodes, because no node exist between p_0 and p_1. Therefore, the surface generated by the triangulation method depends heavily on the choice of the nodes of each contour. We will show here that the problem is solved by using a homotopy model based on the continuous toroidal graph.

There is another problem with the triangulation method, namely the smoothness of the surface generated. Although triangulated surfaces with smooth shading seem smooth, the surface normals are ambiguous and it is hard to understand the exact shape of the objects when the shape is complicated. It is necessary to generate truly smooth surfaces when observation of the exact shape is required. The homotopy model does this.

In our method, the surface connecting consecutive contours is represented as the locus of a homotopy that transforms one of the contours into the other (Figure 10.3). To be precise, the surface is the locus of a homotopy from f to gU, where U is a bijection (correspondence) between the contours. We now discuss this in detail.

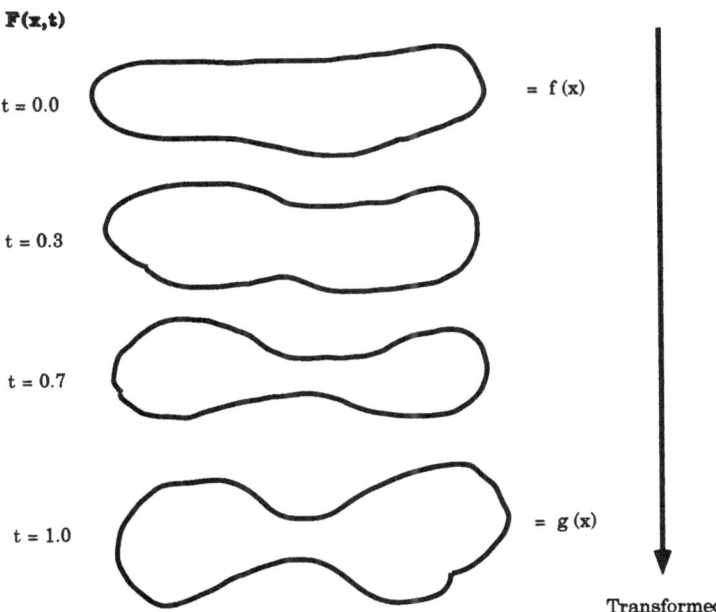

FIGURE 10.3. The transformation of the contours by the homotopy.

Toroidal graph representation. We start by introducing the original discrete version of the toroidal graph representation, proposed in [Fuchs et al. 1977]; we then introduce our continuous version. We assume that a polygonal approximation is used for the contours. Let one contour be defined by a sequence of m distinct contour points $P_0, \ldots, P_{m-1}$, and the other by $Q_0, \ldots, Q_{n-1}$, and assume that the orientation of both loops is the same.

As Christiansen and Sederberg [1978] have pointed out, a triangulation must satisfy two conditions: (1) if two nodes of the same contour are to be defined as the nodes of the same triangle, they must neighbor each other on the contour; and (2) no more than two vertices of any triangle may be recruited from the same contour. Fuchs, Kedem, and Uselton reduced these two rules to one graph-theoretic rule. They represented mutual topological relations of triangles in a toroidal graph, which we adopt as the basic representation in this section. In this graph, vertices correspond to the set of all possible segments between the points $P_0, \ldots, P_{m-1}$ on the one hand and the points $Q_0, \ldots, Q_{n-1}$ on the other, and the arcs correspond to the set of all the possible triangles (see Figure 10.4). The graph of an acceptable surface has exactly one vertical arc in every row of vertical arcs, and exactly one horizontal arc in every column of horizontal arcs. There are two kinds of acceptable surfaces, one homeomorphic to a cylinder and the other to two cones. In this section, the discussion is limited to the former case, unless otherwise noted. Fuchs, Kedem, and Uselton proved that the graph of an acceptable surface homeomorphic to a cylinder is connected and satisfies the property that, for every vertex of the graph, there is one arc going in and one going out.

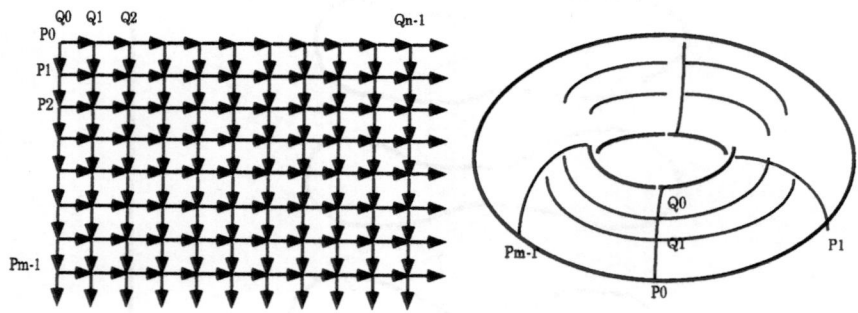

FIGURE 10.4. Toroidal graph representation [Fuchs et al. 1977].

A heuristic approach to finding the optimal path. Based on this result, the same authors provided an algorithm to find the minimum-cost acceptable trail. The algorithm involves a massive search.

By contrast, the shortest diagonal algorithm given by Christiansen and Sederberg is a simple and greedy one. First, adjacent contour lines are mapped onto a unit square. Assume that P_0 and Q_0 are proximate. The shortest diagonal algorithm starts the triangulation by connecting P_0 and Q_0. After vertices P_i and Q_j are connected with an edge, the shorter of the two candidates P_iQ_{j+1} and $P_{i+1}Q_j$ is selected as the next edge in the triangulation. This method produces defective triangles when two consecutive contours differ widely in shape. In Figure 10.5, for example, defective triangles are formed around P_{41} and P_{67}.

Ganapathy and Dennehy [1982] improved the coherence between contour pairs by transforming the contours in such a way that the perimeter of each is equal to one. Then, they proposed selecting P_jQ_i as the next edge in the triangulation when the absolute difference between the length of the polygonal lines $P_0P_1 \ldots P_{j+1}$ and $Q_0Q_1 \ldots Q_i$ is shorter than the absolute difference between the lengths of $P_0P_1 \ldots P_j$ and $Q_0Q_1 \ldots Q_{i+1}$. This algorithm is still based on local constraints only, and cannot avoid defective triangles completely. To remedy this, Kaneda, Harada, Nakamae, Yasuda and Sato [Kaneda et al. 1987b] added another condition to Christiansen's method: for each triangle edge, the shortest of the possible candidates is selected. This condition, however, cannot be always satisfied.

This is obvious when one plots on the toroidal graph every *closest pair vertex*, that is, every pair satisfying the conditions

$$d(P_i, Q_j) = \min_{0 \le k < n} d(P_i, Q_k) \quad \text{and} \quad d(P_i, Q_j) = \min_{0 \le k < m} d(P_k, Q_j),$$

where $d(P, Q)$ is the distance between P and Q. (At most one closest pair vertex exists on each row and on each column of the toroidal graph.) Finding closest pair vertices is a special case of the all nearest-neighbors problem [Preparata and Shamos 1985].

Acceptable paths that connect all the closest pair vertices do not exist when those vertices are distributed as shown in Figure 10.5. Therefore, we next try connecting as many closest pair vertices as possible. First, vertices are weighted 1 when they are closest pair vertices and 0 otherwise. If S is an acceptable path, $\Phi(S)$ is defined as the total weight of the vertices that S passes through. Then the problem can be reduced to finding the acceptable path of maximum total weight. However, a simpler approach, which groups closest pair vertices, is explored in this section. Observation of Figure 10.5 leads to the idea of grouping closest pair vertices into *runs* (P_i, Q_j), (P_{i+1}, Q_{j+1}), $\ldots$, (P_{i+k}, Q_{j+k}). The length $L(R)$ of a run R is the number of vertices in it; if p is a closest pair vertex in the run, we write $p \in R$. Then the following proposition holds.

Proposition 10.2.1. *If there exists an acceptable path S_1 that passes through a closest pair vertex (P_i, Q_j), there exists another acceptable path S_2 that passes through all the closest pair vertices of the run R, where $(P_i, Q_j) \in R$ and $\Phi(S_1) \le \Phi(S_2)$.*

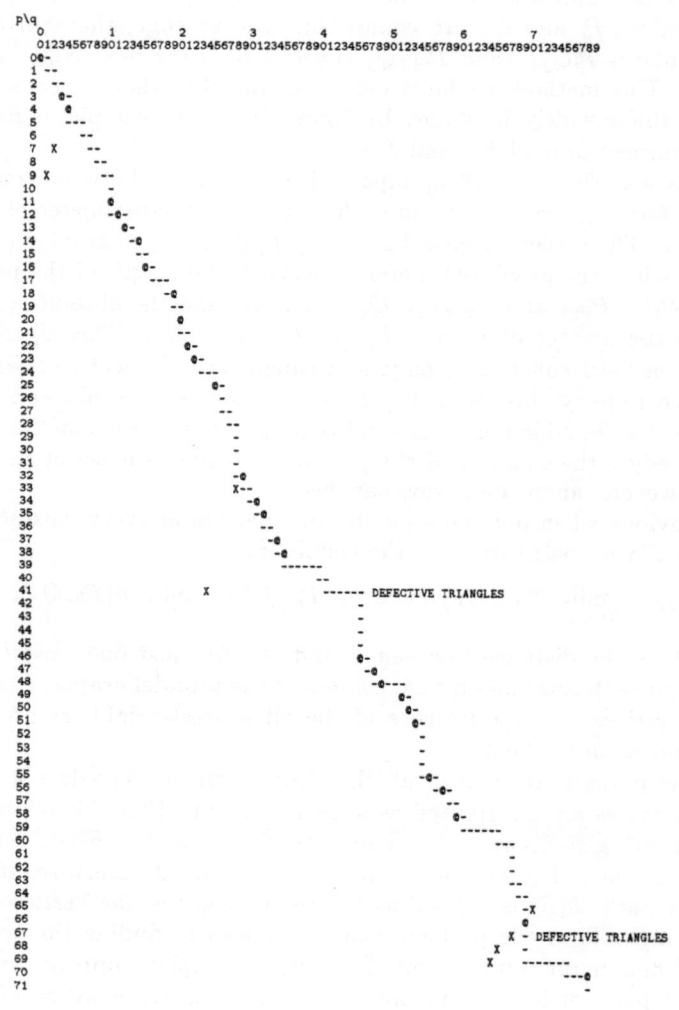

FIGURE 10.5. Example distribution of closest pair vertices and the path chosen by the shortest diagonal algorithm.

Proof. Without loss of generality, it can be assumed that $L(R) \geq 2$.

Case 1: $(P_{i+1}, Q_{j+1}) \in R$ and S_1 *does not pass through this vertex.*

   Case 1a: S_1 *passes through* (P_{i+1}, Q_j). Let (P_k, Q_{j+1}) be the last vertex at which S_1 leaves the row $j+1$. Let $S_1 = S_3 S_4 S_5$, where S_4 is a path from (P_{i+1}, Q_j) to (P_k, Q_{j+1}). $\Phi(S_4) = 0$ holds because at most one closest pair vertex can exist on each row and each column. Obviously there is a path S_6 from (P_{i+1}, Q_{j+1}) to (P_k, Q_{j+1}) and $\Phi(S_6) = 1$. Therefore when S_2 is constructed as the concatenation of S_3, S_6 and S_5, we have $\Phi(S_2) = \Phi(S_1) + 1$.

   Case 1b: S_1 *passes through* (P_i, Q_{j+1}). The discussion is the same as in Case 1a.

Case 2: $(P_{i-1}, Q_{j-1}) \in R$ and S_1 *does not pass through this vertex.* The discussion is the same as in Case 1.

Applying this construction method recursively, the desired path S_2 is obtained. $\qquad \square$

By the Proposition, only runs have to be taken into consideration when we try to maximize Φ. Thus the objective stated previously is reduced to the following one: to choose an acceptable set of runs that maximizes Φ, where we define an *acceptable set of runs* as a set of runs that an acceptable path passes through. Obviously, any set of runs that consists of two runs is acceptable. An unacceptable set of three runs $\{R_1, R_2, R_3\}$ is as follows: suppose $R_1 = (P_{i_1}, Q_{j_1}), \ldots, R_2 = (P_{i_2}, Q_{j_2}), \ldots, R_3 = (P_{i_3}, Q_{j_3}), \ldots,$ and $i_1 < i_2 < i_3 \pmod{m}$. If $j_1 < j_3 < j_2 \pmod{m}$ or $j_2 < j_1 < j_3 \pmod{m}$ or $j_3 < j_2 < j_1 \pmod{m}$, this set is not acceptable. A set of more than three runs is acceptable if and only if any three elements constitute an acceptable set. Therefore, the objective stated above requires an exhaustive search similar to the knapsack problem.

Confronted with this complexity, we decided to explore a heuristic scheme. Since one cause of the complexity lies in the use of modulo m, it is abandoned. This corresponds to cutting open the torus to a plane. With this simplification, acceptability becomes a binary relation between runs instead of a ternary relation. A set of two runs $(P_{i_1}, Q_{j_1}), \ldots,$ and $(P_{i_2}, Q_{j_2}), \ldots,$ is acceptable if $i_1 < i_2$ and $j_1 < j_2$ or $i_1 > i_2$ and $j_1 > j_2$. When a set $\{R_1, R_2\}$ is not acceptable, R_1 is said to be inconsistent with R_2. This is much easier to test, but there is the possibility that a run might be divided into two when we cut the torus to a plane. Finally, to choose the runs that maximize Φ, a heuristic stack algorithm is adopted:

Algorithm 10.2.2. Assume that there are N runs, and that, for each run $R_k = (P_{i_k}, Q_{j_k}), \ldots,$ we have $i_1 < i_2 < \cdots < i_k < \cdots < i_N$. At first, the stack is empty.

Step 1. Push R_1 onto the stack.
Step 2. Set $k = 2$.

Step 3. Let R be the run at the top of the stack.

Step 4. If R_k and R are consistent, push R_k onto the stack. Else if $L(R_k) \leq$ $L(R)$. remove R_k and set $k = k + 1$. Otherwise remove R from the stack.

Step 5. If $k \leq N$, go to step 3.

As for R_k, less than k iterations are needed. Therefore, the time complexity of this algorithm is $O(N^2)$. This algorithm yields an acceptable set of runs that nearly maximizes Φ.

Continuous version of the toroidal graph. Next, it is necessary to select the trail of the path in detail. Although the path must pass through the vertices of the runs chosen by the previous algorithm, the other vertices still remain arbitrary. Here we propose "linear interpolation" of closest pair vertices in order to select the trail. Before discussing this, we must introduce a continuous version of the discrete toroidal graph.

First, the lower and upper contours must be represented by parameters: points on the contours are designated by functions $f, g : I \to \mathbb{R}^3$, where $I = [0, 1] \subset \mathbb{R}$ and $f(0) = f(1)$, $g(0) = g(1)$. In the continuous toroidal graph, the horizontal and vertical distances between two vertices of the continuous toroidal graph represent the differences of parameter values between the two.

Example 10.2.3. Arclength is used as the parameter. Let l_1 and l_2 be the length of the lower and the upper loop. For example, the horizontal distance of the graph between (P_{i_1}, Q_{j_1}) and (P_{i_2}, Q_{j_2}) represents the arclength between P_{i_1} and P_{i_2} over the upper loop and the vertical distance of the graph represents the arclength between Q_{j_1} and Q_{j_2} over the lower loop. The point (x, y) on the graph represents the pair (P, Q), where P is the point whose arclength from P_0 is $x l_1$, and Q is the point whose arclength from Q_0 is $y l_2$.

There are other ways to choose the parameter:

Example 10.2.4. When the contours are close in shape to geometric circles, the argument with respect to a pole near the center of the contour can be used as the parameter. This is essentially a cylindrical coordinate system.

Contours need not be approximated by polygonal lines. Parametric curves such as Bézier curves. B-splines, or cardinal splines can be used instead.

Example 10.2.5. The parameter of the spline basis function can be used as the parameter of the graph. To use the linear interpolation discussed later, arclength representation is desirable. However, the conversion from the curve parameter to the arclength is not computationally easy. One remedy is to approximate the spline curves by line segments and use their length instead of actual arclength. Display examples presented later use this approximation.

When a path passes through (x, y), the corresponding points $P = f(x)$ and $Q = g(y)$ will be *connected by a homotopy*.

As long as the parameter increases monotonically as the point on the
contour moves away from the initial node, it does not matter essentially
what parameter is used in the following discussion. The continuous version
of an acceptable path is represented on this graph as a monotonically increasing (or decreasing) "multi-valued function." Strictly speaking, the path
is represented as the concatenation of the graphs $y = U_i(x)$ and $x = V_i(y)$,
for $i = 0, \ldots, \nu - 1$, where $U_i : [x_{2i}, x_{2i+1}] \to I$ and $V_i : [y_{2i}, y_{2i+1}] \to I$
(with $0 = x_0 \leq x_1 \leq \cdots \leq x_{2\nu-1} < 1$ and $0 = y_0 \leq y_1 \leq \cdots \leq$
$y_{2\nu-1} < 1$) are monotonically increasing (or decreasing) functions, such that
at each joint (x_{2i+1}, y_{2i+1}) of U_i and V_i, we have $U_i(x_{2i+1}) = V_i(y_{2i+1})$ and
$U_1(0) = V_{\nu-1}(1)$, and at each joint (x_{2i+2}, y_{2i+2}) of V_i and U_{i+1}, we have
$V_i(y_{2i+2}) = U_{i+1}(x_{2i+2})$. This representation is a generalization of a monotonically increasing function. Figure 10.6 shows an example where $\nu = 2$. The
discrete toroidal graph is the special case of this continuous version, and the
conversion is straightforward (see Figure 10.7): its graph is the concatenation
of the graphs of $y = q_i$ and of $x = p_i$. The surface represented by this path
is as in Figure 10.8. As for the homotopy, linear interpolation (straight-line
homotopy) is used as an example. It uses triangles, all the points on the
base being connected with the opposite vertex. The heuristic triangulation
method proposed by Ganapathy and Dennehy [1982] can be seen as the approximation of $y = x$ by a step function when we use the continuous toroidal
graph.

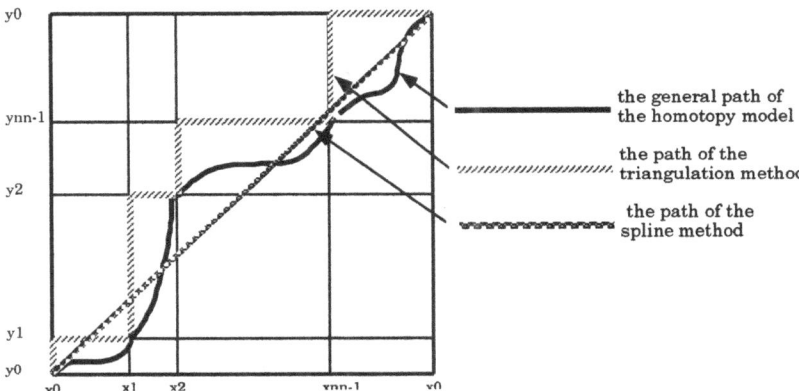

the general path of
the homotopy model

the path of the
triangulation method

the path of the
spline method

FIGURE 10.6. Acceptable path on the continuous toroidal graph.

Finally, the details of the path are determined on this graph. Suppose the
closest pair vertices are (x_i, y_i), where $i = 0, 1, \ldots, k - 1$ and $0 = x_0 < \cdots <$

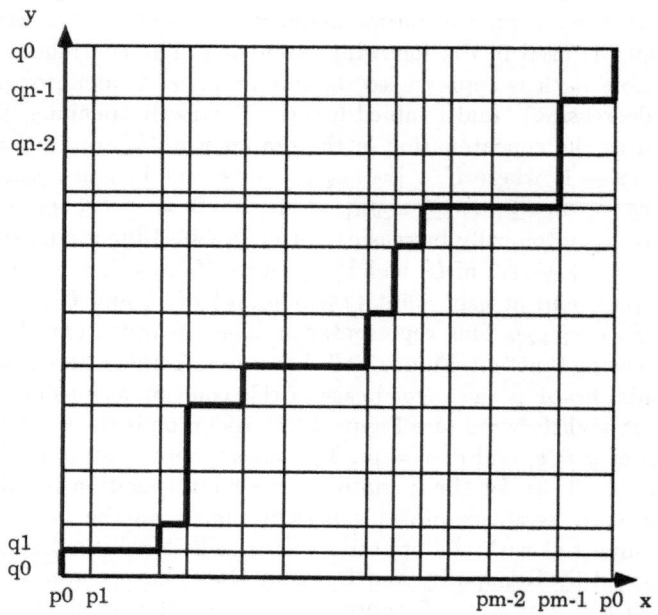

FIGURE 10.7. The discrete toroidal graph as a special case of the continuous toroidal graph.

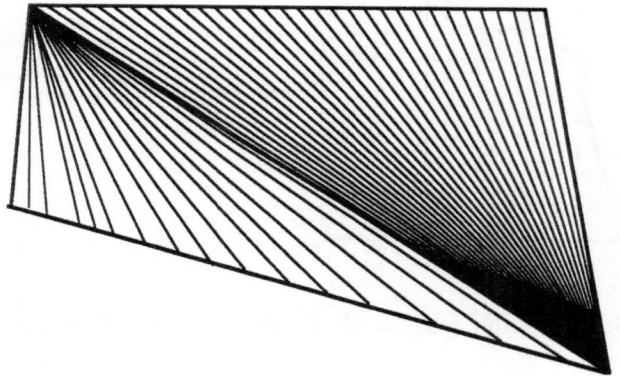

FIGURE 10.8. Correspondence between the two adjacent contours of a triagular mesh.

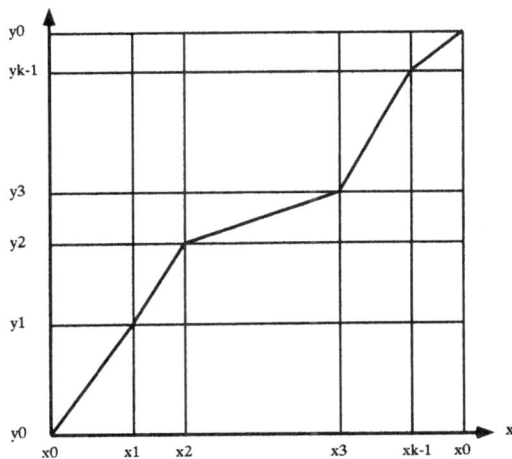

FIGURE 10.9. Linear interpolation of the closest pair vertices.

x_{k-1}. The path is to be represented as a function $U(x) : I \to I$ such that

$$U(x) = (y_{i+1} - y_i)\frac{(x - x_i)}{(x_{i+1} - x_i)} + y_i, \quad \text{for } x_i \leq x < x_{i+1}.$$

As stated earlier, this is the linear interpolation between (x_i, y_i) and (x_{i+1}, y_{i+1}); see Figure 10.9. The surface represented by this path is as in Figure 10.10(a); the contours are the same as in Figure 10.2(a). This example shows a straight-line homotopy. The shaded image is as in Figure 10.10(b), and the surface generated with the contour positions interchanged is shown in Figure 10.10(c). Comparing Figures 10.2(a) and 10.10(a), we see that the surface in the latter figure is smoother, and that defective triangles are avoided. Figure 10.11 shows an example corresponding to Example 10.2.5.

The concept of a closest pair vertex can be also extended to a continuous version: closest pair points can be defined as pair of points on the contours that are mutually closest. We will not discuss this further, however. Since the number of the closest pair points can be infinite, sampling them and choosing a finite number of points among them seems appropriate for current implementation purposes. The homotopy model serves as the basis of the toroidal graph representation expressed by $U(x)$ in order to generate defect-free surfaces between contours.

Overview of the surface generation method. As previously stated, a vertex on the continuous toroidal graph represents a correspondence between a point on the upper and one on the lower contour, and these points are to be connected by a curve, much like a fiber, that is part of the surface we are

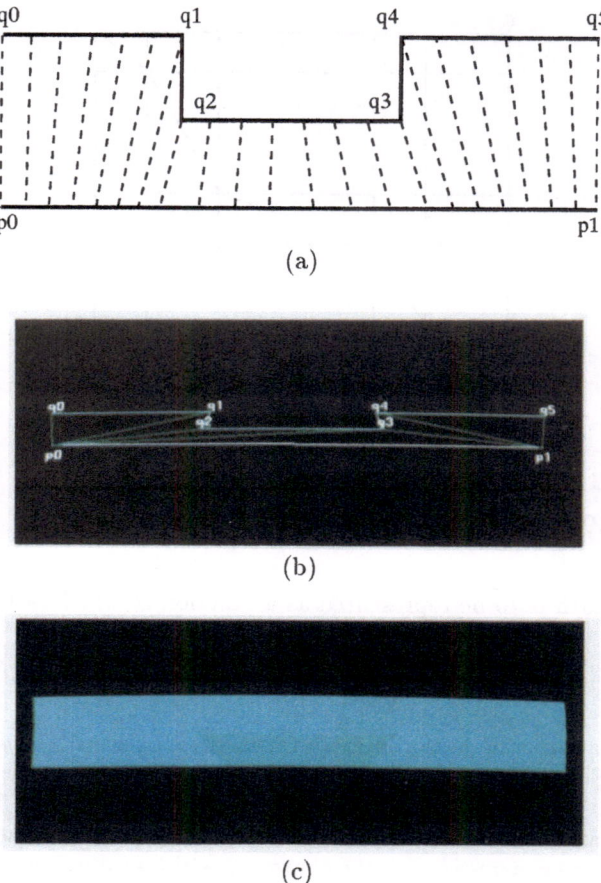

FIGURE 10.10. For the same contours as in Figure 10.2, we show (a) the interpolating surface obtained by a straight-line homotopy, (b) the same surface with shading, and (c) the surface obtained by interchanging the positions of the contours.

generating by means of a homotopy between the contours. For instance, if we choose to use the straight-line homotopy, the two points are connected by a line segment. The surface patch between the contours is regarded as the locus of the homotopy. A simple example to understand this model is a cylinder. The surface between the two circles are generated by connecting corresponding points on each circle. As shown later, this model includes parametric surfaces. Based on this homotopy model, the surface between adjacent contours is generated from the continuous toroidal graph representation.

More precisely, let the lower and upper contours be expressed by the maps $f, g : X \to \mathbb{R}^3$, where $X = [x_0, x_1] \subset \mathbb{R}$. For simplicity, let the lower

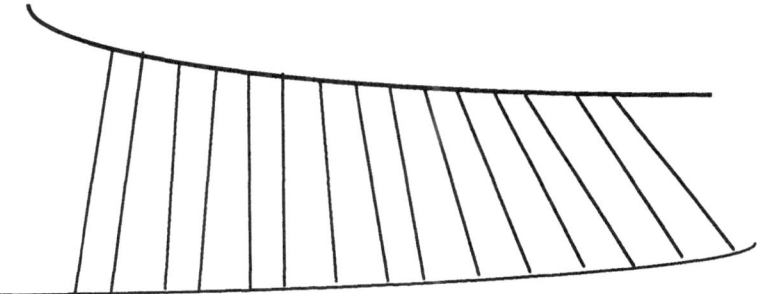

FIGURE 10.11. Correspondence between smoothly curved contours.

contour be on the plane $z = 0$, and the upper one on the plane $z = 1$, and let $X = I$. Then saying that two points $P = (p_1, p_2)$ and $Q = (q_1, q_2)$ are connected by a homotopy F means that there exists $x \in I$ such that $F(x, 0) = f(x) = (p_1, p_2, 0)$ and $F(x, 1) = g(x) = (q_1, q_2, d)$. Then, the cross section at height $z = t$ of the surface being generated (where $0 \le z \le 1$) is given by $F(x, t)$ for $x \in X$. In other words, the surface is represented by the bivariate function $F(x, t)$.

The acceptable path (correspondence between upper and lower contour) must be represented on the toroidal graph by a monotonically increasing continuous function $U : I \to I$, having an inverse U^{-1}. We now give a formula for the surface being generated, based on the homotopy between $f(x)$ and $g(U(x))$. When a straight-line homotopy is used, the formula is

$$\tilde{F}(x, t) = ((1 - t)f(x) + tg(U(x)))$$

Other homotopies can also be used, such as *cardinal splines* [Clark 1981]. In this case we are interpolating between several contours at a time. Let the functions representing a series of four contours be f_{-1}, f_0, f_1, and f_2, and let the respective acceptable paths be U_{-1}, U_0, and U_1. Then the surface between f_0 and f_1 is

$$F(x, t) = w_{-1}(t)f_{-1}(U_{-1}^{-1}(x)) + w_0(t)f_0(x) + w_1(t)f_1(U_0(x))$$
$$+ w_2(t)f_2(U_1(U_0(x))),$$

where
$$\begin{pmatrix} w_{-1}(t) \\ w_0(t) \\ w_1(t) \\ w_2(t) \end{pmatrix} = \begin{pmatrix} -a & 2a & -a & 0 \\ 2-a & -3+a & 0 & 1 \\ -2+a & 3-2a & a & 0 \\ a & -a & 0 & 0 \end{pmatrix} \begin{pmatrix} t^3 \\ t^2 \\ t^1 \\ 1 \end{pmatrix}.$$

When f_i is represented by a cardinal spline function, the surface is referred to as a cardinal spline surface; an example result is displayed later. When f_i is represented by a B-spline, this equation is similar to that of [Wu et al. 1977].

However, the major difference between our approach and theirs is in the use of U_i. Wu's approach can be considered to be the special case where U_i is the identity map. By introducing U_i, our model can reconstruct complicated objects like the one shown in Figure 10.12. The surface generated is not smooth if we take the identity map instead of U_i.

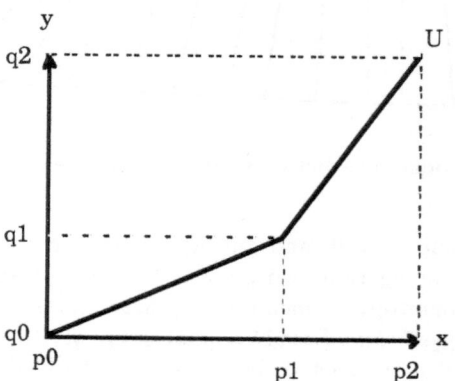

upper contour

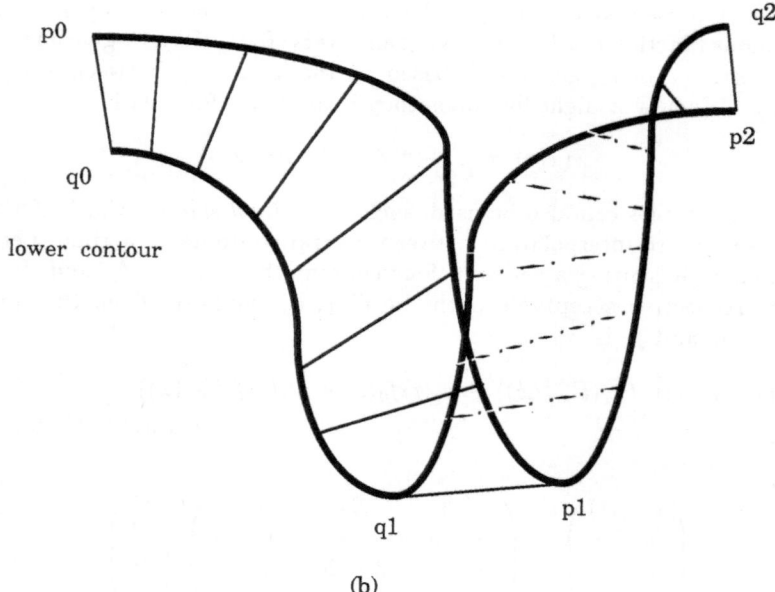

(b)

FIGURE 10.12. (a) Function U; (b) surface generated.

Branching. So far we have assumed that the upper and lower contours are both circles. Sometimes, however, one circle on a contour matches two on an adjacent contour; this corresponds to crossing a saddle singularity in Morse theory (see Figure 8.5). More complicated singularities can occur, as we know, but for simplicity we assume that the upper contour consists of two loops and the lower contour consists of one loop. Interpolating between the two contours means solving the problem of *branching*.

Branch handling can be accomplished by introducing a intermediate contour between the upper and lower contour that contains a singular point where the two loops meet. As illustrated in Figure 10.13, above the crossing the contours consist of two loops, and below they consist of a single loop. The method of [Christiansen and Sederberg 1978] is one way to deal with this branching. If C'^2 continuity is imposed on the surface, the crossing is to be a critical point of the height function.

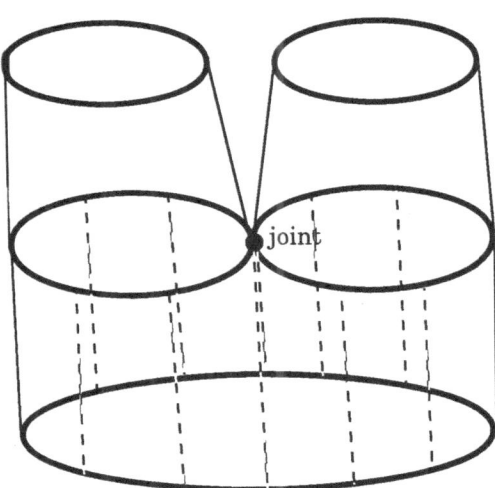

FIGURE 10.13. Branching.

Relationship with other parametric surfaces. The homotopy model of a surface is the generalization of different parametric surfaces. For example, a *suspension*

$$F(x,t) = (1-t)f(x) + tg(x)$$

can be regarded naturally as a straight-line homotopy between f and g.

A Boolean sum surface

$$F(x,t) = (h_0(x), h_1(x)) \cdot (r(t), s(t)) + (h_0(t), h_1(t)) \cdot (f(x), g(x))$$

$$+ (h_0(x), h_1(x)) \begin{pmatrix} f(0) & g(0) \\ f(1) & g(1) \end{pmatrix} \begin{pmatrix} h_0(t) \\ h_1(t) \end{pmatrix},$$

where $F(0, t) = r(t)$, $F(1, t) = s(t)$, and $h_0, h_1 : I \to I$ are blending functions, is another example of a homotopy.

The triangular mesh generated by Fuchs's or Christiansen's method can also be expressed by a straight-line homotopy. For simplicity, we assume that $f(x)$ and $g(x)$ represent single line segments and that between them, triangles $(f(0), f(1), g(0))$ and $(f(0), g(1), g(0))$ are formed. Then the triangular mesh is expressed as

$$\tilde{F}(x, t) = \begin{cases} (1 - t)f(x) + tg(0) & \text{if } x \le t, \\ (1 - t)f(0) + tg(x) & \text{if } x > t. \end{cases}$$

The partial derivative $\partial F / \partial x$ is not defined on the line $x = t$, and the generated surface is not smooth. There is another way of connecting the line segments to produce the triangular mesh. The point $f(0)$ is expanded to the whole lower line segment, and then the whole upper line segments other than $f(0)$ retracts to the point $g(1)$. This is the correspondence that the continuous toroidal graph designates. Although this cannot be expressed by a homotopy, it is not difficult to understand it intuitively as a natural extension of the concept of the homotopy.

The homotopy model has one more degree of freedom. In a boolean sum surface, for example, when x is fixed, the curve $F(x, t)$ is on the plane that contains the points $f(x)$ and $g(x)$, while in the case of homotopy model, the curve $F(x, t)$ is not limited to that plane. In the current implementation, however, this degree of freedom has not yet been exploited yet.

Human ear data.   A human temporal bone with the ear lobe was obtained from a cadaver and subjected to the conventional celloidin processing. A motor-driven drill was used to place reference marks in a celloidin block [Nomura et al. 1989]. A larger celloidin section may be unevenly distended when mounted on a slide, causing movement of reference marks and distortion of parts of the ear structure. The celloidin block was serially sectioned into slices $20\,\mu$m thick. All slices were stained with hematoxylin and eosin (H-E) and mounted on glass slides. A color photograph was taken of each of the H-E stained specimens. The photographs were converted to image data by a Graphica G-225C drum scanner, and outline curves of the objects to be reconstructed were plotted with a stylus pen. The computer used was a Hewlett-Packard HP9000. The contour data were fed into a Silicon Graphics IRIS graphics workstation, where the three-dimensional reconstruction was performed using the homotopy model explained above.

Some of the resulting images are displayed here. Figure 10.14 shows the three auditory ossicles (malleus, incus and stapes). The outline curves are approximated by the cardinal spline and a straight-line homotopy is used

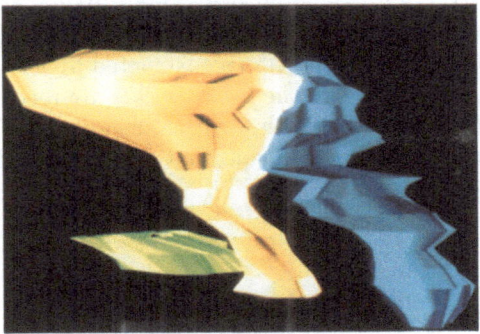

FIGURE 10.14. The three ear ossicles, reconstructed using a straight-line homotopy.

FIGURE 10.15. The ossicles, reconstructed by a cardinal spline homotopy.

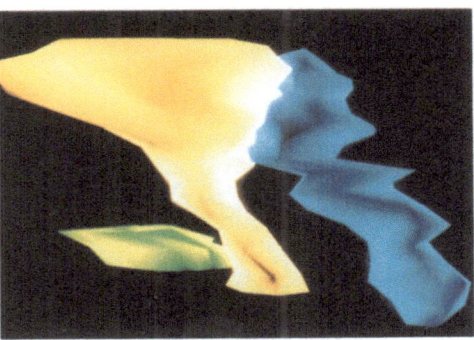

FIGURE 10.16. The ossicles, reconstructed using the triangulation method with Gouraud shading [Nomura et al. 1989].

for surface reconstruction. As noted in Example 10.2.5, the arclength of the cardinal spline curve is approximated by the length of linear line segments. Figure 10.15 shows the same objects reconstructed by using a cardinal spline homotopy, while in Figure 10.16 the same objects are reconstructed by Christiansen's triangulation method with Gouraud shading. As mentioned earlier, the latter technique leads to an image that suffers from ambiguities and is difficult to interpret.

11. Homology

11.1. Simplicial Homology

In this chapter we associate to certain topological spaces a sequence of abelian groups, called *homology groups*, $H_r(K)$, for $r = 0, 1, \ldots$. These groups give us information on the topological structure of K. For instance, $H_0(K)$ measures the number of connected components of the space; the higher homology groups $H_r(K)$, for $r > 0$, intuitively speaking, measure the number of "r-dimensional holes" in K. A one-dimensional hole is a hole in the ordinary sense, a two-dimensional hole is an enclosed space, and so on.

Basic definitions. We start by defining the notion of a *simplicial complex*. An *n-simplex* $\sigma = (p_0 p_1 \ldots p_n)$ is the convex hull of $n + 1$ points $p_0, \ldots, p_n$ in general position in some space $\mathbb{R}^N$, where $N \geq n$; "general position" here means that these $n + 1$ points are not contained in an affine subspace of dimension lower than n (this is the same as saying that $p_1 - p_0$, $p_2 - p_0$, $\ldots, p_n - p_0$ are linearly independent vectors). Thus a 0-simplex is a point, a 1-simplex is a segment, a 2-simplex is a triangle, and a 3-simplex is a tetrahedron.

The points $p_0, \ldots, p_n$ are the *vertices* of the simplex. *We will always consider σ with an orientation*, which can be specified by ordering the vertices; an even permutation of the vertices specifies the same orientation, while an odd permutation specifies the opposite orientation.

An m-dimensional *face* or *subsimplex* of σ is the convex hull of an m-point subset of $\{p_0, \ldots, p_n\}$. In particular, σ is a subsimplex of itself. The zero-dimensional faces are the vertices; the one-dimensional faces are called *edges*. The *interior* of a simplex is what's left when you take away its proper faces; it can also be defined as the topological interior of the simplex considered as a subset of the smallest affine subspace that contains it.

Using simplexes one can compose more complicated objects. A *simplicial complex* is a locally finite collection K of simplexes (in $\mathbb{R}^N$), satisfying the following two conditions: any face of a simplex in K is also in K, and the intersection of two simplexes in K is either empty or a face of both. (*Locally finite* means that any point $p \in \mathbb{R}^N$ has a neighborhood intersecting only finitely many simplexes of K. Most often K will have only finitely many simplexes anyway.) The *dimension* of K is the highest dimension of a simplex appearing in K.

Figure 11.1 shows legal ways in which the simplexes of a simplicial complex can meet.

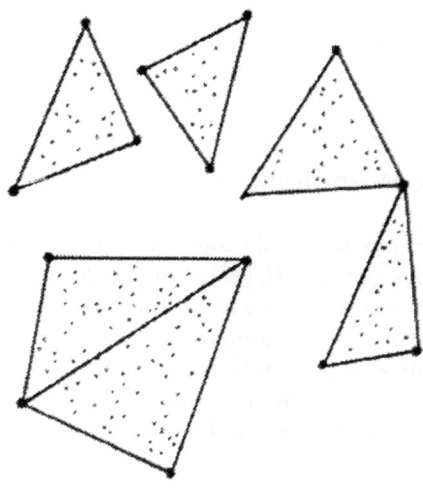

FIGURE 11.1.

The union of all simplexes in K is called the *polyhedron* or *underlying space* of K, and is denoted by $|K|$. The *k-skeleton* of K is the subcomplex consisting of simplexes of dimension k or lower.

A topological space X is *triangulated* if we have a simplicial complex $|K|$ and a homeomorphism between X and $|K|$. We also say in this case that X has a *simplicial subdivision* or *triangulation*: we can think of it as being made up of "curvilinear simplexes" such as the ones in Figure 11.2. This notion of a triangulation coincides with the one laboriously introduced in Definition 4.1.2 for the case of surfaces—it is rather easier to define a triangulation in the context of simplicial complexes!

In the subsequent discussion, all statements about a simplicial complex also apply to an arbitrary topological space X with a simplicial subdivision. Of course, the same X may have many different simplicial subdivisions, so we'd better have a particular subdivision in mind when we apply the statements to X.

Chains and the boundary operator. Assume that K is a simplicial complex of dimension m, and let q be any integer. We denote the q-simplexes of K, *each with a fixed orientation*, by σ_i. We then consider finite linear

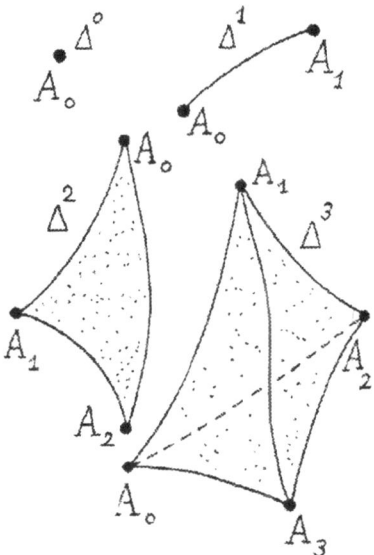

FIGURE 11.2.

combinations of the form
$$c = \sum a_i \sigma_i,$$

where the a_i are integers. This expression is purely formal, that is, an assignment of an integer coefficient to each q-simplex, in such a way that only finitely many coefficients are nonzero—the simplexes are not added in any geometric sense. The set $C_q(K)$ of such linear combinations is the *free abelian group* generated by the q-simplexes of K; it is a direct sum of copies of $\mathbb{Z}$, one for each σ_i. Addition in $C_q(K)$ is carried out by adding the coefficients of terms with the same generator σ_i. The elements of $C_q(K)$ are called (simplicial) q-*chains*; thus $C_q(K)$ is the q-*dimensional chain group* of K.

More precisely, $C_q(K)$ is the group of *integer-valued* chains in K, since its elements are combinations of the generators with *integer* coefficients. When we want to stress this, we write $C_q(K, \mathbb{Z})$ instead of $C_q(K)$. It is possible to define, in an exactly analogous way, groups $C_q(K, G)$ for any abelian group of coefficients G, whose elements are formal combinations

$$c = \sum_i a_i \sigma_i, \quad \text{for } a_i \in G.$$

Such combinations are called G-*valued* q-*chains*. The most interesting groups of coefficients are the groups $\mathbb{Z}$, $\mathbb{Z}_p = \mathbb{Z}/p\mathbb{Z}$, and $\mathbb{R}$. For concreteness, we will develop the theory of simplicial homology with coefficients in $\mathbb{Z}$; the case of other coefficient groups is similar, and differences will be pointed out where relevant.

Since there are no simplexes in dimensions $q > m$ and $q < 0$, the only chain in these cases is 0, and the chain group is the trivial group. We identify the chain $1\sigma_i$ with the simplex σ_i, and the chain $(-1)\sigma_i = -\sigma_i$ with the simplex obtained from σ_i by reversing its orientation. Thus, for example, if $\sigma_i = (p_0 p_1)$, we can write $-\sigma_i = -(p_0 p_1) = (p_1 p_0)$. Here the oriented simplex $(p_1 p_0)$ is not an element of the chosen basis for $C_q(K)$, so the right-hand side of the equality has no prior meaning as a q-chain, but we *define it* to mean the chain $-(p_0 p_1)$. This is a very useful convention; it is used. for example, in the next definition.

Definition 11.1.1. The $(q-1)$-dimensional faces of a q-simplex $\sigma^q = (p_0 p_1 \ldots p_q)$ have an *induced orientation* given by

$$(-1)^i (p_0 \ldots p_{i-1} p_{i+1} \ldots p_q), \quad \text{for } i = 0, \ldots, q.$$

The *boundary* of a q-simplex is the sum of all its $(q-1)$-dimensional faces, taken with the induced orientation:

$$\partial_q (p_0 \ldots p_q) = \sum_{i=0}^{q} (-1)^i (p_0 \ldots p_{i-1} p_{i+1} \ldots p_q).$$

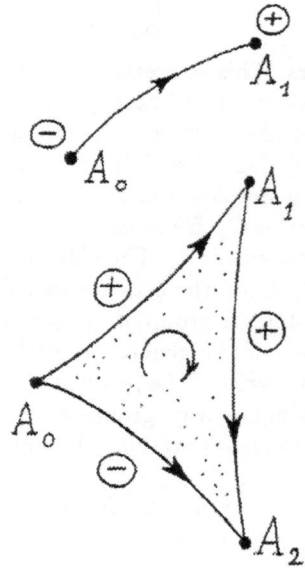

FIGURE 11.3.

Here are some examples (see Figure 11.3). A one-simplex $\sigma^1 = (p_0 p_1)$ is given by its two vertices. Its boundary is

$$\partial(p_0 p_1) = p_1 - p_0.$$

The boundary of a two-simplex $\sigma^2 = (p_0 p_1 p_2)$ is what you get by going around the triangle in the order $p_0 \to p_1 \to p_2 \to p_0$:

$$\partial(p_0 p_1 p_2) = (p_0 p_1) - (p_0 p_2) + (p_1 p_2) = (p_0 p_1) + (p_1 p_2) + (p_2 p_0).$$

The boundary of a three-simplex $\sigma^3 = (p_0 p_1 p_2 p_3)$ is

$$\partial(p_0 p_1 p_2 p_3) = (p_1 p_2 p_3) - (p_0 p_2 p_3) + (p_0 p_1 p_3) - (p_0 p_1 p_2).$$

The boundary of a zero-simplex is 0.

We now extend the notion of boundary from individual simplexes to chains, by linearity. Thus, by definition, the boundary of a q-chain $c = \sum_i n_i \sigma_i$ is $\partial_q(c) = \sum_i n_i \partial_q(\sigma_i)$. This defines a homomorphism $\partial_q : C_q(K) \to C_{q-1}(K)$, called the ($q$-dimensional) *boundary operator*. In fact, we obtain a whole sequence of boundary operators

$$\cdots \longrightarrow C_{q+1}(K) \xrightarrow{\partial_{q+1}} C_q(K) \xrightarrow{\partial_q} C_{q-1}(K) \longrightarrow \cdots,$$

called the *simplicial chain complex* of K. These operators satisfy the important equation

$$\partial_q \partial_{q+1} = 0 \qquad \text{for all } q.$$

Simplicial homology groups.   A q-chain is called a q-*cycle* if its boundary is 0. It is a q-*boundary* if it is the boundary of some $(q+1)$-chain. Thus, the group of q-cycles is

$$Z_q(K) = \ker \partial_q,$$

and the group of q-boundaries is

$$B_q(K) = \operatorname{im} \partial_{q+1}.$$

A q-cycle *homologous to zero* is the same as a q-boundary. Two q-cycles z_1 and z_2 are called *homologous* (to one another) if their difference $z_1 - z_2$ is homologous to zero, that is, if $z_1 - z_2 = \partial h$ for some $h \in C_{q+1}(K)$.

Note that $B_q(K)$ is a subgroup of $Z_q(K)$, since $\partial_q \partial_{q+1} = 0$. In words, *every boundary is a cycle*. However, it is not true that every cycle is a boundary.

Definition 11.1.2. The q-*dimensional homology group* (or simply q-*th homology group*) is the quotient

$$H_q(K) = Z_q(K)/B_q(K).$$

An element of $H_q(K)$ is called a *homology class*. The homology class of a cycle $c \in Z_q(K)$ will be denoted by $[c]$.

Intuitively, the group $Z_q(K)$ counts all q-cycles, and the group of q-boundaries $B_q(K)$ counts all "filled" q-cycles, those that bound a $(q+1)$-cycle. The q-dimensional homology group, therefore, counts only "unfilled" q-cycles, which we may think of as "holes".

Simplicial maps and the invariance of simplicial homology. It turns out that, although simplicial homology groups are defined for a simplicial complex K, any other simplicial structure on the underlying space $|K|$ leads to the same homology groups (though not to the same chain and boundary groups). We shall state this below in more detail; for proofs of this and other statements not proved here, refer to [Munkres 1984], for example.

Definition 11.1.3. Let K and L be simplicial complexes. A map $\psi : K^{(0)} \to L^{(0)}$ from the set of vertices of K to that of L is a *simplicial map* if, for any simplex σ in K, all the vertices of σ are mapped to vertices of some simplex of L (not necessarily all distinct).

We can extend a simplicial map to a continuous map between the underlying polyhedra $|K|$ and $|L|$, as follows. Write $x \in \sigma$ as $x = \sum_{i=0}^{q} a_i p_i$, where each $a_i \geq 0$ and $\sum_{i=0}^{q} a_i = 1$. Then set

$$\psi_\sigma(x) = \psi_\sigma\left(\sum_{i=0}^{q} a_i p_i\right) = \sum_{i=0}^{q} a_i \psi(p_i);$$

this is an affine map, hence continuous. If σ and τ share a face, ψ_σ and ψ_τ agree on that face. Hence we can combine all the ψ_σ's into a map $\psi : |K| \to |L|$, where $\psi|_\sigma = \psi_\sigma$.

The identity map on K and the inclusion map $i : M \to K$, where M is a subcomplex of K, are obviously simplicial maps.

Given a simplicial map $\psi : K \to L$, we can define a homomorphism between chain groups as follows:

$$\psi_q(\sigma) = \psi_q((p_0 p_1 \ldots p_q)) = (\psi(p_0)\psi(p_1)\ldots\psi(p_q))$$

if all the $\psi(p_i)$, for $0 \leq i \leq q$, are distinct; otherwise $\psi_q(\sigma) = 0$. This defines ψ_q on chains consisting of one simplex; we then extend ψ_q to all chains by linearity. This is consistent since $\psi_q(-\sigma) = -\psi_q(\sigma)$. The result is a homomorphism $\psi_q : C_q(K) \to C_q(L)$.

Lemma 11.1.4. *If $\psi : K \to L$ is a simplicial map, we have for all q the important relation*

$$\partial_q \psi_q = \psi_{q-1} \partial_q,$$

that is, the following diagram is commutative.

$$
\begin{array}{ccc}
C_q(K) & \xrightarrow{\ \partial_q\ } & C_{q-1}(K) \\
\downarrow{\psi_q} & & \downarrow{\psi_{q-1}} \\
C_q(L) & \xrightarrow{\ \partial_q\ } & C_{q-1}(L)
\end{array}
$$

This fact implies that the homomorphism $\psi_q : C_q(K) \to C_q(L)$ induces a homomorphism $\psi_{q*} : H_q(K) \to H_q(L)$, defined by

$$\psi_q([c]) = [\psi_q(c)].$$

This is well-defined: we can easily check that, if $c, c' \in Z_q(K)$ satisfy $[c] = [c']$, the right-hand side of the definition above is the same whether we use c or c', since $\psi_q(Z_q(K)) \subset Z_q(L)$ and $\psi_q(B_q(K)) \subset B_q(L)$. The maps $\psi_{q*} : H_q(K) \to H_q(L)$, for all values of q, are known as the homomorphisms *induced by* ψ *in homology* and we often denote them, individually or collectively, by ψ_*.

It is easy to check that the map induced in homology by the identity map id $: K \to K$ is the identity homomorphism $H_q(K) \to H_q(K)$ for each q, and that if $\psi : K \to L$ and $\varphi : L \to M$ are simplicial maps then $(\varphi \circ \psi)_* = \varphi_* \circ \psi_*$.

The following result is very important; if you are not very familiar with the notion of a homotopy equivalence (Definition 10.1.1), simply imagine that it is a weaker version of homeomorphism. In particular, the theorem applies when the polyhedra are homeomorphic. For a proof, see [Munkres 1984].

Theorem 11.1.5. *If two simplicial complexes have homotopically equivalent polyhedra, their homology groups are isomorphic. In particular, the simplicial homology groups of a topological space that admits a simplicial subdivision does not depend on the subdivision.*

11.2. Visualization of Homology Groups

Although homology groups are very useful in grasping the topological structure of an object, the definition of homology group is complicated, and it is useful to be able to make it more visual.

The computation of quotient groups is an essential part of the computation of homology groups. Here we describe a computer technique for visualizing the computation of homology groups; this can help provide insight into such groups [Kunii et al. 1993]. The basic strategy of the animation is to show that a cycle can be transformed into a fixed representative of its homology class by erasing some boundaries. We can see how cycles are classified into their homology classes through the animations of quotient groups.

The pictorial representation: An introductory tool for studying homology groups. We first focus on homology groups of simplicial complexes in lower-dimensional spaces, and draw simplicial complexes accompanied by cycles and boundaries. For such complexes, one can see directly the generators of $Z_r(K)$, $B_r(K)$, and $H_r(K)$, which helps understand the essentials of homology theory.

In order to increase the intuitiveness of the animation, a homotopy technique is employed that enables us to transform a cycle smoothly into another cycle. Figure 11.4 depicts a rough image of this transformation. As the animation proceeds, the cycle C_0 is continuously transformed into the cycle C_1. Although the intermediate object C_t (for $0 < t < 1$) that appears during the transformation does not correspond to any object in the theory, this technique is more informative than an abrupt transformation of erasing a boundary and redrawing the remaining cycle.

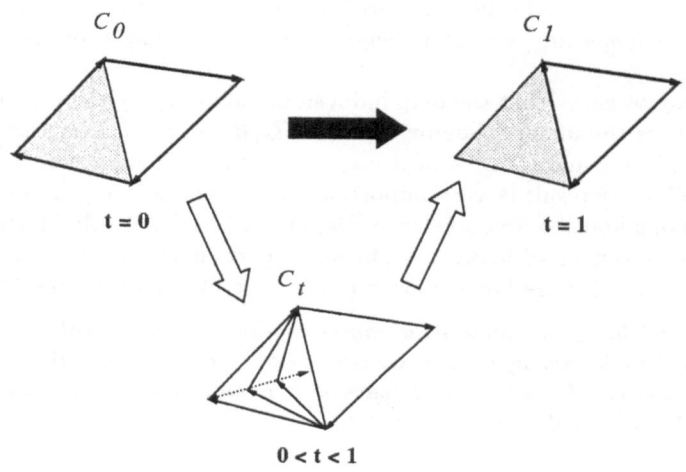

FIGURE 11.4. Smooth transformation of a cycle.

An algorithm for the cycle transformation is given as follows, where we set $Z = Z_r(K)$, $B = B_r(K)$, and $H = H_r(K)$. The idea is to transform one cycle c into another cycle $\bar{c}$, taken as a representative of its homology class $[c]$, by adding or subtracting one simplex at a time. We fix a set of generators b_i of B, with the property that each b_i is the boundary of a single simplex σ_i.

1. For a given cycle $c \in Z$, compute the coefficients m_i needed to go from c to the representative $\bar{c} = c - \sum_i m_i b_i$ of its homology class. Assume that $m_i \in \{-1, 0, 1\}$ for simplicity.
2. For some b_i such that $m_i \neq 0$ and $\gamma_i = b_i \cap c \neq \varnothing$, construct the following retraction $f : \sigma_i \times [0, 1] \to \sigma_i$, where $\partial(\sigma_i) = b_i$:

$$f(p, 0) = p, \quad \text{for all } p \in \sigma_i,$$
$$f(p, 1) \in \bar{\gamma}_i, \quad \text{for all } p \in \sigma_i, \ \bar{\gamma}_i = \gamma_i - m_i b_i$$
$$f(p, t) = p, \quad \text{for all } p \in \bar{\gamma}_i.$$

As the variable t grows from 0 to 1, γ_i is smoothly transformed into $\bar{\gamma}_i$ by the retraction f.

3. At this point, we have a new cycle $c' = c - m_i b_i$. Now let $c = c'$, and go back to the step 2. Repeat steps 2 and 3 until c becomes $[c]$.

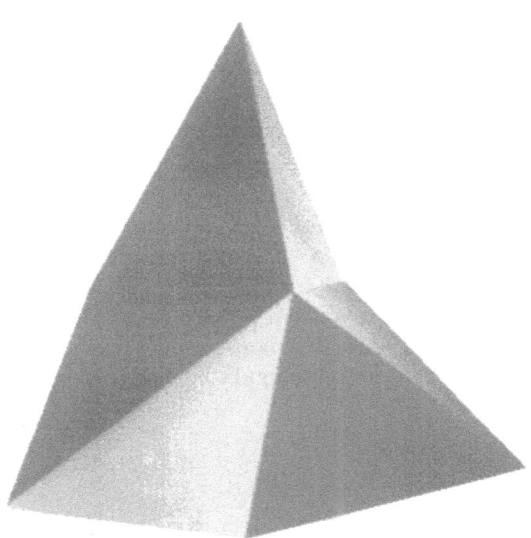

FIGURE 11.5. A tetrapod.

As an example, take the simplicial complex T displayed in Figure 11.5: a "tetrapod", consisting of four solid tetrahedra. It has a two-hole (a closed space) inside. Figure 11.6 shows scenes from the animation for the two-dimensional homology group $H_2(T)$. A two-cycle and a two-boundary are colored by semitransparent blue and solid pink, respectively. Green segments give the one-skeleton of the tetrapod. The animation demonstrates that the two-cycle going around the tetrapod can be homologously transformed into the hull of the two-hole.

The vector representation: a general tool for visualizing finitely generated abelian groups. Although our visualization target is homology groups, the visualization method introduced in this section can be applied to any finitely generated abelian group. A generator of a group generates a finite or an infinite cyclic group. Here we find the analogy between a generator generating a cyclic group and a vector generating a line. This idea leads us to a method visualizing finitely generated abelian groups by sets of vectors.

One difficulty is how to distinguish generators of infinite order from those of finite order. To overcome this problem, we visualize infinite-order generators by a set of vectors in $\mathbb{R}^2$, and visualize finite-order generators by a separate set of vectors on the projective plane $\mathbb{RP}^2$. With this representation, algebraic relations among generators of Z, B, and H can be expressed by geometric relations among vectors. We give a rough sketch of the animation algorithm below.

The homology class of a Z-vector that corresponds to a generator of Z can be represented by the sum of the Z-vector and some B-vectors that correspond to generators of B. Each of the B-vectors is shrunk gradually to zero vectors, in order to show that the elements of B are homologous to the identity element in the quotient group Z/B. The Z-vector follows the shrinking transformation of the B-vectors to maintain the relations among them. The Z-vector consequently becomes the representative of its homology class. We can thus overlap the homologous Z-vectors with each other.

Figure 11.7 shows a torus and a triangulated torus. It also shows how a (triangulated) torus is constructed from a square. If we glue the right-hand and left-hand sides of the triangulated square as in the figure, the square becomes a triangulated cylinder. A triangulated torus can be obtained by identifying both ends of the open triangulated cylinder, as indicated in the figure.

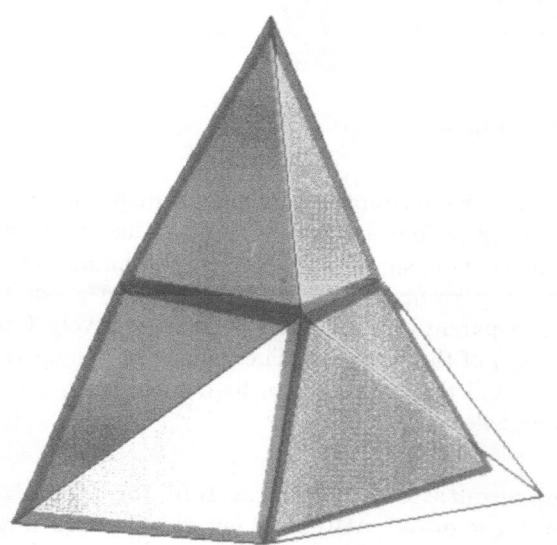

FIGURE 11.6. Animation with pictorial representation.

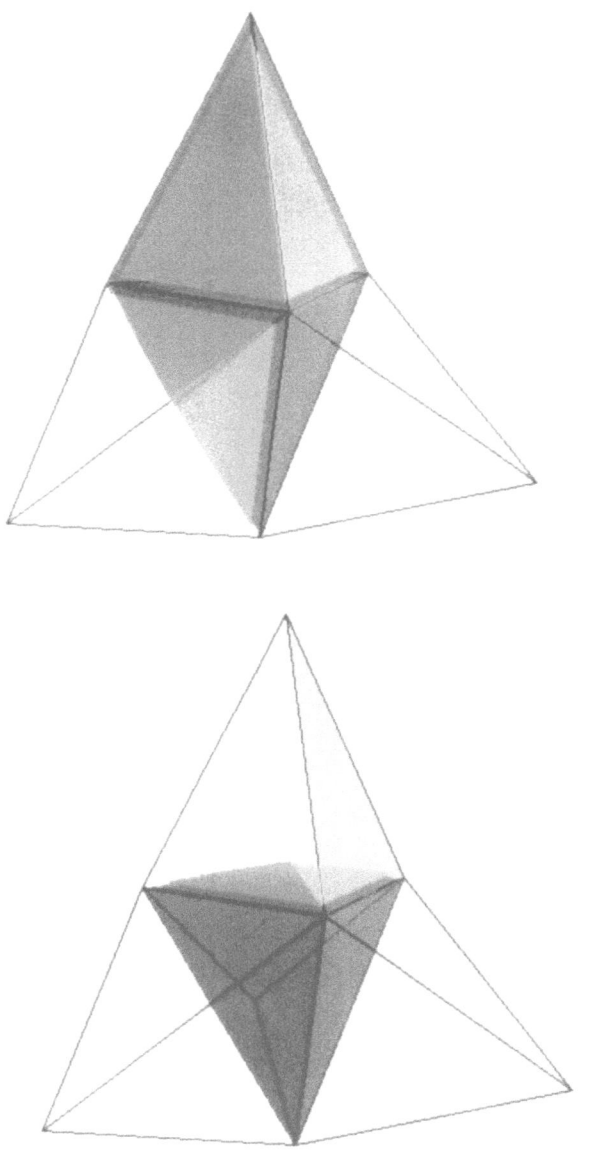

FIGURE 11.6. Animation with pictorial representation (continued).

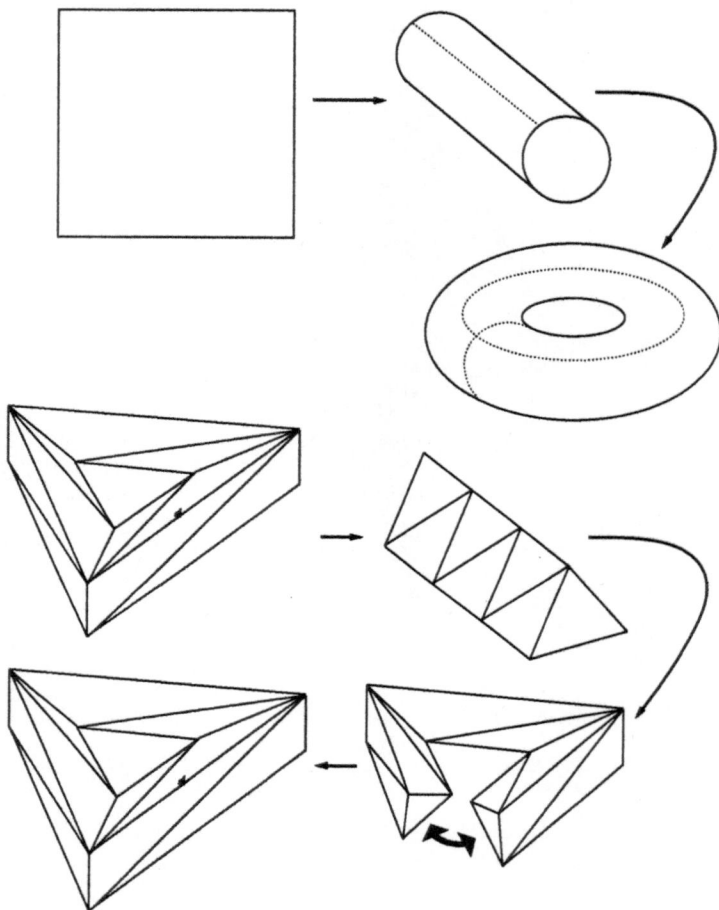

FIGURE 11.7. Construction of a triangulated torus.

Let K be the triangulated torus and L the boundary of the square. The homology groups are:

$$H_0(L) \cong \mathbb{Z}, \quad H_0(K) \cong \mathbb{Z},$$
$$H_1(L) \cong \mathbb{Z}, \quad H_1(K) \cong \mathbb{Z}^2,$$
$$H_2(L) \cong 0, \quad H_2(K) \cong \mathbb{Z}.$$

Figure 11.8 shows animation scenes for $H_0(L)$, using the vector representation. In the center of Figure 11.8, the homology class of a generator of $Z_0(L)$ (the red downward vector) is expressed as a sum of vectors. The green vectors denote generators of $B_0(L)$. These vectors are sequentially shrunk to

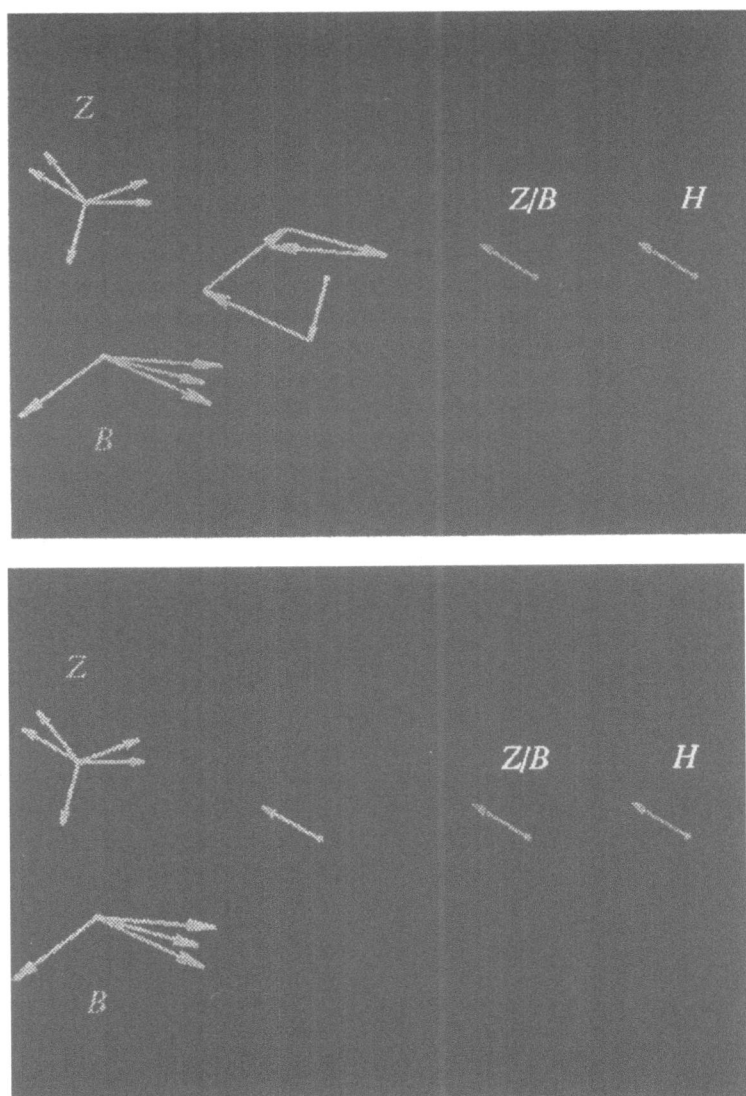

FIGURE 11.8. Animation with vector representation.

zero vectors. At last, the generator becomes its homology class, as shown in the bottom half of the figure.

11.3. Algorithmic Calculation of Homology Groups

This section discusses how one can compute the homology groups of a space by purely algorithmic (mechanical) means, suitable for computer implementation.

Incidence matrices for a simplicial complex. Consider a simplicial complex K. We name its simplexes σ_i^q, where q is the dimension and i is an index ranging from 1 to α_q, the number of q-simplexes in K. We also fix an orientation for each simplex.

For each simplex we have

$$\partial \sigma_j^{q+1} = \sum_{i=1}^{\alpha_q} a_{ij}^q \sigma_i^q,$$

where each a_{ij}^q, for $1 \leq j \leq \alpha_{q+1}$ and $1 \leq i \leq \alpha_q$, equals 0, 1, or -1. More precisely, $a_{ij}^q = 0$ if σ_j^q is not a face of σ_i^{q+1}; otherwise $a_{ij}^q = +1$ if the orientation of σ_j^q coincides with the orientation induced by σ_j^{q+1}; otherwise $a_{ij}^q = -1$. The number a_{ij}^q is called the *incidence coefficient* of σ_i^q and σ_j^{q+1}.

The a_{ij}^q form an $\alpha_q \times \alpha_{q+1}$ matrix, called the *q-dimensional incidence matrix of K*:

$$
I^q = \quad
\begin{array}{c|ccccc}
 & \sigma_1^{q+1} & \cdots & \sigma_j^{q+1} & \cdots & \sigma_{\alpha_{q+1}}^{q+1} \\
\hline
\sigma_1^q & & & & & \\
\vdots & & & & & \\
\sigma_i^q & & & \{a_{ij}^q\} & & \\
\vdots & & & & & \\
\sigma_{\alpha_q}^q & & & & &
\end{array}
$$

Thus we get a sequence of incidence matrices $I^0, \ldots, I^{n-1}$, where $n = \dim K$. The matrix I^{q-1} encodes the action of the boundary operator ∂_q, in the bases for $C_{q-1}(K)$ and $C_q(K)$ consisting of elementary chains in each dimension.

Theorem 11.3.1. *The topology (and combinatorial structure) of a simplicial complex is completely determined by its incidence matrices. There exists an algorithm that computes the homology groups of a simplicial complex from its incidence matrices.*

Indeed, if two simplicial complexes K and L have the same incidence matrices, the map that takes each vertex of K to the vertex of L with the same label is a simplicial map, because the incidences are the same. It is easy to see that the extension of this map to $|K|$ is continuous and one-to-one. An inverse map is defined analogously, also continuous, therefore the $|K|$ and $|L|$ are homeormorphic. As for the assertion that K and L have the same combinatorial structure, this is essentially another way of saying that the incidences are the same.

The algorithm for computing the homology groups will occupy the remainder of this section.

Reducing the incidence matrices. Because $\partial_q \partial_{q+1} = 0$, we obviously have the matrix equation

$$I^{q-1} I^q = 0$$

(see Figure 11.9), since I_{q-1} expresses the action of ∂_q and I_q expresses the action of ∂_{q+1} in the chosen bases.

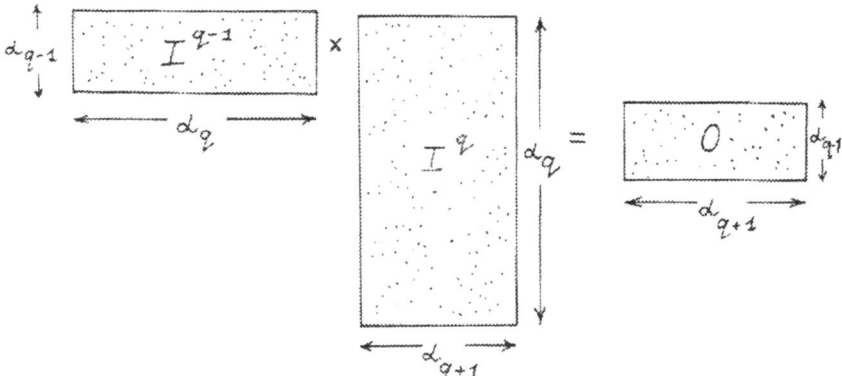

FIGURE 11.9.

Suppose that, instead of expressing our incidence matrices in the basis consisting of elementary chains σ_i^q, we choose new bases for the chain groups, differing from the old by an invertible linear transformation with integer coefficients. Then the matrices I^q will transform into new incidence matrices $\bar{I}^q$. The equation $\bar{I}^{q-1} \bar{I}^q = 0$ will still be satisfied, since the change of basis does not alter the geometrical fact that the boundary of a boundary is zero.

We will use such changes of basis to simplify the incidence matrices to a form that allows the homology groups of K to be read off easily. We achieve this by a series of *elementary transformations*, which are of three types:

(1) Replacing a basis element τ_i^q by $\tau_i^q + \tau_j^q$, where $i \neq j$. The i-th column of I^{q-1} is replaced by the sum of i-th and j-th columns: the j-th row of I^q is replaced by the j-th row minus the i-th row.

(2) Replacing τ_i^q by $-\tau_i^q$. The i-th column of I^{q-1} and the i-th row of I^q are multiplied by -1.

(3) Interchanging τ_i^q by $-\tau_j^q$. The i-th and j-th columns of I^{q-1} are interchanged, as are the i-th and j-th rows of I^q.

(Thus the reduction is similar to Gaussian elimination, with the difference that only transformations involving integer coefficients, and whose inverses also involve integer coefficients. are allowed.)

Any integer-valued matrix A can be reduced in an algorithmic way, using these elementary transformations on rows and columns, to a canonical form where all entries are zero except for the first γ entries along the principal diagonal, where $\gamma = \operatorname{rank} A$. Moreover the nonzero diagonal elements are such that each is a multiple of the next. This canonical form is shown in Figure 11.10, left. In fact we will further modify our matrices, by column permutations. so that the nonzero entries appear in the *top right corner*, as shown in Figure 11.10, right; we call this last configuration the *normalized canonical form*.

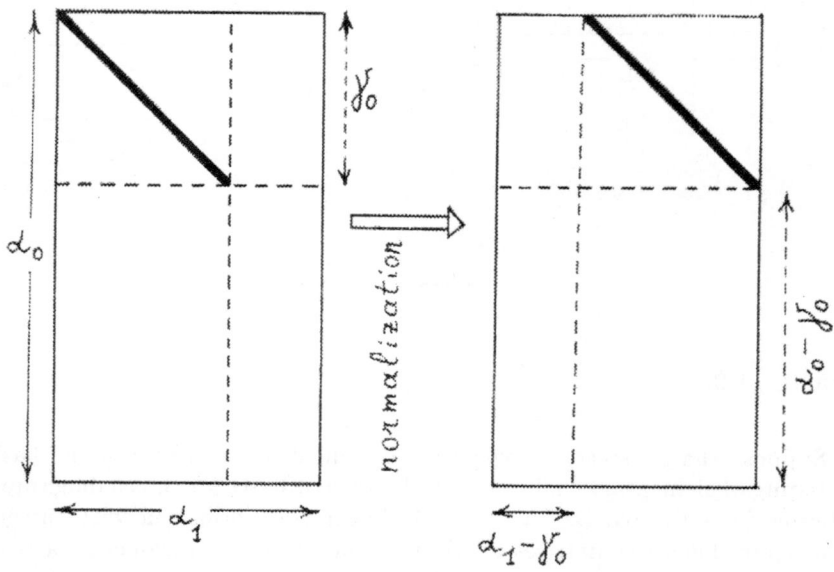

FIGURE 11.10.

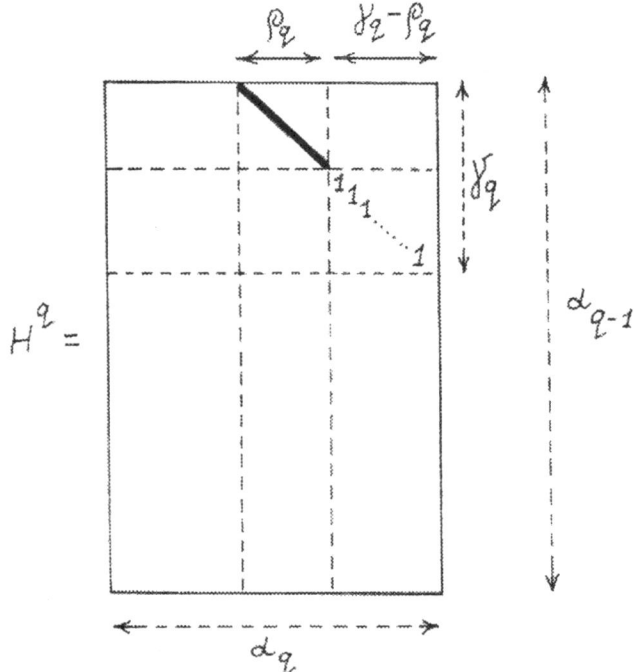

FIGURE 11.11.

We now apply this reduction to the incidence matrices, one at a time. Set $\gamma_i = \operatorname{rank} I^i$, for $i = 0, \ldots, n-1$. We begin by reducing I^0 to normalized canonical form, by changing the bases of $C_0(K)$ and $C_1(K)$ as just explained. This affects I^1 as well, since we're changing the basis of $C_1(K)$. In the end, I^0 has γ_0 diagonal entries on the top right, and all other entries are zero.

Now, because $I^0 I^1 = 0$. *the last γ_0 rows of the transformed I^1 must be zero.* and we needn't worry about them anymore. We apply the reduction to the top $\alpha_1 - \gamma_0$ rows of I^1, by changing the bases of $C_1(K)$ and $C_2(K)$. This affects the first $\alpha_1 - \gamma_0$ columns of I^0, but because these columns already consist entirely of zeros. the matrix doesn't change. Now we have both I^0 and I^1 in normalized canonical form.

Next. because $I^1 I^2 = 0$, the last γ_1 rows of the transformed I^2 are zero, and we can repeat the same process to put I^2 in normalized canonical form. This continues until all the I^q are in the desired form.

Computing the homology groups. Let's look again at the normalized canonical form (Figure 11.11). Let the diagonal entries *different from* 1 be

$\varepsilon_1^q \ldots, \varepsilon_{\rho_q}^q$, each of which is a divisor of a preceeding one. The remaining $\gamma_q - \rho_q$ numbers are equal to 1. Finally, let $\beta_q = \alpha_{q-1} - \gamma_q - \gamma_{q-1}$ for $q = 0, \ldots, n$, with the convention that $\gamma_{-1} = \gamma_n = 0$. Then it can easily be proved that the desired homology groups are

$$H_q(K) = \underbrace{\mathbb{Z} \oplus \mathbb{Z} \oplus \cdots \oplus \mathbb{Z}}_{\beta_q \text{ factors}} \oplus \mathbb{Z}_{\varepsilon_1^q} \oplus \mathbb{Z}_{\varepsilon_2^q} \oplus \cdots \oplus \mathbb{Z}_{\varepsilon_{\rho_q}^q},$$

for $q = 0, \ldots, n-1$, and that $H_n(K)$ is a free abelian group of rank β_n—there are no finite factors in this case.

This concludes the algorithm for computing homology groups. In practice, one should realize that the calculation of homology groups using triangulations is rather cumbersome, since a triangulation of even a fairly simple object contains lots of simplexes. The time needed to reduce an integer-valued matrix to canonical form grows quickly with the size, so it is desirable to make the matrices as small as possible. To this end we will introduce a different type of decomposition of spaces, namely *cell decomposition*, which often requires signficantly fewer elements. Homology theory can be developed for cell decompositions along the same lines we've used above, and the algorithm just described works with virtually no modification apart from the redefinition of the incidence coefficients. As a result, calculation of homology groups is generally much quicker using cell decompositions.

11.4. Cell complexes

The main objects discussed in the preceding chapters were smooth manifolds. But, in many problems of science, more complicated geometrical objects often arise, which can still be grouped into a general category of spaces called *cell spaces* (or *CW-complexes* or *cell complexes*). The simplest examples are the level sets of a smooth function defined on a manifold. These surfaces can have singular points, that is, sometimes they are not manifolds. But they are cell spaces. The concept of a cell space is a natural extension of the class of manifolds; in particular, any smooth manifold is a cell space (Theorem 6.3.3).

We start with an informal description of a cell space. Like manifolds, these spaces are glued from balls; but in the case of manifolds, the balls all have the same dimension, and they are glued by means of homeomorphisms (on common pieces). By contrast, to construct a cell space, each successive ball is glued by its boundary to the already constructed part of the space, using a continuous map from the boundary of the ball to the boundary of the space. This map may not be a homeomorphism. Thus, the cell space is glued from the *images of the balls*. The precise definition is as follows (refer to Figure 11.12):

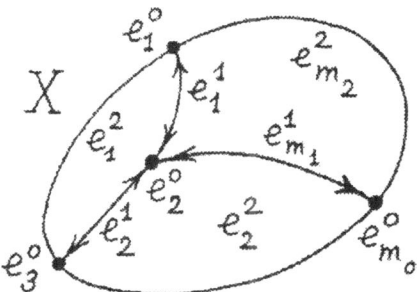

FIGURE 11.12.

Definition 11.4.1. A topological space K is called a *cell space* if it can be represented as the union of nonintersecting sets e_i^q, called *cells* (where q is the *dimension* of the cell e_i^q and i is its number):

$$K = \bigcup_{q=0}^{n} \left(\bigcup_{i=1}^{m_q} e_i^q \right),$$

and the following conditions are satisfied:

1) The closure $\bar{e}_i^q$ of each cell e_i^q is the image of a closed q-dimensional ball (disk) $\bar{D}^q$ under some continuous map

$$\chi_i : \bar{D}^q \to \bar{e}_i^q \subset K,$$

called the *characteristic map* of the given cell.

2) The restriction of χ_i to the *open* ball D^q is a homeomorphism between D^q and e_i^q.

3) The *boundary* of each cell, that is, the set

$$\partial e_i^q = \bar{e}_i^q - e_i^q$$

(where the bar denotes closure in K) is contained in the union of a finite number of cells of lesser dimensions:

$$\partial e_i^q \subset \bigcup_{s \leq q-1} e_j^s.$$

4) A subset Y in K is closed if and only if each inverse image $\chi_i^{-1}(Y)$ is closed in the ball $\bar{D}^q$ for any cell e_i^q.

We will usually write

$$K = e_1^0 + \cdots e_{m_0}^0 + e_1^1 + \cdots + e_{m_1}^1 + \cdots + e_1^n + \cdots e_{m_n}^n,$$

using the symbol $+$ instead of $\cup$.

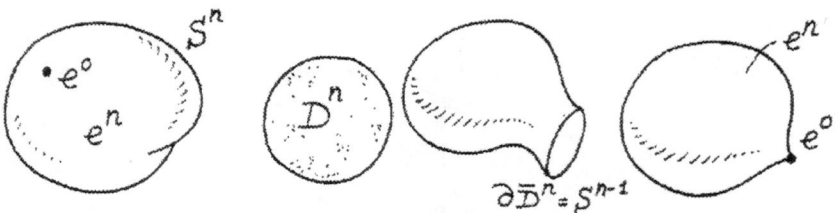

FIGURE 11.13.

Example 11.4.2. Let M be a smooth compact surface. The classification theorem for surfaces (Theorem 4.2.7) says that M can be obtained from a flat fundamental polygon W by means of appropriate identifications in the boundary, given by the standard words

$$W = a_1 b_1 a_1^{-1} b_1^{-1} \ldots a_g b_g a_g^{-1} b_g^{-1} \qquad \text{for } M \text{ orientable of genus } g,$$
$$W = c_1 c_1 c_2 c_2 \ldots c_k c_k \qquad \text{for } M \text{ nonorientable of genus } k.$$

It is natural to take this fundamental polygon as the basis for a cell decomposition. Thus, in the orientable case, the single vertex of the glued-up fundamental polygon is the single zero-cell, the $2g$ loops coming from the boundary of the polygon become the one-cells, and the single two-cell e^2 is the interior of the closed polygon, attached to the set of one-cells by means of a characteristic map that can be defined explicitly by specifying a homeomorphism between the closed disk and the fundamental polygon. Thus the cell decomposition for an orientable surface of genus g is

$$M^2 = e^0 + (a_1 + b_1 + \cdots + a_g + b_g) + e^2;$$

it has one zero-cell, $2g$ one-cells, and one two-cell.

In the nonorientable case the same idea leads to the cell decomposition

$$M^2 = e^0 + (c_1 + \cdots + c_k) + e^2,$$

with one zero-cell, k one-cells, and one two-cell.

Let's consider in detail the case of a torus (Figure 11.14). We start from one point, the 0-cell e^0; this single point is the zero-skeleton $K^{(0)}$ of the torus (in the cell decomposition being considered). Then we glue to this point two one-cells a and b, thus obtaining a bouquet of two circles (the *bouquet* $K \vee Y$ of two topological spaces K and Y is the space obtained from the union $K \cup Y$ by identifying one point of K with one of Y). The result is the one-skeleton $K^{(1)}$. After this we glue the two-cell $W = e^2$ to the bouquet of two circles. This gluing is shown in Figure 11.15.

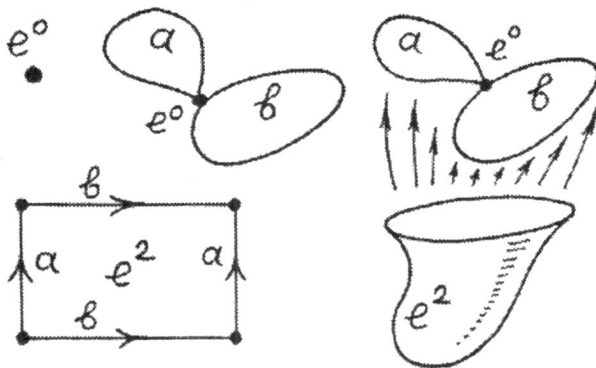

FIGURE 11.14.

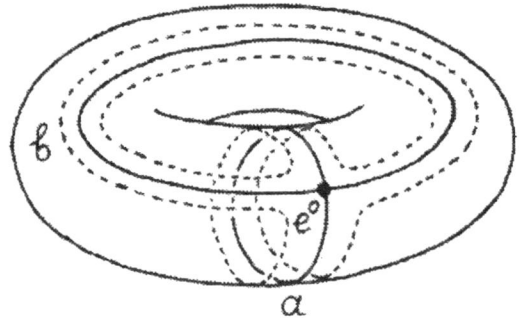

FIGURE 11.15.

11.5. The Degree of a Map

Consider two smooth, compact, connected, orientable n-dimensional manifolds M and P (without boundary), and fix orientations for them. Let $f : M \to P$ be a smooth map from M into P. We will associate to this map an integer, called its *degree*.

Recall that, by Sard's theorem (Theorem 9.2.1), the set of critical values of f has measure zero in P—that is, almost all points y from P are *regular values* of f.

Let $y \in P$ be a regular value and $f^{-1}(y) = \{x_1, \ldots, x_N\}$ the inverse image of y. Because each x_i is a regular point and the manifolds have the same dimension, the differential $df_{x_i} : T_{x_i}M \to T_yP$ is an isomorphism and the determinant $\det(df)$ of Jacobian matrix df with respect to any bases is nonzero. The orientation of both manifolds having been given in advance, we can choose positively oriented bases $\{e_1, \ldots, e_n\}$ for T_yP and $\{f_1, \ldots, f_n\}$ for

$T_{x_i}M$, and record the sign of this determinant as expressed in these bases (Figure 11.16). We put $\varepsilon(x_i) = +1$ if the determinant is positive, and $\varepsilon(x_i) = -1$ if negative.

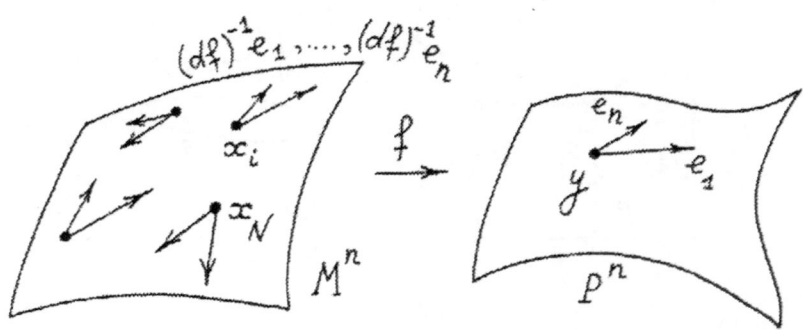

FIGURE 11.16.

We define the degree of f (calculated at y) as the integer

11.5.1.
$$\deg f = \sum_{i=1}^{N} \varepsilon(x_i).$$

The importance of this definition is that it does not depend on the choice of y:

Theorem 11.5.2. *Let $f : M \to P$ be a smooth map between compact, connected, oriented n-manifolds. Then the number $\deg f$ defined in 11.5.1 does not depend on the choice of a regular value $y \in P$, and can be called simply the degree of f. The degree is preserved under homotopy, that is, homotopic maps $f : M \to P$ and $g : M \to P$ have the same degree: $\deg f = \deg g$.*

We consider some visual examples. Figure 11.17 shows the map $f(z) = z^q$ of the complex plane, which maps the unit circle S^1 into the same circle. A basis at the point y reduces here to one tangent vector e_1. We have $\varepsilon(x_i) = +1$ at all points x_i with $f(x_i) = y$, and all these points are regular. Therefore $\deg f = q$, since there are q inverse images.

Figure 11.18 shows the same map in another form: we wind the first circle around the second circle q times in one direction. It is clear that, for all points x_i that are mapped to a point y, this map preserves orientation, so $\varepsilon(x_i) = +1$ for all i.

Figure 11.19 illustrates the invariance of the degree $\deg f$ under a smooth homotopy and its independence from the regular value y. Here we have $\deg f = +1 + 1 - 1 + 1 = +2$.

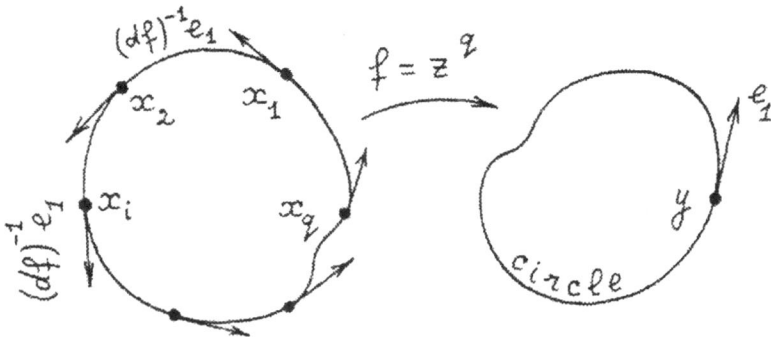

FIGURE 11.17.

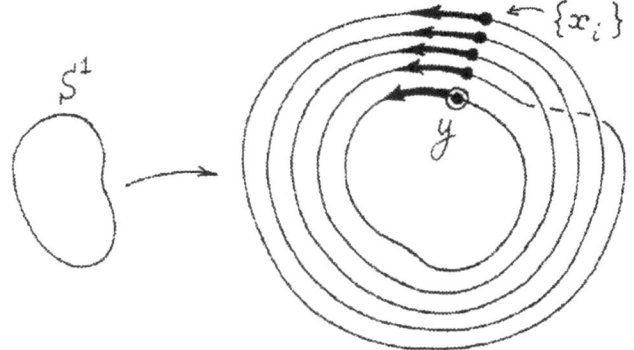

FIGURE 11.18.

Important Remark. We have defined the degree for *smooth* maps between two smooth manifolds. But any *continuous* map $g : M \to P$ can be approximated by a smooth map $f : M \to P$, and f can be chosen homotopic to g. Consequently, if we set $\deg g = \deg f$ for f smooth and homotopic to g, we can extend the definition of the degree to arbitrary *continuous* maps of smooth manifolds. It can be proved that this definition does not depend on the choice of such an approximation.

An important special case is when $f : M \to P$ is surjective and all points $y \in P$ are regular values, or, equivalently, all points $x \in M$ are regular points. Such maps are called *smooth covering maps* (or simply smooth coverings). Using the assumption that P is connected, it is easy to prove that, in this case, for any two points $x, y \in P$, the inverse images $f^{-1}(x)$ and $f^{-1}(y)$ have the same number of points, and that all the ε's are the same. The integer $|\deg f|$ is then called the *multiplicity of the covering f*.

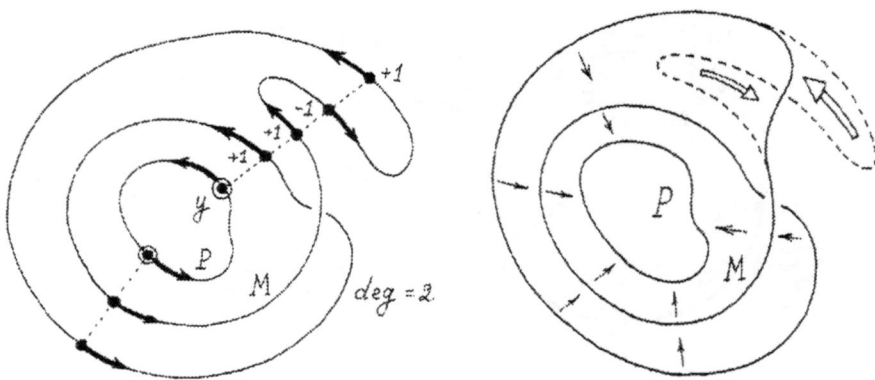

FIGURE 11.19.

Example 11.5.3. As we have seen, the smooth map $f : S^1 \to S^1$ given by $f(z) = z^q$ on the complex plane has degree $\deg f = +q$. This map is a q-fold covering (that is. a covering of multiplicity q).

Example 11.5.4. Consider the *antipodal map* $f : S^n \to S^n$ given by $f(x) = -x$ for each point $x \in S^n$, where S^n is the unit sphere in R^{n+1} (Figure 11.20). Figure 11.21 shows the one-dimensional case, the circle, where it is obvious that $\deg f = +1$. Figure 11.22 is the case of a two-sphere, and shows the basis (e_1, e_2) at the point x and its image (b_1, b_2) at the point $-x$ under the map $x \to -x$. We need to compare the orientation of the basis (b_1, b_2) with the orientation of the initial basis (e_1, e_2). Transporting the basis (e_1, e_2) along the equator from x to $-x$ clearly yields a basis at $-x$ having the opposite orientation of that of (b_1, b_2): For example, the vector e_2 can be made to point along the equator throughout, so it ends up coinciding with b_2, while e_1 is made to point north throughout, so it ends up opposite b_1. Thus in dimension two the degree of the antipodal map is -1. The same reasoning gives $\deg f = (-1)^{n-1}$ in arbitrary dimension n: after sliding along the equator, one basis vector behaves like e_2 above, while all other $n - 1$ vectors behave like e_1.

Example 11.5.5. Consider a torus T^2 given by the parametrization $(e^{i\varphi}, e^{i\psi})$ in $\mathbb{C}^2$, where φ and ψ vary independently from 0 to 2π (this is a direct product of the unit circles in each factor $\mathbb{C}$). Consider the smooth map $f : T^2 \to T^2$ given by

$$f(e^{i\varphi}, e^{i\psi}) = (e^{ip\varphi}, e^{iq\psi}),$$

where p and q are arbitrary fixed integers. It is an easy exercise to prove that $\deg f = pq$. This map is a covering of multiplicity pq.

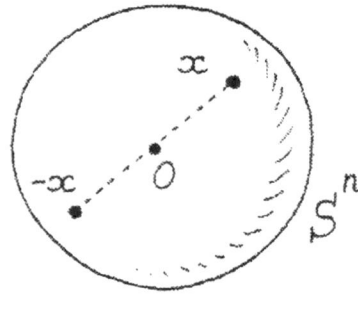

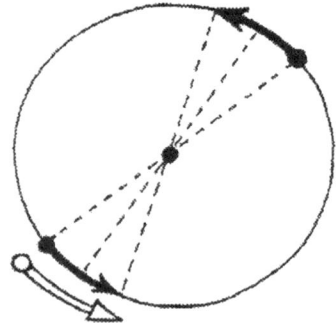

FIGURE 11.20.   FIGURE 11.21.

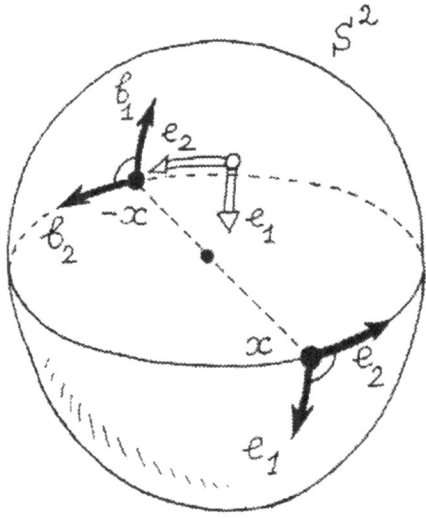

FIGURE 11.22.

11.6. Cell Homology

Cell chains. For the rest of this chapter we make the convention that K is a *finite* cell complex. We look at its q-cells e_i^q, for fixed q, each of which has an orientation induced by the characteristic map $\chi^q : D^q \to K$. We consider the free abelian group $C_q(K)$ generated by the e_i^q, that is, the group of formal integer combinations

$$c = \sum_i a_i e_i^q, \quad \text{for } a_i \in \mathbb{Z}.$$

Because K has finitely many cells, $C_q(K)$ is a finitely generated group. Its generators can be formally written as $1e_i^q$, but we identify them with the e_i^q. We call an element of $C_q(K)$ a (cell) q-chain of K.

More precisely, $C_q(K)$ is the group of *integer-valued* chains in K, since its elements are combinations of the generators with *integer* coefficients. When we want to stress this, we write $C_q(K, \mathbb{Z})$ instead of $C_q(K)$. It is possible to define, in an exactly analogous way, groups $C_q(K, G)$ for any abelian group of coefficients G, whose elements are formal combinations

$$c = \sum_i a_i e_i^q, \quad \text{for } a_i \in G.$$

Such combinations are called *G-valued q-chains*. As in the simplicial case, the most interesting groups of coefficients are the groups $\mathbb{Z}$, $\mathbb{Z}_p = \mathbb{Z}/p\mathbb{Z}$, and $\mathbb{R}$. For concreteness, we will develop the theory of cell homology with coefficients in $\mathbb{Z}$.

The cell boundary operator and cell homology.   Consider two cells e^q and e^{q-1} (we omit their indices for simplicity). Form the continuous map

11.6.1. $S^{q-1} = \partial D^q \xrightarrow{\chi^q} K^{(q-1)}/K^{(q-2)} = \text{bouquet of } (q-1)\text{-spheres},$

where $K^{(q-1)}/K^{(q-2)}$ is the space obtained from the $(q-1)$-skeleton by identifying together all points that belong to the $(q-2)$-skeleton (recall that $K^{(q-2)} \subset K^{(q-1)}$). This quotient space is a bouquet of spheres S^{q-1}, because each closed cell $\bar{e}_i^{q-1}$ is transformed into a sphere when we identify all points in $K^{(q-2)}$—the interior e_i^{q-1} is unchanged but the boundary $\bar{e}_i^{q-1} - e_i^{q-1}$ is collapsed to a point, since it is contained in $K^{(q-2)}$. The top right drawing in Figure 11.23 represents this quotient.

Now, the $(q-1)$-cell we started with corresponds to one of these S^{q-1}'s in the bouquet—denote it S_*^{q-1}. We can project the whole of $K^{(q-1)}/K^{(q-2)}$ into this sphere, as indicated by the second white arrow in Figure 11.23: S_*^{q-1} is mapped onto itself by the identity map, and all other spheres are collapsed to a point.

Composing the map of 11.6.1 with this second collapsing map, we finally obtain a map

$$f : S^{q-1} \to S_*^{q-1}$$

that is completely determined by the cells e^q and e^{q-1}. We denote the degree of this map by $[e^q : e^{q-1}]$, and call it the *incidence coefficient* of the two cells e^q and e^{q-1} in the cell complex K.

Figure 11.24 illustrates the case $[e^q : e^{q-1}] = 2$: the boundary of e^q goes twice (in the same direction) along the boundary cell e^{q-1}. This picture explains the geometric meaning of the incidence coefficient.

Remark. It is clear from the definition that, if e^{q-1} is disjoint from $\bar{e}^q$, their incidence coefficient is zero.

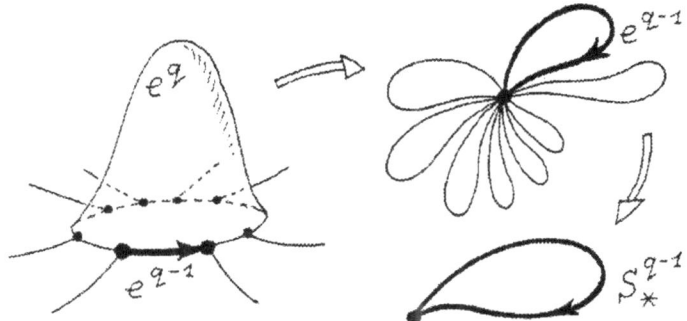

FIGURE 11.23.

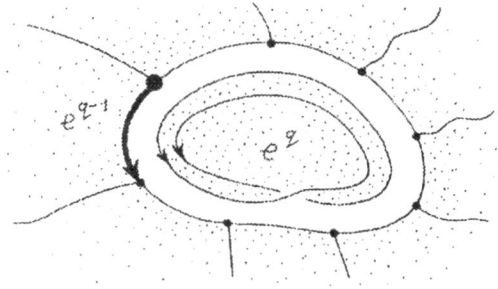

FIGURE 11.24.

Definition 11.6.2. The (cell) *boundary operator* is a homomorphism $\partial_q :$ $C_q(K) \to C_{q-1}(K)$, defined as follows. If e^q is one of the generators of the q-chain group $C_q(K)$, we have

$$\partial e^q = \sum [e^q : e_i^{q-1}] e_i^{q-1},$$

where the sum is over all $(q-1)$-cells of K. For arbitrary chains in $C_q(K)$, the action of ∂_q is extended by linearity; thus the boundary of $c = \sum_k a_k e_k^q$

$$\partial\left(\sum_k a_k e_k^q\right) = \sum_{k,i} a_k \, [e_k^q : e_i^{q-1}] \, e_i^{q-1}.$$

Like its simplicial counterpart, the cell boundary operator satisfies

11.6.3. $\partial_{q-1} \partial_q(c) = 0$ for all $c \in C_q(K)$.

This follows from the definition of ∂_q, and we suggest that the reader check this important topological fact.

A chain $z \in C_q(K)$ having zero boundary, that is, such that $\partial_q z = 0$, is called a *q-cycle*. The set of all cycles is a subgroup of $C_q(K)$ denoted $Z_q(K)$. Thus

$$\ker \partial_q = Z_q(K).$$

A chain $b \in C_q'(K)$ that is the image of some $(q+1)$-chain c under the action of ∂_{q+1}, that is, such that $b = \partial_{q+1}(c)$, is called a *q-boundary*. The set of all q-boundaries form a subgroup of $C_q(K)$ denoted $B_q(K)$. Thus

$$\operatorname{im} \partial_{q+1} = B_q(K).$$

It follows immediately from 11.6.3 that $B_q(K) \subseteq Z_q(K)$, that is, every q-boundary is a q-cycle.

Definition 11.6.4. The q-th *cell homology group* of a cell space K is the quotient

$$H_q(K) = Z_q(K)/B_q(K).$$

An element of $B_q(K)$ is also said to be a chain *homologous to zero*. Two q-chains m and n are called *homologous* if the chain $m - n$ is homologous to zero. Thus an element of $H_q(K)$ is an equivalence class of homologous chains.

When we want to stress that we are using integer coefficients, we denote the homology group by $H_q(K, \mathbb{Z})$. As mentioned above, we can use any other abelian group G as the coefficient group, and in this case we obtain groups $H_q(K, G)$ called *homology groups with coefficients in G*.

Homology groups play an important role in geometry, topology, physics, and many applications.

Remark. The definitions of the boundary operator, cycles, and boundaries are formal analogs of intuitive notions from the theory of smooth manifolds. A compact smooth submanifold Z (without boundary) in some smooth manifold K can be considered as a cycle. The geometric boundary $B = \partial C$ of a smooth manifold C can be considered as a boundary in the sense of homology theory (Figure 11.25).

Some properties of homology groups. We consider the homology group $H_q(K)$ (with integer coefficients). Because $H_q(K)$ is a finitely generated abelian group (K being a finite cell space), the classical structure theorem for finitely generated abelian groups says that

$$H_q(K) = (\mathbb{Z} \oplus \cdots \oplus \mathbb{Z}) \oplus (\mathbb{Z}_{p_1} \oplus \cdots \oplus \mathbb{Z}_{p_N}),$$

where $\mathbb{Z}_{p_i} = \mathbb{Z}/\mathbb{Z}_{p_i}$ is the finite cyclic group of order p_i. Let β_q be the number of free summands $\mathbb{Z}$ in this decomposition, also known as the *rank* of $H_q(K)$; this number is well-defined. We call it the *q-th Betti number* of the cell space K. The number

$$\lambda(K) = \sum_{q=1}^{n} (-1)^q \beta_q$$

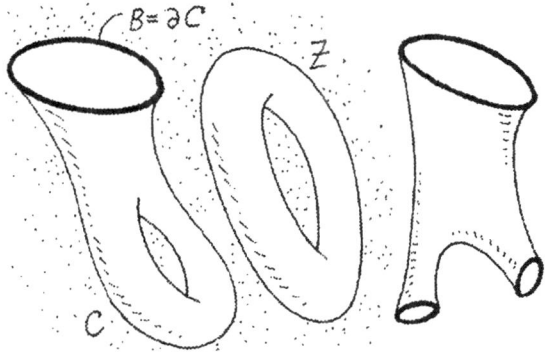

FIGURE 11.25.

is called the *Euler characteristic* of K.

If there are no finite cyclic subgroups in the group $H_q(K)$, that is, if $H_q(K)$ is a free abelian group, we say that $H_q(K)$ is *torsion-free*. (A *torsion element* in an abelian group is one that has finite order.)

It follows immediately from the definitions that, if K is an n-dimensional cell space (that is. n is the highest dimension of a cell in K), we have $H_q(K) = 0$ for $q < 0$ and for $q > n$. It can also be shown that

$$H_0(K) = \mathbb{Z} \oplus \mathbb{Z} \oplus \cdots \oplus \mathbb{Z},$$

where the number of summands is equal to the number of connected components in K; in particular, if K is connected, $H_0(K) = \mathbb{Z}$. (Recall from Section 4.1 that our definition of a connected space is that any two points can be joined by a path.)

The next theorem is extremely important.

Theorem 11.6.5. *Homotopically equivalent spaces have isomorphic homology groups.*

This means that, if K and Y are homotopically equivalent cell spaces (Definition 10.1.1). then for any q and any coefficient group G we have an isomorphism

$$H_q(K, G) = H_q(Y, G).$$

This applies. in particular, when K and Y are homeomorphic. Thus, homology groups are *topological invariants*, and even *homotopy invariants*, of cell spaces.

Corollary 11.6.6. *If two cell spaces K and Y have (for some q) different (non-isomorphic) homology groups $H_q(K)$ and $H_q(Y)$, then K and Y are non-homeomorphic and. in fact. homotopically inequivalent.*

This theorem gives an important tool for comparison of different spaces, because it is generally easier to compare homology groups than to compare topological spaces directly. Thus, if two topological spaces have different homology groups, they are nonhomeomorphic. The converse, of course, is not true: we have already seen that homotopically equivalent spaces (and thus spaces having the same homology groups) need not be homeomorphic; see Figure 10.1, for example.

It is the homotopy invariance of homology groups that is responsible for the important role that they play in geometry and topology.

Corollary 11.6.7. *If a topological space has two distinct cell decompositions, the cell homology groups are the same for both decompositions.*

Relation between simplicial homology and cell homology. The polyhedron $|K|$ of a simplicial complex is a cell space, in an obvious way (fix a homeomorphism between the closed ball and a standard closed simplex in dimension k). Consequently, for any such K we can calculate its simplicial homology groups $H_q^{\text{simp}}(|K|)$ and its cell homology groups $H_q^{\text{cell}}(|K|)$.

Theorem 11.6.8. *If X is the polyhedron of a finite simplicial complex, the simplicial homology groups and the cell homology groups of X are isomorphic.*

For a "nice" space, then, we know two ways of computing homology groups: if we have a cell decomposition, we can compute the cell homology, and if we have a simplicial structure, we can compute the simplicial homology. The result is the same. And although we will talk about cell homology for the remainder of this chapter, the results discussed are equally valid for simplicial homology.

11.7. Example Calculations of Homology Groups

The results above show that computing the homology of (finite) cell spaces is easy. We first consider the chain groups, which are easily derived from the cell structure of K. Then we need to compute the boundary operators for these groups. for which it suffices to compute the incidence coefficients of pairs of cells of consecutive dimensions. After this, it is necessary to compute the quotient groups of the cycles with respect to the boundaries.

Example 11.7.1 (The sphere S^n). The n-sphere admits the simplest cell decomposition

$$S^n = e^0 + e^n,$$

where e^0 is a point and e^n is the complement to e^0 in a sphere. It is clear that, except for $n = 0$. we have $C_0 = \mathbb{Z}$ and $C_n = \mathbb{Z}$, and the boundary operator ∂ is trivial. Consequently, for $n \neq 0$,

$$H_0(S^n) = H_n(S^n) = \mathbb{Z},$$
$$H_i(S^n) = 0 \quad \text{for all other } i.$$

The case $n = 0$ is also very easy: here $C_0 = \mathbb{Z} + \mathbb{Z}$ (the 0-sphere consists of two points), and the only nontrivial homology group is $H_0 = \mathbb{Z} + \mathbb{Z}$.

Example 11.7.2 (Closed orientable surfaces). As we saw earlier, a closed orientable surface has a cell decomposition

$$M^2 = e^0 + (a_1 + b_1 + \cdots + a_g + b_g) + e^2,$$

where e^0 is a point on M, and a_i, b_i are loops starting and ending at e^0. Thus all the loops a_i and b_i are cycles. From the structure of the surface's fundamental polygon, we see that the boundary operator acts on the cell e^2 as follows:

$$\partial e^2 = a_1 + b_1 - a_1 - b_1 + \cdots + a_g + b_g - a_g - b_g = 0.$$

Consequently, e^2 is a cycle in dimension two. Finally, we obtain:

$$H_0(M, \mathbb{Z}) = H_2(M) = \mathbb{Z},$$
$$H_1(M, \mathbb{Z}) = \underbrace{\mathbb{Z} \oplus \mathbb{Z} \oplus \cdots \oplus \mathbb{Z} \oplus \mathbb{Z}}_{2g \text{ times}} = \mathbb{Z}^{2g}.$$

The Euler characteristic $\chi(M)$, therefore, equals $2(1 - g)$. For the sphere we have $\chi = 2$, for the torus $\chi = 0$, and for all other closed surfaces the Euler characteristics is negative.

Corollary 11.7.3. *Orientable closed surfaces of different genus cannot be homeomorphic, or even homotopically equivalent.*

Example 11.7.4 (Closed nonorientable surfaces). A closed nonorientable surface of genus k (sphere with k crosscaps) has a cell decomposition

$$M^2 = e^0 + (c_1 + \cdots + c_k) + e^2,$$

where e^0 is a vertex, the c_i are loops (and therefore cycles), and e^2 is a two-cell. From the structure of the surface's fundamental polygon, we see that the boundary operator acts on e^2 as follows:

$$\partial e^2 = 2(c_1 + \cdots + c_k).$$

Thus e^2 is not a cycle. On the other hand, there exists a relation between one-cycles, given by the preceding formula. A short calculation then gives

$$H_0(M, \mathbb{Z}) = \mathbb{Z},$$
$$H_1(M, \mathbb{Z}) = \underbrace{\mathbb{Z} \oplus \mathbb{Z} \oplus \cdots \oplus \mathbb{Z}}_{k-1 \text{ times}} \oplus \mathbb{Z}_2 = \mathbb{Z}^{k-1} \oplus \mathbb{Z}_2,$$
$$H_2(M, \mathbb{Z}) = 0.$$

Note that we can choose the following basis of generators for the group of one-cycles:

$$c_1, \ldots, c_{k-1}, c_1 + \cdots + c_{k-1} + c_k.$$

It is interesting to compute also the homology groups with coefficients in $G = \mathbb{Z}_2$, because in this case ∂e^2 does vanish (recall that $2 = 0$ in $\mathbb{Z}_2$), so e^2 is a cycle. Thus the $\mathbb{Z}_2$ homology is given by

$$H_0(M, \mathbb{Z}_2) = \mathbb{Z}_2,$$

$$H_1(M, \mathbb{Z}_2) = \underbrace{\mathbb{Z}_2 \oplus \mathbb{Z}_2 \oplus \cdots \oplus \mathbb{Z}_2}_{k \text{ times}} = \mathbb{Z}_2^k,$$

$$H_2(M, \mathbb{Z}_2) = \mathbb{Z}_2.$$

11.8. The Homology Exact Sequence

Although computing homology groups of a finite cell complex is an essentially mechanical task if the incidence coefficients are known, it can be tedious and unenlightening except for the simplest examples. Usually, then, we will want to take a shortcut if possible. One tool that can often be used to compute homology groups more directly is the *exact sequence of a pair*, which we now define.

Suppose that K is a cell space and Y is a cell subspace of K; we say that (K, Y) is a *cell space pair*. Because every cell of Y is also a cell of K, we have $C_q(Y) \subseteq C_q(K)$, and we may consider the group of *relative chains* $C_q(K, Y) = C_q(K)/C_q(Y)$. Since the boundary operator $\partial = \partial_q$ acts as

$$\partial : C_q(K) \to C_{q-1}(K), \qquad \partial : C_q(Y) \to C_{q-1}(Y),$$

it induces an operator (that is, a homomorphism)

$$\partial : C_q(K, Y) \to C_{q-1}(K, Y),$$

which will be denoted, for simplicity, by the same symbol ∂. Now we can construct the new groups of relative cycles $\ker \partial = Z_q(K, Y)$ and relative boundaries $\mathrm{im}\, \partial = B_q(K, Y)$. Then we consider the quotient group

$$H_q(K, Y) = Z_q(K, Y)/B_q(K, Y),$$

called the *relative q-dimensional homology group of K modulo Y*. Like the absolute homology groups defined earlier, these are also finitely generated abelian groups.

We now construct a new operator

$$\partial : H_q(K, Y) \to H_{q-1}(Y).$$

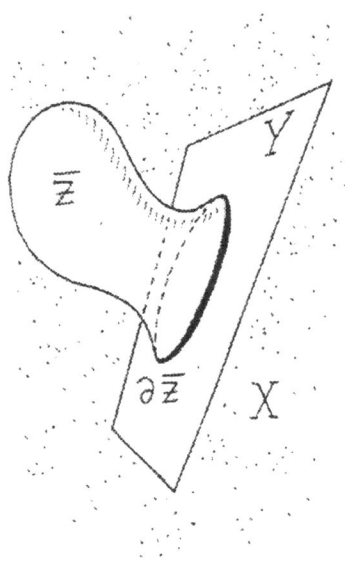

FIGURE 11.26.

Let $z \in C_q(K, Y)$ be a relative cycle and $\bar{z} \in C_q(K)$ be a representative of it. Since $\partial z = 0$, we have $\partial \bar{z} \in C_{q-1}(Y)$ (Figure 11.26). This absolute cycle we denote by $z' \in Z_{q-1}(Y)$; its homology class does not depend on the choice of $\bar{z}$ and on the choice of a representative of the homology class $[z]$, as the reader should check. Thus, if we set $\partial([z]) = [z']$, we thereby define a homomorphism $H_q(K, Y) \to H_{q-1}(Y)$. This homomorphism ∂ is called the *boundary operator for the pair* (K, Y).

On the other hand, if we denote by $i : Y \to K$ the inclusion of Y in K, we have an induced homomorphism

$$i_* : H_q(Y) \to H_q(K),$$

as the reader should check. At the same time, any absolute cycle may be considered as a relative cycle modulo Y, so there is one more natural map

$$j_* : H_q(K) \to H_q(K, Y).$$

Theorem 11.8.1. *The following sequence of groups and homomorphisms is exact, that is, at each group the kernel of the outgoing homomorphism coincides with the image of the incoming homomorphism:*

$$\cdots \to H_{q+1}(K, Y) \xrightarrow{\partial} H_q(Y) \xrightarrow{i_*} H_q(K) \xrightarrow{j_*} H_q(K, Y) \xrightarrow{\partial} H_{q-1}(Y) \to \cdots$$

This is called the *homology exact sequence*. We suggest that the reader prove this theorem by direct analysis of all the definitions.

The following properties of exact sequences are very useful for computations. Proofs are left to the reader as an easy exercise.

(1) The sequence of groups

$$0 \longrightarrow A \longrightarrow 0$$

is exact if and only if $A = 0$.

(2) The sequence of groups

$$0 \longrightarrow A \xrightarrow{\alpha} B \longrightarrow 0$$

is exact if and only if α is an isomorphism.

(3) The sequence of groups

$$0 \longrightarrow A \xrightarrow{i} B \xrightarrow{\pi} C \longrightarrow 0$$

is exact if and only if i is injective (one-to-one with the subgroup $i(A)$ of B). C is isomorphic to the quotient $B/i(A)$, and π is the natural projection onto this quotient.

Proposition 11.8.2. *Let (K, Y) be a pair of finite cell spaces. Then*

$$H_q(K, Y) = H_q(K/Y)$$

for $q > 0$, where K/Y is the cell space obtained from K by collapsing Y to one point.

Given this statement, we can rewrite the homology exact sequence in a slightly different. but equivalent, form, using only *absolute* homology groups (in positive dimensions):

$$\cdots \longrightarrow H_{q+1}(K/Y) \xrightarrow{\partial} H_q(Y) \xrightarrow{i_*} H_q(K) \xrightarrow{j_*} H_q(K/Y) \xrightarrow{\partial} H_{q-1}(Y) \longrightarrow \cdots$$

12. Geodesics

12.1. Definition and basic properties

Consider an n-dimensional Riemannian manifold M (without boundary) with Riemannian metric g_{ij}, expressed in local coordinates $x^1, \ldots, x^n$, so that the (square of) the element of length is given by

$$ds^2 = \sum g_{ij} dx^i dx^j.$$

Denote by g^{ij} the coefficients of the matrix $(g_{ij})^{-1}$ inverse to the matrix (g_{ij}) of Riemannian metric. Define smooth functions Γ^i_{jk} on M by the formulas

$$\Gamma^i_{jk} = \sum_{a=1}^n \tfrac{1}{2} g^{ia} \left(\frac{\partial g_{k\alpha}}{\partial x^j} + \frac{\partial g_{i\alpha}}{\partial x^k} - \frac{\partial g_{jk}}{\partial x^\alpha} \right);$$

these functions are called the *Christoffel symbols* (or *Christoffel functions*) of the metric.

A smooth parametric curve $\gamma(t) = (x^1(t), \ldots, x^n(t))$ on M is called a *geodesic* (for the given metric) if it is a solution of the following system of differential equations on M:

12.1.1.
$$\frac{d^2 x^i}{dt^2} + \sum_{\alpha, k=1}^n \Gamma^i_{\alpha k} \frac{dx^\alpha}{dt} \frac{dx^k}{dt} = 0, \quad \text{for } i = 1, 2, \ldots, n.$$

We call these the *geodesic equations* of the Riemannian manifold. A coordinate change in M changes the Christoffel symbols and the differential equation 12.1.1 in such a way that the solution does not change, when regarded as a curve in M; therefore the definition of geodesics does not depend on the coordinates chosen.

We can apply the classical theorems from the theory of ordinary differential equations to deduce several facts about geodesics. The geodesic equation is a system of second-order ordinary differential equations; therefore it always has a local solution, determined uniquely by the initial position and velocity data. In coordinates, if we choose a point $(x_0^1, \ldots, x_0^n)$ and a vector

$(a^1, \ldots, a^n)$, there is exactly one function $\gamma(t) = (x^1(t), \ldots, x^n(t))$, for $|t|$ small enough, satisfying 12.1.1 and

$$x^i(0) = x_0^i \quad \text{and} \quad \frac{dx^i}{dt} = a^i, \quad \text{for } i = 1, 2, \ldots, n.$$

In coordinate-free language, we can write

$$\gamma(0) = P \quad \text{and} \quad \dot{\gamma}(0) = a,$$

where $P = (x_0^1, \ldots, x_0^n)$ and $a \in T_P M = (a^1, \ldots, a^n)$ in the given coordinate system (Figure 12.1). Thus, we have $2n$ constants in all, n for the initial point and n for the initial velocity.

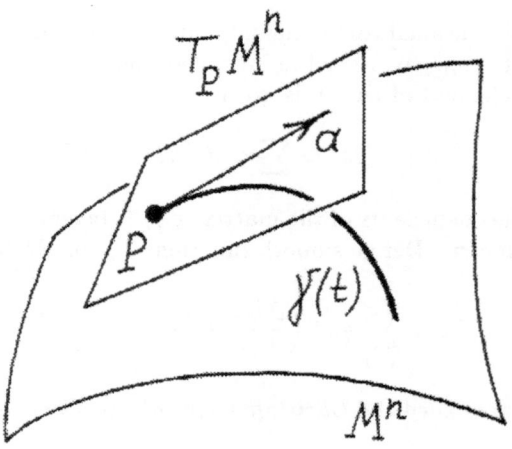

FIGURE 12.1.

We restate what we have said in the form of a thorem.

Theorem 12.1.2. *Let M be a Riemannian manifold, and choose $P \in M$ and $a \in T_M P$. Then there exists a unique geodesic $\gamma(t)$ (defined for $|t|$ small enough) such that $\gamma(0) = P$ and $\dot{\gamma}(0) = a$.*

Consequently, two geodesics that are tangent at a point coincide.

The velocity of a geodesic is constant, that is, $\langle \dot{\gamma}(t), \dot{\gamma}(t) \rangle = \langle a, a \rangle$ for all t. In particular, if the initial velocity is a unit vector, the parameter t along γ is arclength.

If $f : M \to N$ is an isometry between Riemannian manifolds, the image $f \circ \gamma$ of a geodesic γ in M is also a geodesic. This follows because (in the appropriate coordinate systems) f preserves g_{ij}, therefore also the Christoffel symbols.

Theorem 12.1.3. *For each point S on a Riemannian manifold M there exists a neighborhood U of S and a number $\varepsilon > 0$ such that the following statements hold* (refer to Figure 12.2):

1. *Any two points P and Q in U can be connected by a unique geodesic γ of length smaller than ε and belonging to U.*
2. *Any other continuous curve $\alpha(t)$ connecting the same points P and Q has length no less then that of γ. In this sense the geodesic γ is the shortest path between the two sufficiently close points P and Q. Thus, a geodesic locally minimizes distance.*
3. *This geodesic segment γ depends smoothly on the initial and terminal points.*

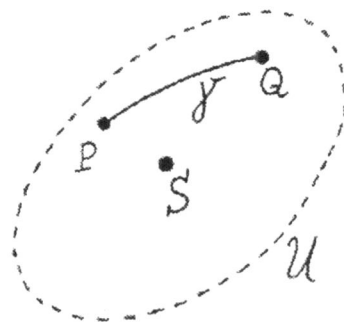

FIGURE 12.2.

For compact manifolds there is a global version of statement (a):

Theorem 12.1.4. *Any two points of a compact connected Riemannian manifold can be connected by a geodesic.*

Generally speaking, this geodesic is not unique (see examples below).

For *noncompact* manifolds the analog of Theorem 12.1.4 can be wrong. Consider the manifold M obtained by removing from the Euclidean plane a closed disk (Figure 12.3). This is a noncompact manifold with locally Euclidean metric (and without boundary, because we are removing a closed set). Then two points P and Q as shown in Figure 12.3 cannot be connected by a geodesic. Indeed, assume that there exists some geodesic $\alpha : [0, t] \to M$ connecting P and Q. Since the image $\alpha([0, t])$ of the geodesic is a compact set, we can choose the number ε of Theorem 12.1.3 in such a way that it will be the same for all points on the geodesic. But we know that within a Euclidean ball the minimal distance between two points is provided by a straight line; thus the geodesic must be a straight line between any two points that are less

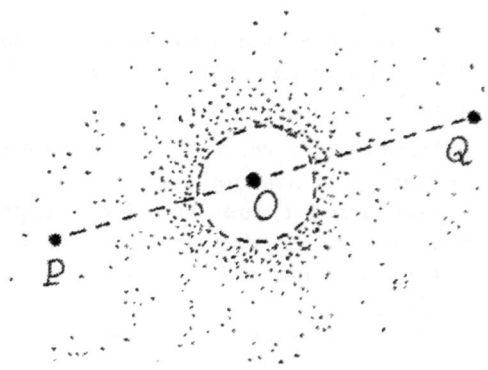

FIGURE 12.3.

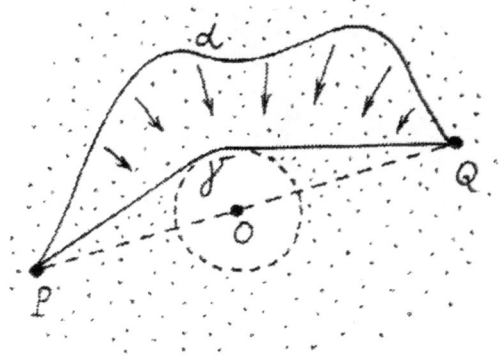

FIGURE 12.4.

than ε apart, and this is only possible if the whole geodesic is a line segment. But a line segment is ruled out because it goes outside M.

(Intuitively, we can think of tying a string at P, making it go through Q, and trying to shorten it by pulling it. In Figure 12.4 we see what would happen: the desired final configuration is shown as γ, but it can never be reached, since the boundary of the disk has been removed. The same situation takes place if we remove from the plane *only one point*, instead of a closed disk.)

A geodesic $\gamma(t)$, where $a \leq t \leq b$, is called *minimal* if it is shorter than any other path (continuous, or smooth) connecting its extreme points $\gamma(a)$ and $\gamma(b)$.

Theorem 12.1.3 asserts that any sufficiently small segment of a geodesic curve is minimal. At the same time, a rather long geodesic may not be minimal, as we shall see below.

12.2. Geodesics on Simple Surfaces

Example 12.2.1 (Euclidean space). In Euclidean space we have $g_{ij} = 0$, so $\Gamma^i_{jk} = 0$. Thus the geodesic equation 12.1.1 reduces to

$$\frac{d^2 x^i}{dt^2} = 0,$$

and the solutions are $x^i(t) = a^i t + b^i$, where the a^i and b^i are constants. In other words, *in Euclidean space the geodesics are the straight lines, and only they.*

Example 12.2.2 (Cone and cylinder). Consider in Euclidean three-space a cone M (Figure 12.5), and remove its vertex O to obtain a smooth, non-compact, surface, with the induced Riemannian metric from $\mathbb{R}^3$. How can we describe geodesics on M? First we cut the cone along any of its generators (lines through the vertex). Now, it can be proved that the metric on M induced from $\mathbb{R}^3$ is locally Euclidean, that is, any small piece of the cone can be isometrically unwrapped onto the Euclidean plane, as shown in Figure 12.5. This experiment can be done with a real sheet of paper. Because this transformation is locally an isometry, sufficiently short geodesics on the cone are transformed into geodesics on the plane, that is, straight lines. Thus, we obtain a very simple local description of geodesics on the cone: they are the lines that yield Euclidean line segments under the unwrapping map.

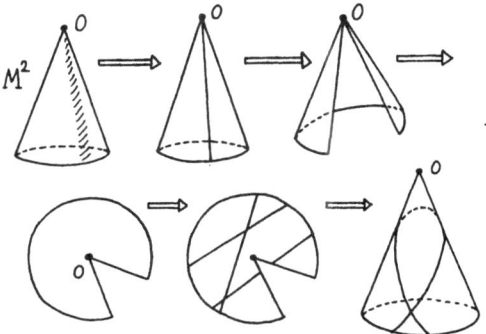

FIGURE 12.5.

To get a description applicable to *global* geodesics, we should remember that a curve in the Euclidean plane that is locally straight is globally straight, so we can try to extend the straight line on the opened-up cone (sector of the plane) in both directions. It it hits the cut, it comes back in from the other side of the cut in a different direction, since the angle it makes with the cut has to be the same on both sides. Thus we see that, if the cone is narrow—that is, if it makes a wedge of less than 180° when opened up—a geodesic extending in both directions will be self-intersecting, because the corresponding line in the wedge will hit the cut sooner or later.

A cylinder, too, can be developed onto the plane by an isometry (Figure 12.6). Thus, the same description of geodesics is valid for cylinders: they are images of straight lines from the plane. For the cylinder of revolution, a geodesic is either a spiral, a generator (straight lines parallel to the axis), or circles orthogonal to the axis.

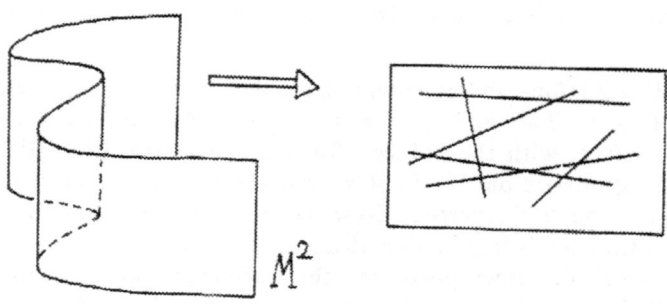

FIGURE 12.6.

Example 12.2.3 (The sphere). Let S^2 be a standard (round) sphere in $\mathbb{R}^3$. Then it is easy to show that *geodesics are exactly the great circles*, that is, the plane sections of the sphere through its center (Figure 12.7). To see this, consider a point P and a tangent vector a at P; we know there is a geodesic $\gamma(t)$ through P with initial velocity a. Let μ be the great circle through P and having a as a tangent vector (Figure 12.8). Then $\gamma(t)$ must be entirely contained in μ! If not, the reflection of γ in the plane containing μ would be a different geodesic with the same starting position and velocity, which is impossible. This shows any geodesic must be a great circle. It follows also that any great circle (parametrized by arclength) is a geodesic: we choose a starting point on that circle and a starting velocity along it, and use the fact that a geodesic with those initial data exists.

We can, of course, prove the same result using explicit formulas. In spherical coordinates (θ, φ), the metric has the form $ds^2 = d\theta^2 + \sin^2\theta d\varphi^2$; setting $x^1 = \theta$ and $x^2 = \varphi$ we have $g_{11} = 1$, $g_{12} = g_{21} = 0$, and $g_{22} = \sin^2\theta$. Using the formulas for the Christoffel symbols, we obtain (verify!)

$$\Gamma^1_{22} = -\tfrac{1}{2}\sin 2\theta, \qquad \Gamma^2_{12} = \cot\theta,$$

and $\Gamma^i_{jk} = 0$ for all other combinations of indices (i, j, k). Thus the geodesic equations become

$$\theta'' - \tfrac{1}{2}(\varphi')^2 \sin 2\theta = 0, \qquad \varphi'' + \varphi'\theta' \cot\theta = 0.$$

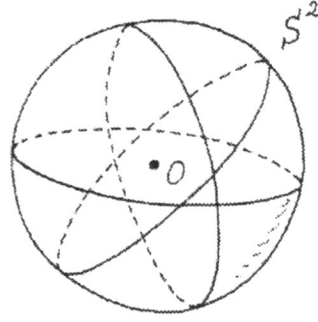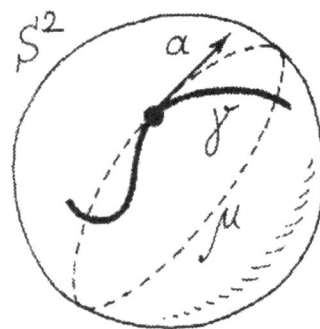

FIGURE 12.7.   FIGURE 12.8.

One solution is $\varphi = \text{const}$ and $\theta = t$, that is, a meridian μ emerging from the north pole of the sphere. Thus, any great circle going through the north pole is a geodesic, if for the parameter θ we take the arclength. Moreover, any initial point is equivalent to the north pole, since we can apply a rotation of space to put it there, so great circles through other points are also geodesics. Finally, since we know a geodesic for every point and tangent vector, we know all geodesics.

Example 12.2.4 (The hyperbolic plane). The open upper half-plane (Figure 12.9), with the Riemannian metric

$$ds^2 = \frac{dx^2 + dy^2}{y^2},$$

is called the *hyperbolic plane*, or *Lobachevskian plane*. Its geodesics are described as follows (Figure 12.10):

1. straight lines orthogonal to the x-axis, and
2. upper halves of circles with center on the x-axis.

These geodesics can be considered as *straight lines* on the hyperbolic plane.

We prove this by direct integration of the geodesic equation. The Christoffel symbols here are (verify!):

$$\Gamma_{12}^1 = \frac{1}{y}, \qquad \Gamma_{11}^2 = -\frac{1}{y}, \qquad \Gamma_{22}^2 = \frac{1}{y};$$

the remaining symbols vanish. The geodesic equation becomes

$$\frac{d^2 x}{dt^2} = \frac{2\dot{x}\dot{y}}{y}, \qquad \frac{d^2 y}{dt^2} = \frac{\dot{y}^2 - \dot{x}^2}{y},$$

where the dot indicates differentiation with respect to t. It follows that

$$\frac{d^2 y}{dx^2} = -\frac{1}{y}\left(\frac{\dot{y}^2}{\dot{x}^2} + 1\right) = -\frac{1}{y}({y'_x}^2 + 1),$$

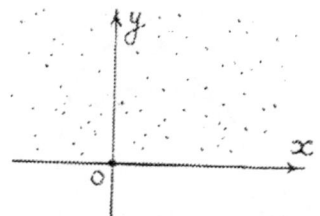

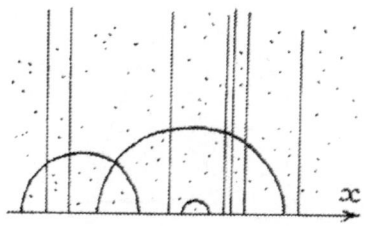

FIGURE 12.9.   FIGURE 12.10.

hence successively $y'' = -(1/y)(y_x'^2 + 1)$, $yy'' + y'^2 = -1$, $(yy')' = -1$, $yy' = -x + C$, $ydy = (-x + C)dx$, $y^2 = -x^2 + 2Cx + D$, $x^2 - 2Cx + y^2 = D$, $(x - C)^2 + y^2 = C^2 + D$, which is a circle orthogonal to the real axis.

In this calculation it was assumed that $\dot{x} \neq 0$. If $\dot{x} = 0$, we obtain instead straight lines orthogonal to the real axis. We can consider these lines as circles of infinite radius.

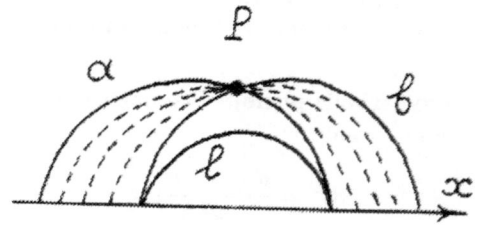

FIGURE 12.11.

Figure 12.11 shows that if a point P does not belong to a hyperbolic straight line l, there exist infinitely many hyperbolic straight lines passing through P and *parallel* to l in the sense that they don't intersect l. The lines a and b in the figure are limit cases.

Example 12.2.5 (The torus). The torus T^2 can be given a locally Euclidean metric, as follows. Think of T^2 as a direct product of two circles, and consider the standard angular coordinates φ and ψ on these circles (Figure 12.12); they determine a point of the torus as the pair (φ, ψ). Then write the metric on the torus as

$$ds^2 = d\varphi^2 + d\psi^2.$$

One can also represent the torus as the quotient $\mathbb{R}^2/\mathbb{Z}\oplus\mathbb{Z}$ of the Euclidean plane by the square lattice

$$\{(m, n) : m, n \in \mathbb{Z}\}$$

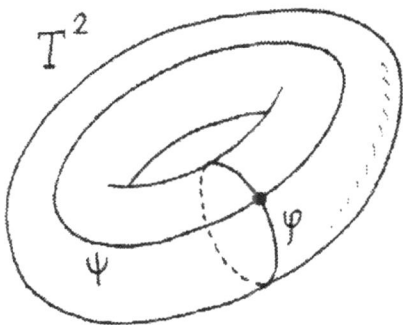

FIGURE 12.12.

(Figures 12.13 and 12.14). The projection map, associating to $(x, y) \in \mathbb{R}^2$ the point $(\varphi, \psi) \in T^2$ with $\varphi = x$ modulo 1 and $\psi = y$ modulo 1, is a local isometry. Consequently, the geodesics on the torus are exactly the projections of straight lines on the Euclidean plane. There are no other geodesics on the torus.

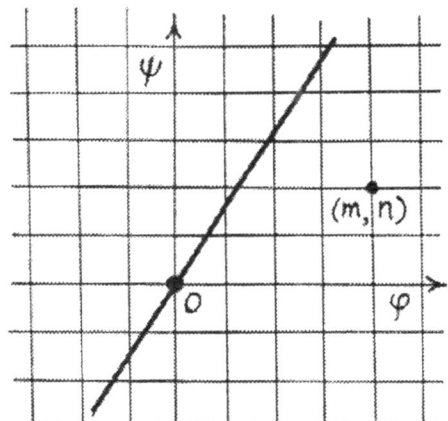

FIGURE 12.13.

The geodesics are divided into two classes: *closed* and *nonclosed*. Note that *closed* indicates that the geodesic not only returns to the same point, but also with the same velocity vector. It is convenient to depict geodesics on a torus as straight lines on $\mathbb{R}^2$, the lattice $\mathbb{Z} \oplus \mathbb{Z}$ being fixed (Figure 12.13).

Consider a geodesic l passing through the point $(0,0)$ on the plane. Obviously, the corresponding geodesic on the torus is *closed* if and only if l

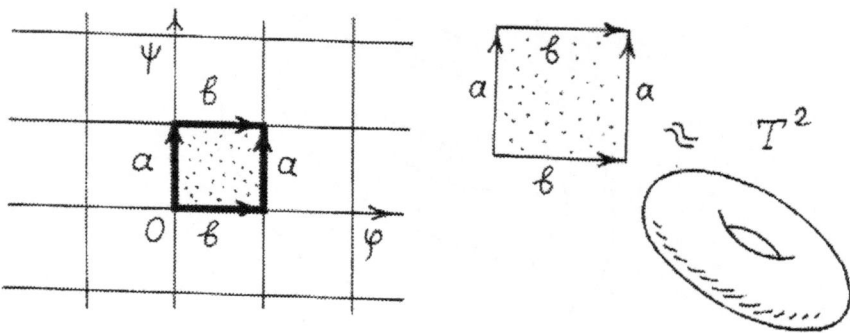

FIGURE 12.14.

goes through some lattice point (m, n). We can always assume that m and n are relatively prime. The pair (m, n) is called the *type* of the given closed geodesic.

Conversely, the geodesic on T^2 coming from l is nonclosed if and only if l is a straight line that does not meet lattice points apart from $(0, 0)$.

This criterion can be expressed in other terms. Consider the angle α between l and the x-axis (Figure 12.15). Then the geodesic on the torus is closed if and only if $\tan \alpha$ is rational. Thus geodesics on the torus are sometimes classified as *rational* and *irrational*, or as *periodic* and *nonperiodic*.

A closed geodesic on the torus is homeomorphic to a circle (this is true of a non-self-intersecting closed geodesic in any manifold). Figures 12.16 and 12.17 show a line of slope $\frac{2}{3}$ in the plane, and the corresponding geodesic of type $(3, 2)$ in the torus. This curve goes 3 times around the meridian and 2 times around the parallel of latitude of the torus. In general, a geodesic of type (m, n) goes m times around the meridian and n times around the parallel. (This presupposes a conventional homeomorphism between the abstract torus $S^1 \times S^1$ and the torus of revolution in $\mathbb{R}^3$, where the first S^1 maps to a parallel of latitude and the second to a meridian.)

Now consider the irrational case (Figure 12.18). Here the geodesic is dense on the torus (that is, its closure is all of T^2); it passes arbitrarily close to any point infinitely many times.

A periodic geodesic determines a *periodic motion* on the torus, and a nonperiodic geodesic determines an *almost periodic motion*.

Example 12.2.6 (Liouville metrics). Consider a surface M with local coordinates u, v. A Riemannian metric on M is called a *Liouville metric* (in the given coordinates u and v) if it has the form

$$ds^2 = (f(u) + g(v))(du^2 + dv^2),$$

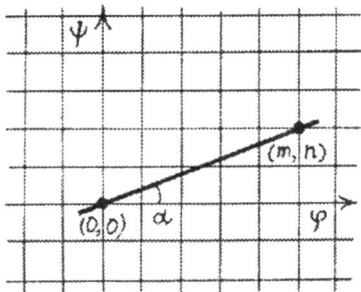

FIGURE 12.15.

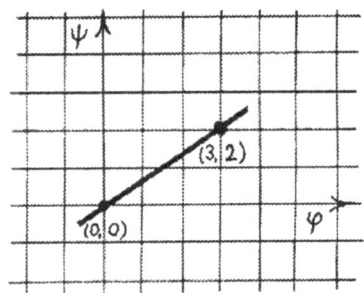

FIGURE 12.16.

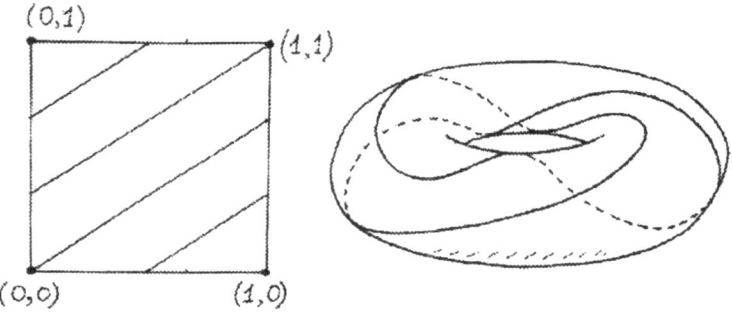

FIGURE 12.17.

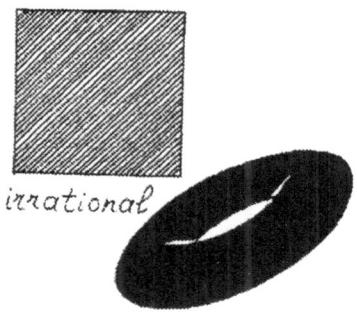

FIGURE 12.18.

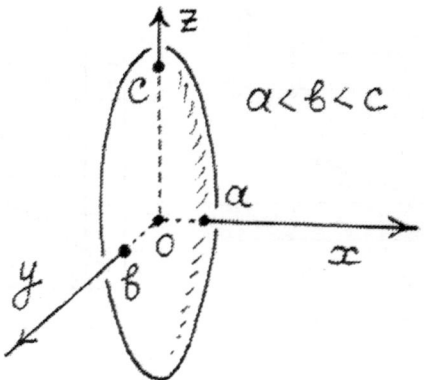

FIGURE 12.19.

where f and g are arbitrary smooth positive functions. The level curves of the function

$$z(u, v) = \int \frac{du}{\sqrt{f(u) - a}} \pm \int \frac{dv}{\sqrt{g(v) + a}}$$

are the geodesics of this metric.

Example 12.2.7 (The ellipsoid). We take the ellipsoid

$$\frac{x^2}{a} + \frac{y^2}{b} + \frac{z^2}{c} = 1,$$

where the positive numbers $a < b < c$ are the squares of the semiaxes, and x, y, z are Euclidean coordinates in $\mathbb{R}^3$ (Figure 12.19).

For the given values of a, b, c, we introduce a system of orthogonal coordinates in $\mathbb{R}^3$ classically known as *ellipsoidal coordinates*. For any $P = (x, y, z) \in \mathbb{R}^3$, the three roots of the cubic equation

12.2.8.
$$\frac{x^2}{a + \lambda} + \frac{y^2}{b + \lambda} + \frac{z^2}{c + \lambda} = 1$$

are real, and each lies in one of the intervals $(-a, \infty)$, $(-b, -a)$, and $(-c, -b)$. Let $\lambda_1 > \lambda_2 > \lambda_3$ be these roots. Then

$$x^2 = \frac{(a + \lambda_1)(a + \lambda_2)(a + \lambda_3)}{(a - b)(a - c)},$$

$$y^2 = \frac{(b + \lambda_1)(b + \lambda_2)(b + \lambda_3)}{(b - a)(b - c)},$$

$$z^2 = \frac{(c + \lambda_1)(c + \lambda_2)(c + \lambda_3)}{(c - b)(c - a)}.$$

Fixing λ_1 and letting λ_2, λ_3 vary gives a parametrized surface. the same as the implicit surface obtained by fixing $\lambda \in (-a, +\infty)$ in 12.2.8. It is easy to see from the implicit representation that this surface is an ellipsoid. Similarly, fixing λ_2 and letting λ_1, λ_3 vary is the same as fixing $\lambda \in (-b, -a)$ in 12.2.8. and gives a one-sheeted hyperboloid, while λ_3 gives a two-sheeted hyperboloid. These three surfaces intersect orthogonally at (x, y, z), as shown in Figure 12.20. and they are each confocal with the original ellipsoid. We call $(\lambda_1, \lambda_2, \lambda_3)$ the *ellipsoidal coordinates* of (x, y, z). (There are up to eight points with the same ellipsoidal coordinates, obtained by negating each Euclidean coordinate independently.)

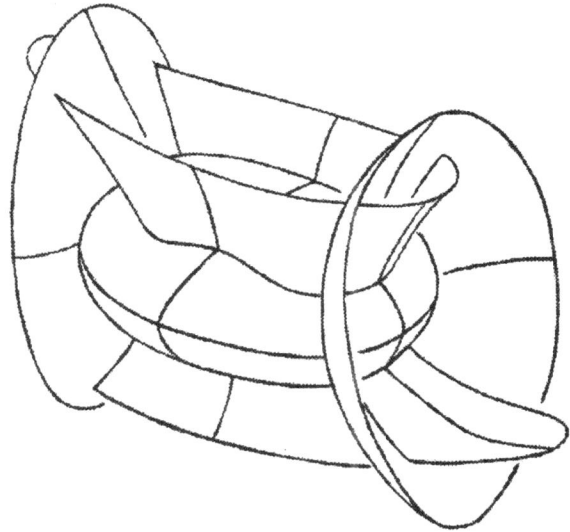

FIGURE 12.20.

The original ellipsoid E has equation $\lambda_1 = 0$ in these coordinates. We now give a description of the geodesics on it, without proofs; for details, see [Knörrer 1980]. for example.

Consider a parameter t varying in the interval $[a, c]$. If $t \in [a, b]$, consider the region G_t of E defined by

$$a \le -\lambda_2 \le t.$$

This region is homeomorphic to an annulus (Figure 12.21), except when $t = a$, in which case it reduces to the ellipse in the yz-plane. and when $t = b$, in which case it is the whole ellipsoid. Likewise, if $t \in [b, c]$, consider the closed

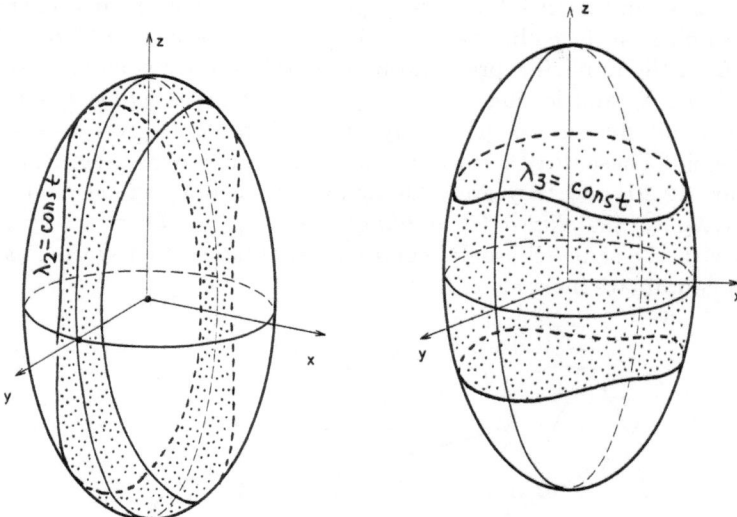

FIGURE 12.21.   FIGURE 12.22.

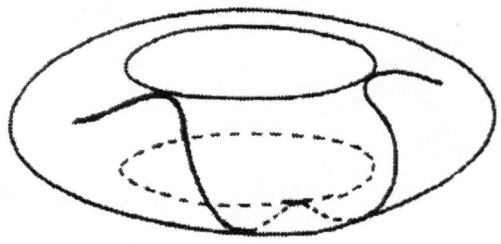

FIGURE 12.23.

region G_t

$$c \geq -\lambda_3 \geq t,$$

also homeomorphic to an annulus (Figure 12.22) unless $t = c$ or $t = b$. It can be proved that *each geodesic of E occupies some region G_t, for some $t \in [a, c]$*. The geodesic oscillates around the central ellipse of G_t; it alternately touches either boundary curve of G_t, then returns to the interior (Figure 12.23). Except for $t = a, c$ and in one case for $t = b$, the geodesic is nonclosed and everywhere dense in G_t, and determines an almost periodic motion there. The exceptional cases are the ellipses determined by the symmetry planes, which are obviously geodesics, by the same symmetry argument used in Example 12.2.3.

There is an interesting geometric description for the geodesics that fill G_t. Let H be the confocal hyperboloid (one-sheeted or two-sheeted) that determines the boundary of G_t—that is, the surface $\lambda_2 = -t$ or $\lambda_3 = -t$ as the case may be. Then consider all the straight lines in $\mathbb{R}^3$ that are tangent to both H and E (not necessarily at the same point); the tangency points with E can be shown to lie in G_t. Such a line tangent to x at E determines a direction in $T_x E$; for each x in the interior of G_t there are two such directions, and for each $x \in \partial G_t$ there is one, tangent to ∂G_t. A smooth unit vector field on G is coherently defined (locally) by one of these sets of directions. It turns out that the *the integral trajectories of this field are geodesics on the ellipsoid*—exactly those that fill G_t. To obtain *almost all* the geodesics, apply this construction for all $t \in (a, b) \cup (b, c)$. That leaves out the ellipses in the xy- and yz- planes (corresponding respectively to the limits $t \to c$ and $t \to a$), and the geodesics that go through the umbilic points

$$\left(\pm\sqrt{\frac{a\,(a - b)}{a - c}},\, 0,\, \pm\sqrt{\frac{c\,(b - c)}{a - c}} \right)$$

(A point on a surface is called an *umbilic* if the principal curvatures are equal there.) The four umbilics are always outside G_t, for $t \notin b$; as $t \in (b, c)$ approaches b from above the complement of G_t becomes two narrow strips joining umbilics in pairs, and likewise as $t \in (a, b)$ approaches b from below. A geodesic going through an umbilic also goes through the opposite one, and back, but when it comes back it has a different direction (unless it happens to be the ellipse in the xz-plane), so it is not a closed geodesic.

Example 12.2.9 (Surfaces of revolution). Now we describe all geodesics on a surface of revolution in $\mathbb{R}^3$. Let $r(z)$ be any positive smooth function. Take the smooth curve $y = r(z)$ in the yz-plane and rotate it around the z-axis, thus creating a surface M in $\mathbb{R}^3$ (Figure 12.24). The circles traced by each point of the curve $y = r(z)$ are called the *parallels* of M. Consider an arbitrary smooth curve γ on M. For a point $P \in M$ with ordinate z, let $\alpha(z)$ be the angle between γ and the parallel passing through P. Then it can be proved that, *if γ is a geodesic,*

12.2.10. $\qquad\qquad r(z) \cos \alpha(z) = \text{const}$

as P moves along γ. This is therefore a necessary (but not sufficient) condition γ to be a geodesic.

We take some examples. For $r(z) = \text{const}$, the surface of revolution is a *cylinder* (Figure 12.25). Then it follows from the equation that $\alpha(z) = \text{const}$ and the geodesics are the *spirals*, the straight lines parallel to the z-axis, and the parallels. As we saw in Example 12.2.2, these are the images of straight lines in the Euclidean plane when the plane is rolled into a cylinder. In this case 12.2.10 is necessary and sufficient for γ to be a geodesic.

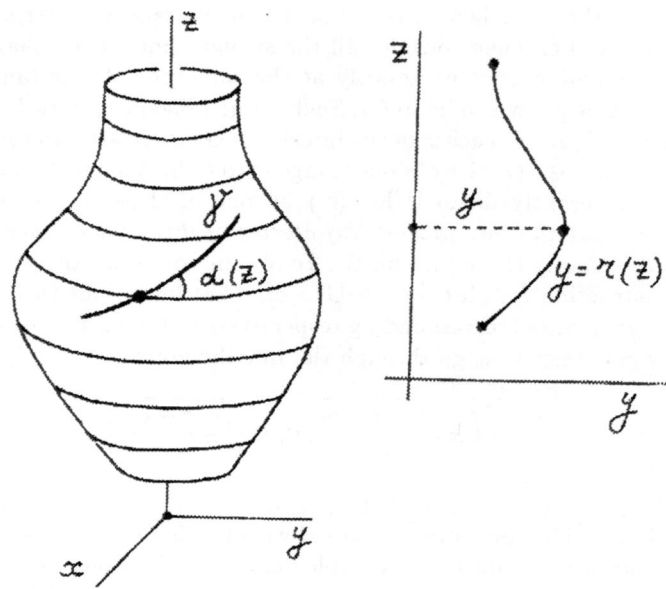

FIGURE 12.24.

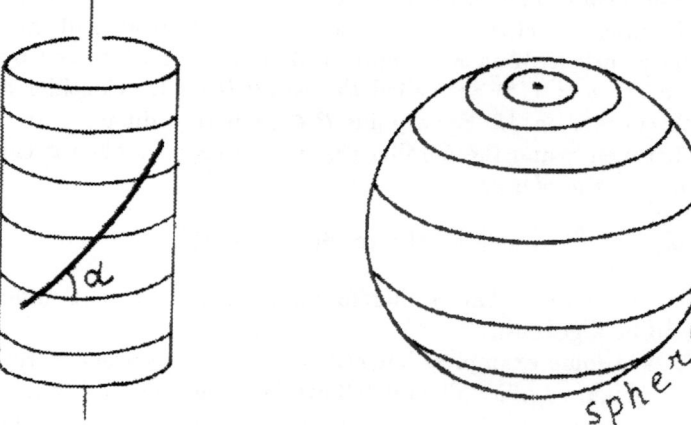

FIGURE 12.25.   FIGURE 12.26.

In the case of a *sphere* (Figure 12.26), the geodesics are great circles, as we saw earlier. The reader should verify that 12.2.10 is satisfied for great circles. If γ is a parallel of latitude, 12.2.10 is satisfied as well, since r and α are both constant, but γ is *not* a geodesic unless it is the equator.

This phenomenon occurs for all other surfaces of revolution apart from the cylinder: Figure 12.27 shows a *cone* with its parallels, which again are not geodesics. A parallel γ of a surface of revolution is a geodesic if and only if the normals to the surface at γ are perpendicular to the axis of revolution.

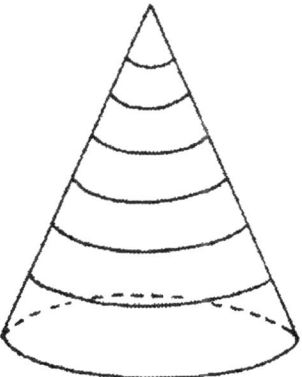

FIGURE 12.27.

Figure 12.80

13. Transformation Groups

13.1. Lie Groups and Lie Algebras

In geometry and its applications there often arise groups of transformations that have the structure of a smooth manifold.

Definition 13.1.1. A smooth manifold G is called a *Lie group* if there is a group operation defined on it, generally denoted multiplicatively, such that the maps

$$k : (x, y) \mapsto xy \qquad \text{and} \qquad n : x \mapsto x^{-1},$$

where $x, y \in G$, are smooth. The identity element for the group law is called the *identity of G*, and is usually denoted by 1. The connected component of the identity in G is labeled G_0; it is easy to see that G_0 is a normal subgroup of G. The dimension of a Lie group is its dimension as a manifold.

For example, the space of invertible $n \times n$ real matrices is a Lie group: it is a manifold (being an open subset of the vector space of all $n \times n$ matrices, which is isomorphic to $\mathbb{R}^{n^2}$), and the multiplication and inversion maps are smooth (multiplication is a linear function of the entries and inversion is a rational function). This group is known as $\mathrm{GL}(n, \mathbb{R})$, and will be studied in more detail in Examples 13.2.4, 13.3.1, and 13.4.1.

For the same reason, any subset of the set of nondegenerate $n \times n$ matrices (real or complex) that is a smooth manifold and is closed under multiplication is a Lie group. Such Lie groups are called *matrix groups*. Note that for us matrix groups will always be finite-dimensional. Multiplication in a matrix group is given by standard matrix multiplication, and the identity element is the identity matrix, denoted I.

Usually we will consider matrix groups of positive dimension, that is, not a collection of isolated points.

Definition 13.1.2. A vector space V is called a *Lie algebra* if it is equipped with a skew-symmetric bilinear operation $[\ ,\ ]$, usually called the *commutator*, that satisfies the *Jacobi identity*

$$[X, [Y, Z]] + [Z, [X, Y]] + [Y, [Z, X]] = 0$$

for all elements $X, Y, Z \in V$.

Theorem 13.1.3. *Let G be a matrix group and let $V = T_1G$ be the tangent space to G at 1. Then V can be identified with some vector space of matrices, and has a natural Lie algebra structure, with commutator given by*

$$[X, Y] = XY - YX.$$

We call V the *Lie algebra of G.*

There is a natural correspondence between *connected* matrix Lie groups and their Lie algebras. Let G be any matrix Lie group (not necessarily connected), and consider a smooth curve $g(t)$ passing through the unit element I in the group G, that is, $g(0) = I$ (Figure 13.1). If the value of the time parameter t is small, the matrix $g(t)$ can be written as

$$g(t) = I + tX + \cdots,$$

where the ellipses represent terms of higher order in t, and $X = \dot{g}(0)$ is the velocity of $g(t)$ at $t = 0$. The set of all such velocity vectors is the tangent space to G at I, that is, the Lie algebra of G.

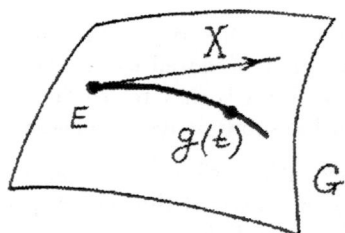

FIGURE 13.1.

A converse procedure allows us to construct a group from the Lie algebra V of G. Fix some element $X \in V$, and take the family of matrices $g(t)$ given by

13.1.4.
$$\exp(tX) = e^{tX} = \sum_{k=0}^{\infty} \frac{(tX)^k}{k!};$$

this is a convergent series for all t, so the map $t \mapsto \exp(tX)$ defines a smooth curve in the space of matrices. By construction,

$$\frac{d}{dt} \exp(tX)\Big|_{t=0} = X.$$

The images of all these curves, for all $X \in V$, form a subset G' of the space of matrices. It can be proved that G' is a Lie group, and that its tangent space at I is V; in fact G' is exactly G_0, the component of the identity of the group G we started with. In particular, if G is connected, $G' = G$.

Setting $t = 1$ in 13.1.4 we get a map $\exp : V \to G$, called the *exponential map*, from the Lie algebra of G into G.

Theorem 13.1.5. *For any matrix group, the exponential map* $\exp : V \to G$ *gives a diffeomorphism between a small neighborhood of zero in V and a small neighborhood of I in G.*

Theorem 13.1.6. *For any compact connected matrix group the exponential map is surjective.*

In other words, in compact matrix groups any element g has a *logarithm*, that is, an element X of the Lie algebra such that $\exp X = g$. Such an element is obviously not unique: the Lie algebra is noncompact and the Lie group is compact, so $\exp$ cannot be injective.

For noncompact Lie group the analogous stamement is not valid in general (there are counterexamples).

13.2. Some Transformation Groups in the Plane

Example 13.2.1 (Plane rotations around the origin). Consider the Euclidean plane $\mathbb{R}^2$ and the group G of transformations that preserve the metric, the orientation, and the origin—in other words, the group of rotations around the origin. This is a one-dimensional Lie group, denoted by $SO(2)$ and identified with the following group of matrices:

$$g = \begin{pmatrix} \cos\varphi & \sin\varphi \\ -\sin\varphi & \cos\varphi \end{pmatrix}.$$

If we relax the condition that the transformation preserves orientation. we get also those of the form

$$g = \begin{pmatrix} \cos\varphi & \sin\varphi \\ \sin\varphi & -\cos\varphi \end{pmatrix},$$

which are reflections in lines through the origin (also called by some authors "improper rotations"). Together with the elements of $SO(2)$, these transformations make up the group $O(2)$ of orthogonal transformations of the plane. $SO(2)$ is commutative and is homeomorphic to a circle, with φ as the angle parameter; $O(2)$ is homeomorphic to the disjoint union of two circles, and $SO(2)$ is the component of the identity in it (and in particular a normal subgroup).

The Lie algebra V of $SO(2)$ and $O(2)$ can be identified with $\mathbb{R}$; it consists of the matrices

$$X = \begin{pmatrix} 0 & \varphi \\ -\varphi & 0 \end{pmatrix},$$

where $\varphi \in \mathbb{R}$. It is clear that

$$g(\varphi) = \begin{pmatrix} \cos \varphi & \sin \varphi \\ -\sin \varphi & \cos \varphi \end{pmatrix} = \exp \begin{pmatrix} 0 & \varphi \\ -\varphi & 0 \end{pmatrix}.$$

Example 13.2.2 (Translations of the plane). For an arbitrary vector a on the Euclidean plane $\mathbb{R}^2$, the distance-preserving transformation

$$g_a(P) = P + a$$

is called a *translation* by a (Figure 13.2). When two translations g_a and g_b are composed, we get another translation g_{a+b}. Thus the set of all plane translations forms a two-dimensional, noncompact, abelian Lie group T, the elements of which can be identified with the points of $\mathbb{R}^2$. Under this identification, the group operation is vector addition. The Lie algebra of T is also identified with $\mathbb{R}^2$ (as a vector space).

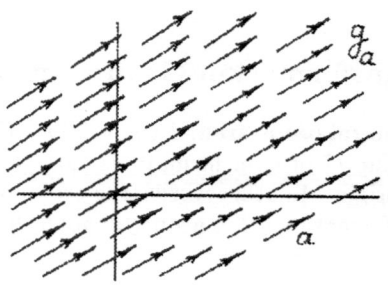

FIGURE 13.2.

Example 13.2.3 (Rigid motions of the plane). A *rigid motion* of the Euclidean plane is an orientation-preserving transformation preserving the Euclidean metric. Any plane rigid motion has the form

$$P \mapsto g(P) + a,$$

where $g \in \mathrm{SO}(2)$ is a rotation around the origin and a is some vector on the plane. That is, a rigid motion is the composition of a rotation around the origin and a translation.

Let $E(2)$ denote the group of plane rigid motions. It follows from the preceding paragraph that $E(2)$ is a connected three-dimensional Lie group, having $\mathrm{SO}(2)$ and T as subgroups; together, these subgroups generate $E(2)$. Topologically, $E(2)$ is homeomorphic to $S^1 \times \mathbb{R}^2$, but the group structure is *not* a direct product $\mathrm{SO}(2) \times \mathbb{R}^2$. The group of translations is normal, but $\mathrm{SO}(2)$ is not: if we take the conjugate of an element of $\mathrm{SO}(2)$ by a nontrivial translation g_a, the result is a rotation about a point different from the origin, namely. the image of the origin under g_a.

All elements of $E(2)$ apart from translations arise in this way, as rotations about arbitrary points of the plane.

We can represent $E(2)$ as a matrix group *in three dimensions*. In cartesian coordinates x and y, a transformation $g(P) + a$ from $E(2)$ has the form

$$\begin{pmatrix} x' \\ y' \end{pmatrix} = \begin{pmatrix} \cos\varphi & \sin\varphi \\ -\sin\varphi & \cos\varphi \end{pmatrix} \begin{pmatrix} x \\ y \end{pmatrix} + \begin{pmatrix} \alpha \\ \beta \end{pmatrix},$$

where α and β are the coordinates of the vector a. We now identify our $\mathbb{R}^2$ with the plane $\Pi = (x, y, 1)$. It can easily be seen that the transformation just given is the restriction to Π of the following linear transformation of $\mathbb{R}^3$ (Figure 13.3):

$$\begin{pmatrix} x' \\ y' \\ z' \end{pmatrix} = \begin{pmatrix} \cos\varphi & \sin\varphi & \alpha \\ -\sin\varphi & \cos\varphi & \beta \\ 0 & 0 & 1 \end{pmatrix} \begin{pmatrix} x \\ y \\ z \end{pmatrix}.$$

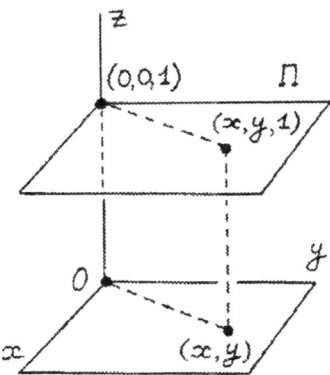

FIGURE 13.3.

Thus, $E(2)$ is isomorphic to the group of matrices of the form

$$\begin{pmatrix} \cos\varphi & \sin\varphi & \alpha \\ -\sin\varphi & \cos\varphi & \beta \\ 0 & 0 & 1 \end{pmatrix}.$$

Its Lie algebra is isomorphic to the three-dimensional vector space of matrices of the form

$$\begin{pmatrix} 0 & \varphi & \alpha \\ -\varphi & 0 & \beta \\ 0 & 0 & 0 \end{pmatrix}.$$

Example 13.2.4 (The plane linear group). At the beginning of this chapter we defined the *linear group* $GL(n, \mathbb{R})$ as the group of invertible $n \times n$ real matrices. Naturally, $GL(2, \mathbb{R})$ can also be regarded as the group of all invertible linear transformations of the plane $\mathbb{R}^2$. It consists of all matrices

$$g = \begin{pmatrix} a & b \\ c & d \end{pmatrix}, \qquad \text{for } a, b, c, d \text{ with } ad - bc \neq 0.$$

It is, therefore, a noncompact four-dimensional Lie group. We can visualize its topology as follows. Let (a, b, c, d) be cartesian coordinates in $\mathbb{R}^4$, so points of $\mathbb{R}^4$ are identified with 2×2 matrices. The condition $ad - bc = 0$ gives a three-dimensional set $C^3 \subset \mathbb{R}^4$: some cone with vertex at the origin (Figure 13.4). Then $GL(2, \mathbb{R})$ is the complement of C^3 in $\mathbb{R}^4$, and therefore an open submanifold of $\mathbb{R}^4$. It has two connected components, defined by $ad - bc > 0$ and $ad - bc < 0$, respectively.

The Lie algebra V is isomorphic to the four-dimensional vector space of matrices of the form

$$\begin{pmatrix} p & q \\ r & s \end{pmatrix}, \qquad \text{for any real } p, q, r, s.$$

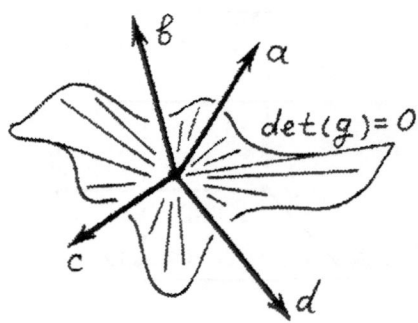

FIGURE 13.4.

Example 13.2.5 (The plane affine group). The group of transformations of the form

$$\begin{pmatrix} x' \\ y' \end{pmatrix} = \begin{pmatrix} a & b \\ c & d \end{pmatrix} \begin{pmatrix} x \\ y \end{pmatrix} + \begin{pmatrix} \alpha \\ \beta \end{pmatrix},$$

where all the constants are real and the matrix $\begin{pmatrix} a & b \\ c & d \end{pmatrix}$ is invertible, is called the two-dimensional *affine group*. It can be expressed as a matrix group by means of the same trick we used in the case of the group of rigid motions (Example 13.2.3): if we embed $\mathbb{R}^2$ as the plane $\Pi \subset \mathbb{R}^3$, the affine transformation above is realized by the matrix

$$\begin{pmatrix} a & b & \alpha \\ c & d & \beta \\ 0 & 0 & 1 \end{pmatrix}.$$

The dimension of the plane affine group is 6. The Lie algebra V is isomorphic to the six-dimensional vector space of matrices of the form

$$\begin{pmatrix} p & q & \gamma \\ r & s & \delta \\ 0 & 0 & 0 \end{pmatrix}, \qquad \text{where } p, q, r, s, \gamma, \delta \in \mathbb{R}.$$

Example 13.2.6 (The one-dimensional affine group). The counterpart of the preceding example in one dimension is the group of affine transformations of the line, which have the form

$$x' = ax + b,$$

where $a \neq 0$ and b is arbitrary. Using again the trick of embedding the space of interest in a space one dimension higher, we see that the aforementioned affine transformation is realized by the matrix

$$\begin{pmatrix} a & b \\ 0 & 1 \end{pmatrix},$$

so the group is two-dimensional and homeomorphic to the plane with the line $a = 0$ removed. The Lie algebra consists of matrices $\begin{pmatrix} p & q \\ 0 & 0 \end{pmatrix}$, where p and q are arbitrary.

The group of isometries of the Euclidean real line consists of the maps $x' = \pm x + c$, and is therefore one-dimensional, with two connected components, each homeomorphic to $\mathbb{R}$.

More generally, any isometry of Euclidean n-space is an affine map, and can be expressed as an element of the n-dimensional orthogonal group $O(n)$ followed by a translation in $\mathbb{R}^n$. The reader is encouraged to verify this.

Example 13.2.7 (The special linear group). The special linear group in two dimensions, $SL(2, \mathbb{R})$, consists of the elements of $GL(2, \mathbb{R})$ having determinant 1. Recalling the notation of Example 13.2.4, we see that $SL(2, \mathbb{R})$ is defined by a single polynomial equation $ad - bc = 1$ in four-dimensional space. To prove that the space thus defined is a three-manifold in $\mathbb{R}^4$, and therefore a Lie group, it is enough to prove that the map $f : (a, b, c, d) \mapsto ad - bc$ from $\mathbb{R}^4$ to $\mathbb{R}$ has 1 as a regular value, for then we can apply Theorem 3.4.3. The differential of this map has coordinates

$$\left(\frac{\partial f}{\partial a}, \frac{\partial f}{\partial b}, \frac{\partial f}{\partial c}, \frac{\partial f}{\partial d} \right) = (d, -c, -b, a),$$

and therefore can only vanish if $a = b = c = d = 0$. But $f(0,0,0,0) \neq 1$, so 1 is a regular value.

The elements of $\mathrm{SL}(2,\mathbb{R})$ are called *unimodular transformations*, because their determinant is 1. Any unimodular transformation g of the plane can be represented (in a unique way) as the composition of a rotation around the origin and a transformation of the form

$$\begin{pmatrix} a & b \\ 0 & a^{-1} \end{pmatrix},$$

where $a > 0$ and b is arbitrary. Thus, $\mathrm{SL}(2,\mathbb{R})$ is homeomorphic to the direct product of a circle and the open half-plane $\{(a,b) : a, b \in \mathbb{R}, \ a > 0\}$.

The Lie algebra V is the three-dimensional vector space of matrices of the form

$$\begin{pmatrix} p & q \\ r & -p \end{pmatrix},$$

where p, q, r are arbitrary. This is because of the well-known relation

$$\det e^X = e^{\operatorname{trace} X}.$$

Example 13.2.8 (Hyperbolic rotations of the plane). Consider on the plane the *indefinite* nondegenerate scalar product $\langle A, B \rangle = -a^1 b^1 + a^2 b^2$, where $A = (a^1, a^2)$ and $B = (b^1, b^2)$ are vectors. The matrix C of this scalar product equals $\begin{pmatrix} -1 & 0 \\ 0 & 1 \end{pmatrix}$. The plane with this scalar product is called the *pseudo-Euclidean plane* and is denoted $\mathbb{R}^{1,1}$. Hyperbolic rotation(an *orthogonal transformation* of $\mathbb{R}^{1,1}$) is a linear transformation preserving this indefinite scalar product.

To find all such transformations, we need to solve the matrix equation $C = gCg^t$, where ^{t} denotes transposition. If we set $g = \begin{pmatrix} a & b \\ c & d \end{pmatrix}$, this condition is equivalent to the following three scalar equations for a, b, c, d (verify!):

$$a^2 - b^2 = 1, \quad ac = bd, \quad d^2 - c^2 = 1.$$

This system has the solution

$$g = \begin{pmatrix} \pm \cosh \psi & \pm \sinh \psi \\ \pm \sinh \psi & \pm \cosh \psi \end{pmatrix}.$$

alternatively, we can write

$$g = \frac{1}{\sqrt{1 - \beta^2}} \begin{pmatrix} \pm 1 & \pm \beta \\ \pm \beta & \pm 1 \end{pmatrix}, \quad \text{where } \beta = \frac{b}{a} = \tanh \psi.$$

When both diagonal entries are positive in these matrices, we say that the hyperbolic rotation is *proper*. In this case we lose nothing by taking all $\pm$ signs to be $+$, since $\sinh \psi$ can have either sign. Thus, the group of proper hyperbolic rotations is a one-dimensional, commutative, noncompact Lie group.

A Euclidean rotation is determined by the angle of rotation φ. The analogous parameter for hyperbolic rotations is ψ. Consider the action of such a rotation on an orthogonal basis $e_1 = (1,0)$, $e_2 = (0,1)$. The orthogonal transformation

$$g = \begin{pmatrix} \cosh\psi & \sinh\psi \\ \sinh\psi & \cosh\psi \end{pmatrix}$$

transforms this frame as is shown in the Figure 13.5: the endpoint of the vector e_1 moves along the first hyperbola, and that of e_2 moves along the second hyperbola. These hyperbolas can be thought of as a pseudocircle, analogous to the actual circle in the Euclidean case. Under a hyperbolic rotation each pseudocircle is invariant.

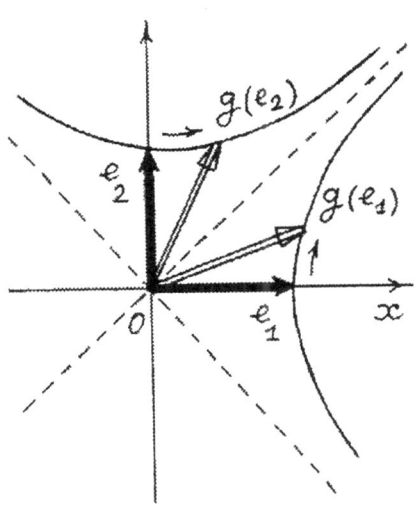

FIGURE 13.5.

Example 13.2.9 (The plane symplectic group). A linear transformation of the plane that preserves area is called *symplectic*. Such a transformation can also be defined by the property that it preserves the *skew-symmetric form*

$$(A, B) = a^1 b^2 - a^2 b^1,$$

where $A = (a^1, a^2)$ and $B = (b^1, b^2)$ are vectors in $\mathbb{R}^2$. The group $\mathrm{Sp}(1, \mathbb{R})$ of such transformations turns out to be the same as the group $\mathrm{SL}(2, \mathbb{R})$ of Example 13.2.7. To see this, observe first that (A, B) is the area of the par-

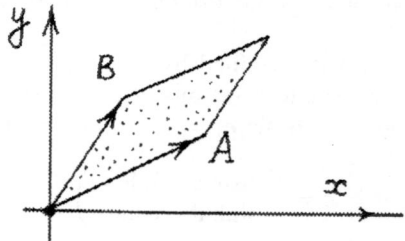

FIGURE 13.6.

allelogram $\Pi(A, B)$ spanned by A and B (Figure 13.6). Let

$$g = \begin{pmatrix} a & b \\ c & d \end{pmatrix}$$

be a symplectic transformation. Consider the images gA and gB of A and B, and the parallelogram $\Pi(gA, gB)$ they span. By elementary linear algebra we have

$$\text{area } \Pi(gA, gB) = \text{area } \Pi(A, B) \det g.$$

Because g preserves area, we have $\det g = 1$. Thus, g preserves the skew-symmetric scalar product if and only if g is unimodular.

Example 13.2.10 (Conformal transformations of the plane). We now regard the Euclidean plane as a complex line, by introducing a complex coordinate $z = x + iy$ and its conjugate $\bar{z} = x - iy$. The Euclidean metric can then be written in the form

$$ds^2 = dx^2 + dy^2 = dz\, d\bar{z},$$

because $dx^2 + dy^2 = (dx + i\, dy)(dx - i\, dy) = dz\, d\bar{z}$, where $dz = dx + i\, dy$ and $d\bar{z} = dx - i\, dy$. We will study the remarkable transformations $g : z \to w$ of the plane given by

$$w = \frac{az + b}{cz + d},$$

where a, b, c, d are arbitrary complex numbers such that $ad - bc \neq 0$. We may as well assume that $ad - bc = 1$, because we can multiply $az + b$ and $cz + d$ by the same scalar factor λ without changing g. In order for the map to be defined for all z, we allow the value $z = \infty$, and set $g(-d/c) = \infty$, $g(\infty) = a/c$. (In effect, then, the map is defined on the Riemann sphere.)

Such transformations are called *linear-fractional* maps, and they form a group G. Linear-fractional maps are complex analytic. Consider the matrix group $SL(2, \mathbb{C})$ consisting of all complex matrices

$$\begin{pmatrix} a & b \\ c & d \end{pmatrix} \quad \text{where } ad - bc = 1.$$

Proposition 13.2.11. *The group G of linear-fractional maps is isomorphic to the quotient group $\mathrm{SL}(2,\mathbb{C})/\mathbb{Z}_2$, where the normal subgroup $\mathbb{Z}_2$ consists of the two elements $I = \left(\begin{smallmatrix} 1 & 0 \\ 0 & 1 \end{smallmatrix}\right)$ and $-I = \left(\begin{smallmatrix} -1 & 0 \\ 0 & -1 \end{smallmatrix}\right)$. Thus, G is a Lie group of (real) dimension six.*

Proof. We can construct a homomorphism $h : \mathrm{SL}(2,\mathbb{C}) \to G$ by associating to $\left(\begin{smallmatrix} a & b \\ c & d \end{smallmatrix}\right)$ the linear-fractional map

$$w = \frac{az + b}{cz + d}.$$

The reader is encouraged to check that h is indeed a homomorphism, that it is surjective, and that its kernel is $\pm I$. Consequently, G is the quotient $G = \mathrm{SL}(2,\mathbb{C})/\ker h = \mathrm{SL}(2,\mathbb{C})/\mathbb{Z}_2$, as desired. $\qquad\square$

We now show that any linear-fractional map g is *conformal*, that is, preserves angles between smooth curves at intersection points. An alternative characterization of conformal maps is that they transform the Riemannian metric (here the Euclidean metric) at a point by a scalar factor, the same in all directions. It is sufficient to calculate the Euclidean metric $dw\,d\bar{w}$ in the new coordinates $z = x + iy$. We have:

$$dw = \frac{a(cw + d)\,dz - c(az + b)\,dz}{(cz + d)^2} = \frac{ad - bc}{(cz + d)^2}dz,$$

and consequently

$$dw\,d\bar{w} = \frac{|ad - bc|^2}{|cz + d|^2}dz\,d\bar{z}.$$

In terms of the real coordinates (u, v) and (x, y) we have:

$$du^2 + dv^2 = \frac{|ad - bc|^2}{|cz + d|^2}(dx^2 + dy^2).$$

Thus g multiplies the Euclidean metric by a positive factor. Moreover g preserves the orientation of the plane.

Another important property of linear-fractional maps is left to the reader to show: Such a map transforms each straight line of the Euclidean plane into a straight line or a circle, and transforms each circle into a circle or straight line.

Example 13.2.12 (Isometries of the hyperbolic plane). Recall from Example 12.2.4 that the hyperbolic plane is the upper half-plane with the metric

$$ds^2 = \frac{-4\,dz\,d\bar{z}}{(z - \bar{z})^2} = \frac{dx^2 + dy^2}{y^2}.$$

We now study the isometries of this Riemannian manifold.

Proposition 13.2.13. *Let g be the linear-fractional transformation given by $g(z) = (az + b)(cz + d)$, where $ad - bc = 1$. Then g maps the upper half-plane into itself if and only if a, b, c, d are real. In this case g acts on the hyperbolic plane as an isometry. Every orientation-preserving isometry of the hyperbolic plane is of this type. In particular, the group of orientation-preserving isometries of the hyperbolic plane is isomorphic to $\mathrm{SL}(2, \mathbb{R})/\mathbb{Z}_2$, where $\mathbb{Z}_2 = \{I, -I\}$.*

We first assertion is very easy. We prove the second. Let g be as in the proposition, with a, b, c, d real, and set $w = g(z)$. Then

$$dw = \frac{ad - bc}{(cz + d)^2}\, dz = \frac{1}{(cz + d)^2}\, dz.$$

Direct calculation shows that

$$\frac{-dw\, d\bar{w}}{(w - \bar{w})^2} = \frac{-dz\, d\bar{z}}{(z - \bar{z})^2},$$

that is, g really is an isometry. We will not prove that this exhausts all orientation-preserving isometries.

13.3. Some Transformation Groups in Space

Example 13.3.1 (The three-dimensional linear group). The group $\mathrm{GL}(3, \mathbb{R})$ consists of all (real) nondegenerate 3×3 matrices. Its dimension is 9 and topologically this group is obtained from the 9-dimensional vector space of all 3×3 matrices by removing all matrices of determinant zero. The Lie algebra is the vector space of all 3×3 matrices.

Example 13.3.2 (The three-dimensional special linear group). The group $\mathrm{SL}(3, \mathbb{R})$ is the subgroup of $\mathrm{GL}(3, \mathbb{R})$ determined by the condition $\det X = 1$. It is easy to see that $\dim \mathrm{SL}(3, \mathbb{R}) = 8$, and that the Lie algebra consists of all 3×3 matrices with trace zero.

Example 13.3.3 (Three-dimensional rotations). The group $\mathrm{SO}(3)$ consists of all (real) 3×3 matrices g that preserve the Euclidean scalar product in $\mathbb{R}^3$ and preserve orientation. Consequently, the corresponding matrix g must satisfy $I = gIg^t$, or $I = gg^t$, or $g^t = g^{-1}$, and $\det g = 1$. Such matrices are called *proper orthogonal*. $\mathrm{SO}(3)$ can also be thought of as the group of orientation-preserving isometries of a sphere.

We now show that each orthogonal matrix $g \neq I$ in $\mathrm{SO}(3)$ uniquely determines an axis $l(g)$ through the origin and an angle φ (determined modulo 2π). The action of g is a rotation through the angle φ about the line $l(g)$. Conversely, an axis l and an angle φ determine a rotation $g \in \mathrm{SO}(3)$.

To see this, observe first that, by linear algebra, an orthogonal transformation in n-space for n odd always has a real eigenvalue. An eigenvector

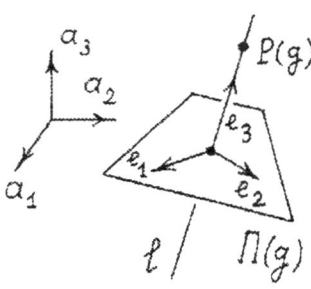

FIGURE 13.7.

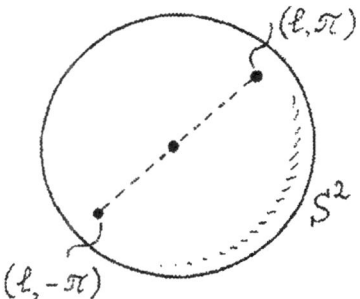

FIGURE 13.8.

with this eigenvalue generates a line l invariant under g, and the plane Π orthogonal to l is also invariant, because g is an orthogonal transformation. For the same reason, g has eigenvalue ± 1 on l. If the eigenvalue is -1, the restriction of g to Π is an orientation-reversing isometry, and therefore is a reflection in a line (see Example 13.2.1). By interchanging this line with l, if necessary, we may assume that l is fixed pointwise, and that g acts on Π by a rotation. The axis is uniquely defined when $g \neq I$, because if l and l' are both fixed pointwise by g, the whole plane spanned by l and l' is fixed, and moreover the line normal to that plane is also fixed (since g preserves orientation); therefore we would have $g = I$. The angle is determined (always modulo 2π) by the action of g on Π.

It follows from this characterization, by choosing an orthogonal matrix A that maps l to the z-axis, that we can conjugate g to the canonical form

$$AgA^{-1} = \begin{pmatrix} \cos\varphi & \sin\varphi & 0 \\ -\sin\varphi & \cos\varphi & 0 \\ 0 & 0 & 1 \end{pmatrix} .$$

Since there are two degrees of freedom for the choice of an axis and one for the choice of an angle, $SO(3)$ is a three-dimensional connected Lie group. We now show that it is homeomorphic to three-dimensional projective space $\mathbb{RP}^3$, by defining a correspondence $SO(3) \to \mathbb{RP}^3$. To do this we take $g \neq I$ and consider its axis l and the orthogonal plane Π. Assuming for the moment that $0 \leq |\varphi| \neq \pi$, choose in Π an arbitrary vector e_1 and let e_2 be the vector obtained from e_1 by the rotation g (Figure 13.7). Complete e_1 and e_2 to a positively oriented coordinate frame e_1, e_2, e_3. Then l becomes a real axis *with orientation* (determined by e_3). Mark the point $P(g)$ on the oriented line l at a positive distance $|\varphi| < \pi$ from the origin. This associates with $g \in SO(3)$ a point $P(g) = P(l, \varphi)$ in $\mathbb{R}^3$. Also set $P(I) = 0$. This gives, for $|\varphi(g)| \neq \pi$, a continuous one-to-one correspondence between matrices g and points $P(g)$.

When $|\varphi| = \pi$ there is no unique way to mark $P(g)$, since the rotations about $l(g)$ through π and $-\pi$ coincide; in this case we associate with g both points $P(l, \pi)$ and $P(l, -\pi)$, later to be identified (Figure 13.8). After we

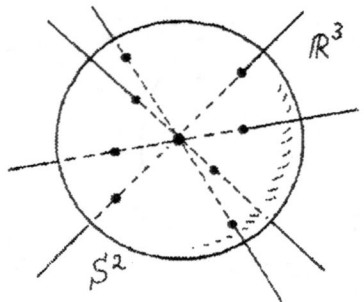
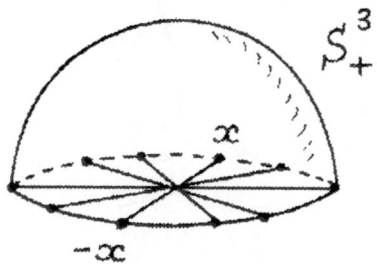

FIGURE 13.9.   FIGURE 13.10.

perform the identification of these antipodal points on the sphere of radius π, we get a one-to-one correspondence, and in fact a homeomorphism, between SO(3) and the space obtained from the three-dimensional closed ball in $\mathbb{R}^3$ by identification of antipodal points of the boundary.

Now it remains to prove that this object is $\mathbb{RP}^3$. According to the standard definition of $\mathbb{RP}^3$, this is the space of all straight lines in $\mathbb{R}^4$ through the origin (Figure 13.9). Because each straight line is uniquely determined by two antipodal points of its intersection with a sphere S^3, it is clear that $\mathbb{RP}^3$ can be represented as the set of all pairs $(x, -x)$, where x runs over all the points of a three-sphere. Thus, we need to identify the antipodal points of a sphere S^3 to obtain $\mathbb{RP}^3$.

The same result is obtained if we take only half of a three-sphere, for example, the upper hemisphere S_+^3, and identify antipodal points on its boundary (which is a two-sphere); see Figure 13.10. But S_+^3 is homeomorphic to a three-ball. This concludes the proof that SO(3) is homeomorphic to $\mathbb{RP}^3$.

We conclude this section by finding the Lie algebra of the orthogonal group SO(3), that is, by determining the tangent space at the identity. Any orthogonal matrix g close to I can be written in the form $g = I + \varepsilon X + \cdots$, where ε is a small parameter and X is a matrix in the Lie algebra. Because $g^{-1} = g^t$. we have

$$I - \varepsilon X + \cdots = I + \varepsilon X^t + \cdots,$$

that is, $X^t = -X$, so X is skew-symmetric. Thus, the Lie algebra of SO(3) is the three-dimensional space of all real skew-symmetric 3×3 matrices.

13.4. Some Transformation Groups in Higher Dimensions

Example 13.4.1 (The complete linear groups). We collect here some facts about the (complete) *linear group* GL$(n, \mathbb{R})$ in arbitrary dimension, and its complex counterpart GL$(n, \mathbb{C})$, consisting of all $n \times n$ complex matrices with nonzero determinant (or, equivalently, of all invertible linear transformations of $\mathbb{C}^n$).

Both $GL(n, \mathbb{R})$ and $GL(n, \mathbb{C})$ are noncompact Lie groups, of dimensions n^2 and $2n^2$ respectively. This can be proved by observing that $GL(n, \mathbb{R})$, for example, is an open topological subspace of the space $n \times n$ matrices, which can be regarded as $\mathbb{R}^{n^2}$; it is open because it is the inverse image of $\mathbb{R} - \{0\}$ under the determinant map, which is continuous as a function of the entries of a matrix.

The Lie algebra of $GL(n, \mathbb{R})$ is the linear space $gl(n, \mathbb{R})$ consisting of all $n \times n$ matrices with real coefficients. The Lie algebra of $GL(n, \mathbb{C})$ is the vector space $gl(n, \mathbb{C})$ consisting of all $n \times n$ matrices with complex coefficients.

Example 13.4.2 (The special linear groups). The *special linear group* is defined as the subset $SL(n, \mathbb{R})$ of $GL(n, \mathbb{R})$ determined by the single polynomial equation $\det g = 1$. The complex group $SL(n, \mathbb{C})$ is defined in the same way as a subset of $GL(n, \mathbb{C})$.

Using the same reasoning as in Example 13.2.7, we see that $SL(n, \mathbb{R})$ and $SL(n, \mathbb{C})$ are Lie groups of dimensions $n^2 - 1$ and $2n^2 - 2$ respectively. They are noncompact.

The Lie algebra of $SL(n, \mathbb{R})$ is the vector space $sl(n, \mathbb{R})$ consisting of all real $n \times n$ matrices of trace zero. Similarly, the Lie algebra of $SL(n, \mathbb{C})$ is the vector space $sl(n, \mathbb{C})$ of all complex matrices with trace zero.

Example 13.4.3 (The orthogonal groups). Let $\langle a, b \rangle$ be the Euclidean scalar product in $\mathbb{R}^n$. The *orthogonal group* $O(n)$ is defined as the group of all real matrices g that preserve this scalar product, namely those for which

$$\langle ga, gb \rangle = \langle a, b \rangle$$

for any vectors a and b. The group $O(n)$ contains a subgroup $SO(n)$, called the *special orthogonal group*, determined by the equation $\det(g) = 1$. The subgroup $SO(n)$ is the connected component of the identity in $O(n)$. The matrix of an element $g \in O(n)$ is characterized by the equation

$$g\, g^t = I.$$

Thus $SO(n)$ is a compact Lie group of dimension $\frac{1}{2}n(n-1)$. The Lie algebra of $SO(n)$ is the vector space $so(n)$ of all real skew-symmetric $n \times n$ matrices.

Example 13.4.4 (The unitary groups). Let $\langle a, b \rangle$ be the Hermitian scalar product in $\mathbb{C}^n$ defined by

$$\langle a, b \rangle = \operatorname{Re} \sum_{i=1}^{n} a^i \bar{b}^i,$$

where Re is the real part of a complex number and the bar denotes complex conjugation. The *unitary group* $U(n)$ is the group of all complex matrices in

$\mathbb{C}^n$ that preserve this scalar product, namely those for which $\langle ga, gb \rangle = \langle a, b \rangle$. Thus, unitary matrices are determined by the matrix equation

$$g \, \bar{g}^t = I.$$

The group $U(n)$ contains a subgroup $\mathrm{SU}(n)$ consisting of all unitary matrices with unit determinant. $U(n)$ is a compact n^2-dimensional Lie group, while $\mathrm{SU}(n)$ is a compact $(n^2 - 1)$-dimensional Lie group. The Lie algebra $u(n)$ of $U(n)$ is the space of all skew-Hermitian complex matrices, while the Lie algebra $su(n)$ of $\mathrm{SU}(n)$ is the space of all skew-Hermitian matrices with zero trace.

Example 13.4.5 (The symplectic groups). Consider a $2n$-dimensional vector space with a skew-symmetric nondegenerate scalar product (a, b), that is, one for which

$$(a, b) = -(b, a)$$

for any vectors a and b. Such a space is called a *symplectic space*. An example is given on $\mathbb{R}^{2n}$ (with its standard basis) by the scalar product

$$(a, b) = \sum_{i,j} \omega_{ij} a^i b^j,$$

$\Omega = (\omega_{ij})$ being the matrix

$$\begin{pmatrix} 0 & -I \\ I & 0 \end{pmatrix},$$

where I is the $n \times n$ identity matrix. This example is typical in the sense that, for any symplectic structure in a $2n$-dimensional vector space (that is, any scalar product making the space into a symplectic space), we can find a basis in which the scalar product is expressed by

$$(a, b) = \sum_i p^i q'^i - p'^i q^i,$$

where $a = (p^1, \ldots, p^n, q^1, \ldots, q^n)$ and $b = (p'^1, \ldots, p'^n, q'^1, \ldots, q'^n)$: this corresponds exactly to the matrix Ω.

A linear transformation $g : \mathbb{R}^{2n} \to \mathbb{R}^{2n}$ is *symplectic* if it preserves the symplectic structure, that is, if $(ga, gb) = (a, b)$ for any vectors a and b. The set of all symplectic homogeneous transformations of $\mathbb{R}^{2n}$ is called the *real symplectic group* and denoted by $\mathrm{Sp}(n, \mathbb{R})$. It is a noncompact Lie group of dimension $n(2n + 1)$. Assuming the standard symplectic structure given by Ω, the Lie algebra $sp(n, \mathbb{R})$ of $\mathrm{Sp}(n, \mathbb{R})$ consists of the real matrices

$$\begin{pmatrix} X & Y \\ Z & -X^t \end{pmatrix},$$

where X is an arbitrary matrix of order n, while the matrices Y and Z are of order n and symmetric.

Consider the simple and important example of $Sp(1, \mathbb{R})$, which was introduced in Example 13.2.9 under a different guise. The connection of the present definition with the earlier one is that the skew-symmetric product of a pair of vectors a and b in $\mathbb{R}^2$ is equal to the area of the parallelogram spanned by a and b. As we have seen, $Sp(1, \mathbb{R})$ is isomorphic to $SL(2, \mathbb{R})$.

We say that a vector subspace Π of a symplectic space is *isotropic* if it is skew-orthogonal to itself, that is, if the skew-symmetric scalar product of any two vectors in Π is zero. If $\dim \Pi = n$, the isotropic plane is called a *Lagrangian plane*.

The dimension of an isotropic plane in the symplectic space $\mathbb{R}^{2n}$ never exceeds n. A symplectic transformation takes any isotropic plane into an isotropic plane. In particular, the image of a Lagrangian plane under a symplectic transformation is a Lagrangian plane. The determinant of any symplectic transformation is 1.

13.5. Discrete Transformation Groups in Two and Three Dimensions

Lattices.   Lattices are the models for the description of real *crystals*. We shall consider a crystal lattice of points on the plane or in three-space. The crystal is regarded as consisting of a few types of atoms fixed in space and distributed in a regular fashion. We consider a special class of lattices, namely those invariant under certain translations. Suppose that the crystal lattice L contains as a subset the set of all points of the form

$$n_1 A_1 + n_2 A_2 + n_3 A_3,$$

where n_1, n_2, n_3 are arbitrary integers and the A_1, A_2, A_3 are linearly independent vectors, called *primitive vectors* of the lattice. (In the plane case we have only two primitive vectors.)

Consider the basic translations T_1, T_2, T_3 along the vectors A_1, A_2, A_3. Then the general translation generated by lattice can be written as

$$T = n_1 T_1 + n_2 T_2 + n_3 T_3.$$

Definition 13.5.1. The lattice L is called *translation-invariant* if there exist primitive translations T_1, T_2, T_3 such that L is invariant under any translation of the form $T = n_1 T_1 + n_2 T_2 + n_3 T_3$, and under no other translation. The parallelepiped Π generated by the vectors A_1, A_2, A_3 is called a *primitive cell* or *fundamental region* of the lattice (Figure 13.11).

We assume that the primitive translation vectors are such that the volume of the parallelepiped Π is smallest possible.

Figure 13.12 shows a plane lattice. A primitive cell is shaded. In general, there may be other atoms of the lattice inside a fundamental region.

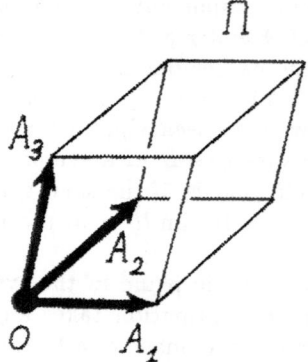

FIGURE 13.11.

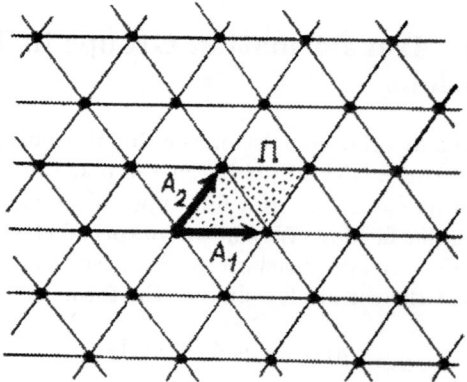

FIGURE 13.12.

Consider the group $E(3)$ of all isometries of a Euclidean three-space, and define $E(2)$ analogously for the plane. Recall that every element g of $E(3)$ or $E(2)$ can be expressed in exactly one way as a product $g = T\alpha$, where T is a translation and α is an orthogonal transformation. Translations do not commute with rotations. Translations form an abelian subgroup, normal in $E(3)$ or $E(2)$, and isomorphic to $\mathbb{R}^3$ or $\mathbb{R}^2$ respectively.

Let L be a lattice. The subgroup $E_3(L) \subset E(3)$ consisting of all isometries preserving L is called the *crystallographic space group* of L. The subgroup $T_3(L)$ consisting of all translations of L is called the *translation group* of L. The *stabilizer* $H_3(T)$ of L is the subgroup of $E_3(L)$ consisting of all motions of L that do not move the origin O. Thus the elements of $H_3(L)$ are orthogonal transformations, and $H_3(L) \subset O(3)$. We denote by $H_3(T)_0$ the

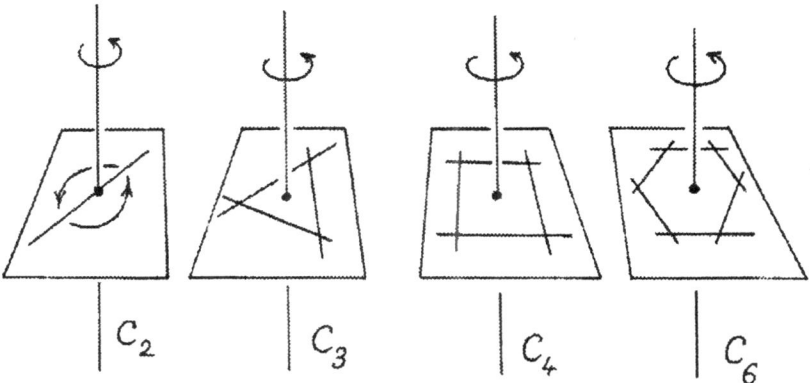

FIGURE 13.13.

subgroup of $H_3(T)$ consisting of rotations (orientation-preserving orthogonal transformations).

Analogous definitions for $E_2(L)$, $T_2(L)$, $H_2(L)$, and $H_2(L)_0$ apply in the plane case.

Two-dimensional lattices. We now list all the possibilities for the stabilizer $H_2(L)_0$, where L is a planar, translation-invariant lattice.

Let C_n be the cyclic group of order n acting on the plane by rotations through multiples of $2\pi/n$ (see Figure 13.13). Note that C_1 is the trivial group. Let D_n be the dihedral group of order $2n$ generated by C_n and a reflection in a line l. Thus D_n is the semidirect product of C_i with the cyclic group $\mathbb{Z}_2$ of order 2.

Theorem 13.5.2 (Classification of stabilizers of plane lattices). *Suppose L is a planar, translation-invariant lattice.*

1. *The possibilities for the proper stabilizer $H_2(L)_0$ are*

$$C_1, C_2, C_3, C_4, C_6.$$

2. *The possibilities for the stabilizer $H_2(L)$ are*

$$C_1, C_2, C_3, C_4, C_6; \quad D_1, D_2, D_3, D_4, D_6.$$

Note that C_5 cannot occur.

It is not difficult to construct for each of these ten groups a plane lattice having that group as its stabilizer. Figure 13.14 gives examples where $H_2(L)_0$ is each of the groups C_2, C_3, C_4, C_6 (and $H_2(L)$ is the corresponding dihedral group).

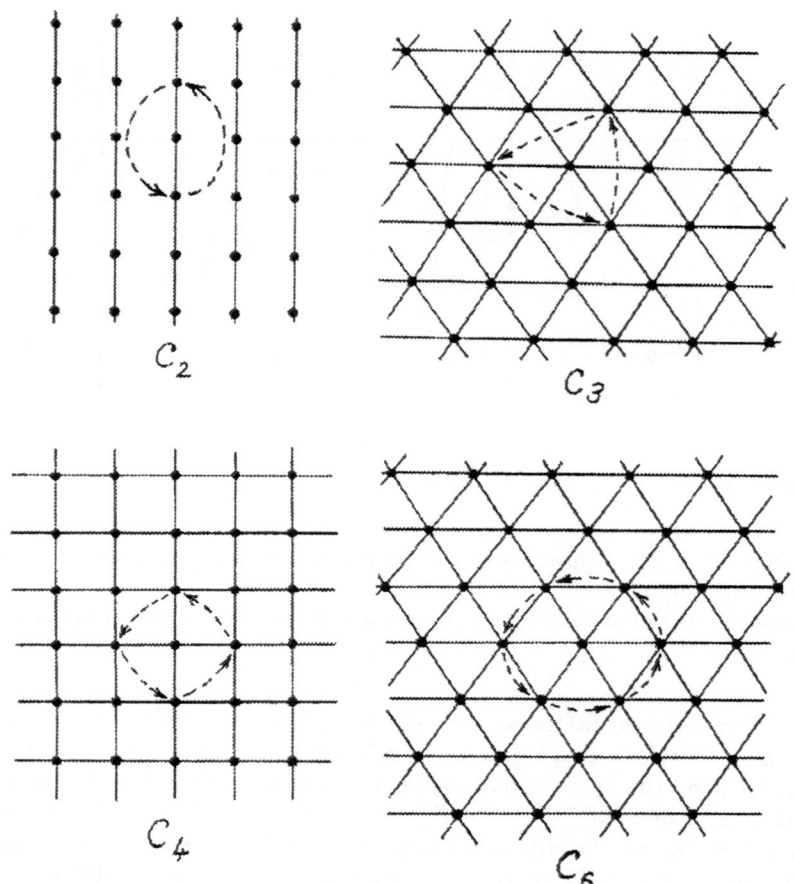

FIGURE 13.14.

Three-dimensional lattices. We now turn to three-dimensional lattices L. We start from the complete list of all possible finite subgroups of rotations of Euclidean three-space, that is, the finite subgroups of $SO(3)$.

C_n. Fix some axis in Euclidean 3-space through the origin O, and denote by Π the plane through O orthogonal to l. As in the two-dimensional case, we denote by C_n the cyclic subgroup generated by a rotation in Π through an angle $2\pi/n$.

D_n. D_n is generated by C_n and a reflection in a line q contained in Π and passing through O (Figure 13.15). A reflection in a line in space is the same as a rotation through π about that line. Thus, orientation-reversing orthogonal transformations of Π are induced by orientation-preserving orthogonal transformations of three-space. D_n consists of the rotations in C_n, together with n reflections in axes lying in Π and passing through O; these axes are obtained from q by π/n rotations.

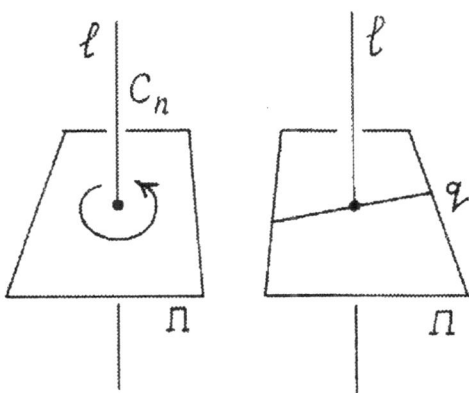

FIGURE 13.15.

The three exceptional finite groups T, W and P. Apart from these two infinite sequences of subgroups, there are a few more exotic finite subgroups of SO(3). They arise is connection with the five regular, or *platonic*, polyhedra (Figure 13.16), namely, the tetrahedron, the cube, the octahedron, the dodecahedron, and the icosahedron.

The geometrical structure of these polyhedra is described in the following table, where the column "Valency" indicates how many faces or edges meet at a vertex:

Name	Type of face	Vertices	Edges	Faces	Valency
Tetrahedron	Triangle	4	6	4	3
Octahedron	Triangle	6	12	8	4
Icosahedron	Triangle	12	30	20	5
Cube	Square	8	12	6	3
Dodecahedron	Pentagon	20	30	12	3

It is interesting to look at these polyhedra *from inside*. If we remove one of the faces and look at the interior of the polyhedron through this hole, we obtain the pictures shown in Figure 13.17.

Each of these polyhedra (imagined with its center at O) has a *symmetry group*, the group of isometries of $\mathbb{R}^3$ that take the polyhedron to itself. The cube and octahedron have isomorphic symmetry groups, as do the dodecahedron and the icosahedron. To see this, for example in the case of the cube

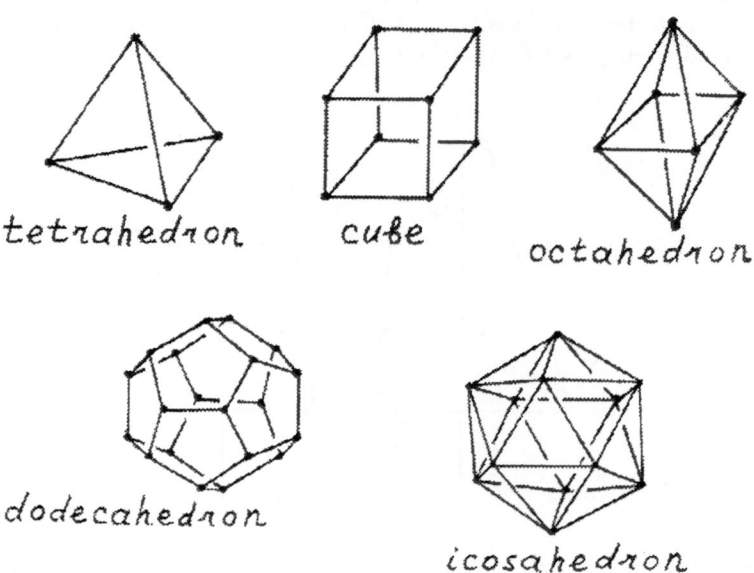

FIGURE 13.16.

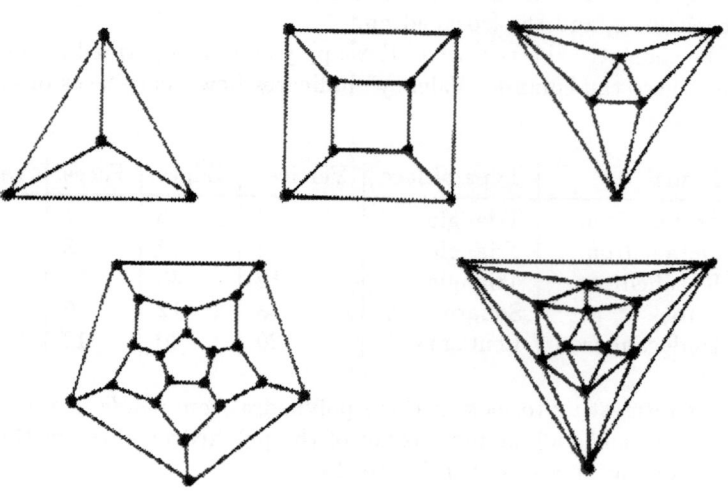

FIGURE 13.17.

and octahedron, consider a sphere inscribed in a cube, and inside the sphere an octahedron whose vertices are at the tangency points of the sphere with the cube—that is, at the centers of the faces of the cube (Figure 13.18). It is clear that any symmetry of the cube is also a symmetry of the octahedron, and conversely. Entirely similar considerations show that the icosahedral and dodecahedral groups are isomorphic.

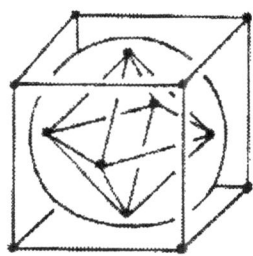

FIGURE 13.18.

We set

T = group of orientation-preserving symmetries of the tetrahedron,
W = group of orientation-preserving symmetries of the octahedron,
P = group of orientation-preserving symmetries of the icosahedron.

These groups are called the (proper) *tetrahedral group, octahedral group*, and *icosahedral group*. Their orders are 12, 24, and 60, respectively. They are not isomorphic to any C_n or D_n.

It turns out that the above groups of rotations of Euclidean space exhaust the possibilities for finite rotation groups in $\mathbb{R}^3$: *Every finite subgroup of the rotation group* SO(3) *of Euclidean three-space is isomorphic to one of C_n. D_n, T. W. and P.*

We now turn to finite subgroups $O(3)$, that is, groups of orthogonal transformations of three-space, regardless of orientation. If H is one of C_n, D_n, T, W, and P, we define $\bar{H}$ as the group obtained from H by adjunction of the reflection in the origin. This reflection is given by the matrix $-I$, where I is the identity, and is therefore orientation-reversing. $\bar{H}$ is a semidirect product of H with $\mathbb{Z}_2 = \{I, -I\}$. The groups $\bar{T}$, $\bar{W}$, and $\bar{P}$ are called the *full tetrahedral, octahedral*, and *icosahedral groups*. respectively. They can alternatively be obtained by adjoining to T, W, and P a plane reflection that fixes the relevant polyhedron. Their orders are 24, 48 and 120.

One more construction is needed to conclude the list of possible finite subgroups of $O(3)$. Let Γ be a subgroup of index 2 in a group $\Phi \subset$ SO(3). Let $S = \Phi - \Gamma$, and replace each element of S by its product with $-I$, the reflection in the origin. The result is a new group $\Phi\Gamma$ of orthogonal transformations in $\mathbb{R}^3$, isomorphic as an abstract group to the initial group Φ. Half of its transformations come from Γ, and the other half is made up of plane reflections. For example, we can form a new group WT since the proper tetrahedral group T occurs as a subgroup of index two in the proper octahedral group W.

Theorem 13.5.3. *Here is the complete list of all possible stabilizers of three-dimensional, translation-invariant lattices:*

$$C_1, C_2, C_3, C_4, C_6,$$
$$\bar{C}_1, \bar{C}_2, \bar{C}_3, \bar{C}_4, \bar{C}_6,$$
$$D_2, D_3, D_4, D_6,$$
$$\bar{D}_2, \bar{D}_3, \bar{D}_4, \bar{D}_6,$$
$$C_2 C_1, C_4 C_2, C_6 C_3,$$
$$D_4 D_2, D_6 D_3. D_2 C_2, D_3 C_3, D_4 C_4, D_6 C_6,$$
$$T, W, \bar{T}, \bar{W}, WT.$$

Each of these groups is realized as the stabilizer of some translation-invariant, three-dimensional lattice. Other finite subgroups of $O(3)$, such as P, cannot occur.

Part II

Advanced Subjects

List of Contributors

Elena V. Anoshkina
University of Aizu, Graduate School
Tsuruga Ikki-machi, Aizu-Wakamatsu City,
Fukushima, 965-80 Japan

Alexander G. Belyaev
Associate Professor
University of Aizu, Graduate School
Tsuruga Ikki-machi, Aizu-Wakamatsu City,
Fukushima, 965-80 Japan

Alexey V. Bolsinov,
Differential Geometry and Applications,
Faculty of Mathematics and Mechanics
Moscow State University
Moscow 119 889, Russia

Anatolij T. Fomenko
Professor of Differential Geometry and Applications
Faculty of Mathematics and Mechanics
Moscow State University
Moscow 119 889, Russia

Tosiyasu L. Kunii
Director, Laboratory of Digital Art and Technology
1-25-21-602, Hongo, Bunkyo-ku, Tokyo, 113 Japan
and
Chairman of Digital Planet Institute
Senior Partner of MONOLITH Co. Ltd.
1-7-3 Azabu-Juban, Minato-ku, Tokyo, 106 Japan

Sergei V. Matveev
Department of Mathematics
Chelyabinsk State University
ul. Molodogvardeitsev, 70B
454136 Chelyabinsk
Russia

Olga E. Orël
Differential Geometry and Applications
Department of Mathematics and Mechanics
Moscow State University
Moscow 119 889, Russia

Shigeo Takahashi
Assistant Professor
Department of Computer Science
Gunma University
Kiryu, Gunma, 376 Japan

14. Computers and Visualization in Hyperbolic Three-Dimensional Geometry and Topology

A. T. Fomenko and S. V. Matveev

14.1. Introduction

This chapter is devoted to a new branch of modern geometry that is the result of interaction between hyperbolic topology, computer geometry, the theory of three-dimensional manifolds, and the new theory of topological classification of integrable Hamiltonian systems of differential equations. This activity was initiated with the work of S. V. Matveev and A. T. Fomenko.

We give here a brief survey of these results to show the new possibilities of application of the modern "computational thinking" to some deep problems in modern and classical geometry and topology. We hope to give a simple description of these ideas, and to suggest their further development to students and professional mathematicians in the area of computer geometry who want to extend the field of application of their computational methods. As a result, they may be able to obtain new theoretical results in a nontrivial field of mathematics with applications to mathematical physics. We will formulate several open problems that can certainly be solved by modern powerful methods of computer geometry.

The reader who wants to penetrate more deeply into this field can refer to [Fomenko 1988; 1989; 1992; Matveev and Fomenko 1988; Thurston 1981; 1982; Seifert and Weber 1933; Adams 1983; Matveev 1975].

Here are the basic problems of computer geometry that we will discuss. Details are given in the following sections.

1) Continue the computer calculation of the volumes of closed three-dimensional hyperbolic manifolds.
2) Continue the algorithmical classification and description of three-dimensional manifolds. Visualize the "first simplest three-manifolds."
3) Solve the problem of minimal hyperbolic volume for closed hyperbolic three-manifolds. Visualize the three-manifolds of a minimal hyperbolic volume.
4) Continue the computer classification of integrable Hamiltonian systems of differential equations with two degrees of freedom.

14.2. Two-Dimensional Hyperbolic Geometry

We start with some elementary notions of two-dimensional hyperbolic geometry, to recall some necessary ideas.

Consider the circle of radius 1 in the Euclidean plane $\mathbb{R}^2$, and the open disc bounded by this circle (Figure 14.1). We will consider a new geometry in the disc. given as follows: points of this geometry are ordinary points of the disc (except boundary points), and "straight lines", henceforth called *hyperbolic lines*, are circular arcs intersecting the unit circle at right angles. In particular, all diameters of the disc are hyperbolic lines, for these diameters can be considered as arcs of infinite radius. The geometry thus defined is called *Lobachevskian* or *hyperbolic geometry*, and the space is called the *hyperbolic plane*. Its model as a unit disc in $\mathbb{R}^2$, as just given, is called the *Poincaré model* of the hyperbolic plane. The boundary of the Poincaré model, that is, the circle $x^2 + y^2 = 1$, is called the *absolute*, or the *circle at infinity*; a point on it is not properly part of the hyperbolic plane, but is often said to be "at infinity", that is, infinitely distant from the points of the hyperbolic plane.

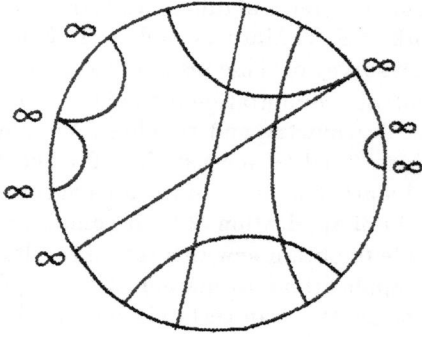

FIGURE 14.1.

Using the Poincaré model, we can easily verify (see Figure 14.1) that, given a hyperbolic line and a point not on it, there are infinitely many other hyperbolic lines going through the point and not intersecting the given line. These lines are said to be *parallel* to the given line.

Consider the standard polar coordinates (r, φ) in the unit disk (Figure 14.2). Then the Lobachevskian metric has the form

$$ds^2 = \frac{dr^2 + r^2 d\varphi^2}{(1 - r^2)^2}.$$

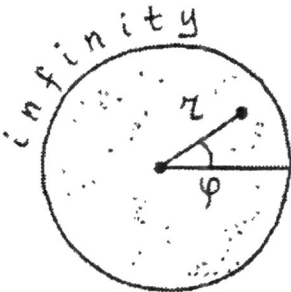

FIGURE 14.2.

If we introduce the complex coordinate $z = u + iv$ on the unit Euclidean disk, this metric can be rewritten in the equivalent form

$$ds^2 = \frac{dz\,d\bar{z}}{(1 - |z|^2)^2}.$$

There exists another useful representation of the Lobachevskian plane, namely, as the upper half-plane. Consider another copy of the Euclidean plane, introduce a new complex coordinate $w = x + iy$; the upper half-plane is the set of points satisfying $y > 0$. Now take the following mapping from the upper half-plane onto the unit disk:

$$z = \frac{1 + iw}{1 - iw}.$$

This mapping is a diffeomorphism (Figure 14.3). Consequently, we can consider the hyperbolic metric in the new coordinates (x, y) on the upper-plane using this mapping as a change of regular coordinates. It is easy to calculate that the Lobachevskian metric on the upper half-plane has the form

$$ds^2 = \frac{dx^2 + dy^2}{y^2}.$$

It is an important property of both the Poincaré and the upper half-plane models that they are *conformal*, that is, the hyperbolic angle between two tangent vectors at a point is the same as the Euclidean angle in the model.

We can now describe the group of isometries of the hyperbolic plane. Consider, for simplicity, only isometries that preserve orientation. It turns out that all such isometries can be represented as transformations on the upper half-plane having the form

$$w \to \frac{aw + b}{cw + d},$$

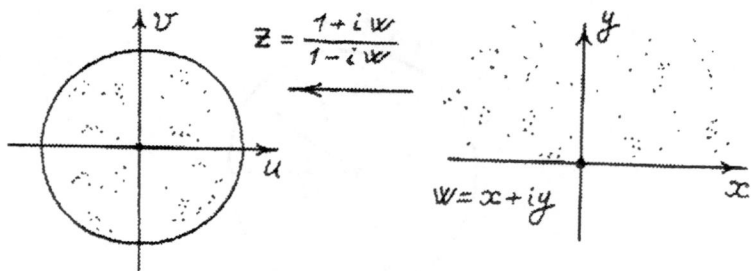

FIGURE 14.3.

where a, b, c. and d are real numbers and $ad - bc = 1$. (You can check that all these transformations map the upper half-plane onto itself). We can represent all these transformations by the matrices

$$\begin{pmatrix} a & b \\ c & d \end{pmatrix},$$

where $ad - bc = 1$. These matrices form the matrix group SL(2, R). The group of orientation-preserving isometries of the hyperbolic plane is isomorphic to the quotient group $SL(2, \mathbb{R})/\mathbb{Z}_2$, where the normal subgroup $\mathbb{Z}_2$ consists of I (the identity matrix) and $-I$.

Some essential properties of the hyperbolic plane can be concretely modeled on a certain surface in three-dimensional Euclidean space, called the *Beltrami surface*, or *pseudosphere*. To define this surface, we proceed as follows.

Consider on the plane (x, y) a *tractrix*, that is, a smooth curve γ characterized by the property that the segment of the tangent lying between the point of tangency and the x-axis has *constant length* a (Figure 14.4). When the point A slides along the curve γ, the point B slides along the x-axis. and the segment AB has length a. The mechanical interpretation: Think of connecting A and B by an inelastic thread of length a, and move the point B along the x-axis. starting from the position $O = B_0$. As it moves, B drags the point A along, starting from the initial position A_0 shown in Figure 14.4. The point A will then draw a certain curve tangent to the y-axis at the point A_0. having the x-axis as an asymptote. The segment AB is tangent to this curve (at the point A. which slides along the curve).

We can easily find the differential equation for γ. From the triangle ABx (Figure 14.4) we have: $\tan \varphi = -y'_x$, where $y = y(x)$ is the graph of γ and $a \sin \varphi = y$. From this we obtain

$$\sin \varphi = \frac{y'_x}{(1 + (y'_x)^2)^{1/2}} \quad \text{or} \quad x'_y = -\frac{(a^2 - y^2)^{1/2}}{y},$$

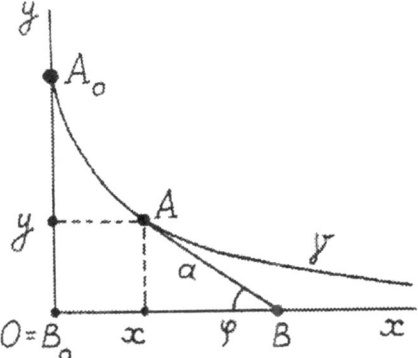

FIGURE 14.4.

where $x = x(y)$ is the graph of γ. Therefore,

$$x(y) = -(a^2 - y^2)^{1/2} + \frac{a}{2} \ln\left(\frac{a + (a^2 - y^2)^{1/2}}{a - (a^2 - y^2)^{1/2}} \right).$$

Thus, we have derived an explicit expression for the curve $x = x(y)$. We now consider the surface of revolution formed by rotating γ about the horizontal x-axis. We obtain a surface M, which is the promised *Beltrami surface* or *pseudo-sphere* (Figure 14.5).

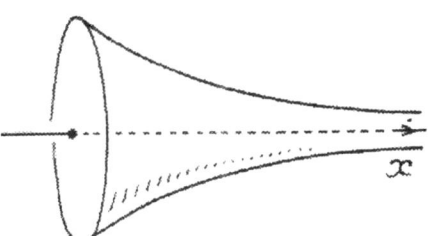

FIGURE 14.5.

Recall that any Riemannian metric on a two-dimensional manifold has a remarkable scalar invariant K, called *Gaussian curvature*. This curvature is completely determined only by the metric itself, and does not depend on the concrete isometric embedding of the surface in ambient Euclidean three-space. But sometimes it is convenient to calculate the Gaussian curvature K in terms of such an embedding. If λ_1 and λ_2 are the principal curvatures at the point P on M, then $K = \lambda_1 \lambda_2$.

It can be easily proved that *the Beltrami surface has a constant negative Gaussian curvature $K = -1/a^2$.*

The Beltrami surface has the boundary, a circle with radius a. It can be shown (a nontrivial fact!) that the surface cannot be continued beyond this boundary circle without violating the condition $K = -1/a^2 < 0$.

It can be proved that *the Riemannian metric on the Beltrami surface induced by the ambient Euclidean metric coincides with the hyperbolic metric.* In other words, *the Beltrami surface is locally isometric to the Lobachevskian plane.*

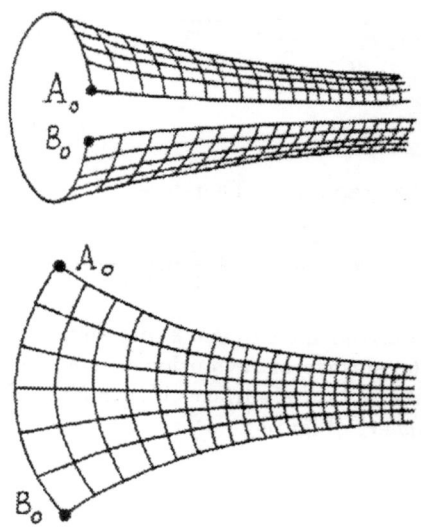

FIGURE 14.6.

This means that we have constructed an isometric embedding of a certain region of the hyperbolic plane into three-dimensional Euclidean space. What particular part of the hyperbolic plane admits such an isometric embedding? To answer this question, cut a Beltrami surface along any of its generators (Figure 14.6). The result is a surface that admits an isometric embedding in the hyperbolic plane, whose image is the region illustrated in Figure 14.7. This region has the form of a curvilinear rectangle with vertices ∞, A_0, B_0. The sides of this triangle are formed by two parallel hyperbolic lines coming from one point at infinity, which we denote by ∞. The third side of the triangle is the arc $A_0.B_0$ with length equal to $2\pi a$. This arc is a portion of the Euclidean circle tangent to the boundary of the Poincaré model at the point ∞. Consequently, the region (∞, A_0, B_0) is an infinite band lying between two parallel hyperbolic lines on the hyperbolic plane, and limited on one side by the arc A_0, B_0. We say it is an *ideal triangle* (this means a triangle where one or more vertices are at infinity).

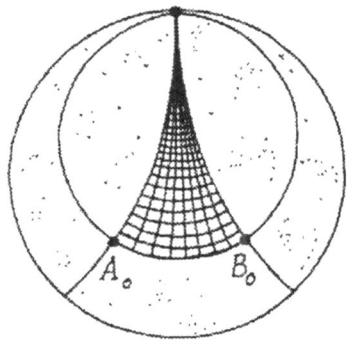

FIGURE 14.7.

The considerations above give rise to the natural question of whether or not the whole hyperbolic plane (and not only a part of it, such as the band described above) can be isometrically realized in three-dimensional Euclidean space in the form of a smooth two-dimensional surface of constant negative curvature. The answer is negative, as was proved by D. Hilbert. Later, this theorem was generalized by N. V. Efimov, who showed that a plane endowed with an arbitrary *complete* smooth Riemannian metric with a negative curvature restricted from above to a negative number has no global isometric embedding into three-dimensional Euclidean space.

The hyperbolic plane is closely connected with closed orientable surfaces of genus $g > 1$ (Figure 14.8). Any such surface can be endowed with a hyperbolic Riemannian metric, and can thus be made *locally* isometric to the hyperbolic plane. Equivalently, any such surface can be represented as the quotient space of the hyperbolic plane by some discrete group of orientation-preserving isometries.

Definition 14.2.1. Let Γ be a discrete group of orientation-preserving isometries of the hyperbolic plane. A subset D of the hyperbolic plane is called a *fundamental region* for Γ if:

1. D is a closed set;
2. the sets $\gamma(D)$, for $\gamma \in \Gamma$, cover the whole hyperbolic plane;
3. this covering of the hyperbolic plane by the sets $\gamma(D)$ is such that a sufficiently small neighborhood of an arbitrary point intersects only a finite number of sets $\gamma(D)$;
4. the images $\gamma(\overset{\circ}{D})$, for $\gamma \in \Gamma$, are pairwise disjoint, where $\overset{\circ}{D}$ is the interior of D.

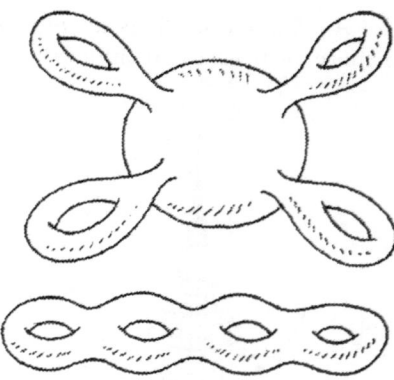

FIGURE 14.8.

We now give an example of a discrete isometry group of the hyperbolic plane whose fundamental region D is a regular $4g$-gon centered at the center of the unit disc in the Poincaré model, having interior angles equal to $\pi/(2g)$ (Figure 14.9). That such a polygon exists is clear, because a very small regular $4g$-gon around the center of the Poincaré disk has interior angles almost the same as those of its Euclidean counterpart, namely $\pi(1 - 1/(4g))$, whereas a very large polygon, with vertices almost at infinity, has angles very close to zero. For an appropriately chosen size in between, therefore, the interior angles equal $\pi/(2g)$.

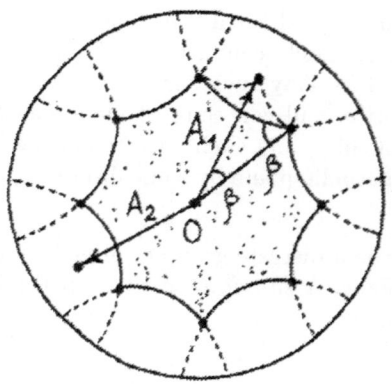

FIGURE 14.9.

We divide the sides of the $4g$-gon into pairs of opposite sides. Suppose $A_1, \ldots, A_{2g}$ are "translations" of the hyperbolic plane under which one member of a pair is mapped to the opposite (Figure 14.9). All these translations are conjugate: A_{i+1} is obtained from A_i by conjugating with a rotation through $\pi(1 - /(2g))$ about the origin. It can be verified that the transformations $A_1, \ldots, A_{2g}$ are linked by the relation

$$A_1 A_2 \ldots A_{2g} A_1^{-1} A_2^{-1} \ldots A_{2g}^{-1} = 1.$$

From this we can derive explicit formulas for the matrices of $A_1, \ldots, A_{2g}$ in the upper half-plane, that is, for the corresponding elements of $SL(2, \mathbb{R})$. We do this as follows. We may assume that A_1 (in the upper half-plane model) sends the imaginary semi-axis to itself. Then it has the form $w \to \lambda w$ for $\lambda = e^a$, where a is twice the leg of an isosceles right triangle with angles $\pi/2$, $\pi/(4g)$, $\pi/(4g)$; see Figure 14.9. Computation shows that

$$A_1 = \begin{pmatrix} \dfrac{\cos \beta + (\cos \beta)^{1/2}}{\sin \beta} & 0 \\ 0 & \dfrac{\sin \beta}{\cos \beta + (\cos \beta)^{1/2}} \end{pmatrix}, \quad \text{where} \quad \beta = \frac{\pi}{4g}.$$

The matrices $A_2, \ldots, A_{2g}$ are obtained from A_1 by conjugation, and $A_k = B_g^{-k+1} A_1 B_g^{k-1}$, where B_g is the matrix of rotation through an angle $\pi(2g-1)/2g$ around the point i (on the upper half-plane). Explicitly,

$$B_g = \begin{pmatrix} \cos \alpha & \sin \alpha \\ -\sin \alpha & \cos \alpha \end{pmatrix}, \quad \text{where} \quad \alpha = \frac{\pi(2g - 1)}{4g}.$$

Denote by Π_g the discrete transformation group generated by the matrices $A_1, \ldots, A_{2g}$, and by L^2 the hyperbolic plane. Consider the quotient space L^2/Π_g of L^2 by the action of Π_g, which is a closed smooth surface M_g. This surface is closed, orientable, and has genus g, because of the way the sides of the fundamental polygon are identified (Figure 14.8).

Theorem 14.2.2. *The quotient $M_g = L^2/\Pi_g$ is a surface of genus g, whose fundamental group is isomorphic to Π_g. The projection $L^2 \to L^2/\Pi_g$ induces on M_g a Riemannian metric of constant negative curvature. (When M_g is given this metric, the projection is by definition a local isometry; it is also a covering map, of infinite multiplicity).*

In particular, any oriented surface of genus $g > 1$ has a Riemannian metric of constant negative curvature. Conversely, any closed smooth compact surface with a complete Riemannian metric of constant negative curvature has genus $g > 1$, and can be obtained as a quotient of the hyperbolic plane L^2 by a group Γ of orientation-preserving isometries, where Γ is isomorphic to Π_g.

We now calculate the *area* of a compact hyperbolic surface. We can assume that its Gaussian curvature is normalized, that is, equal to -1. We can use the following celebrated result of Gauss and Bonnet, valid for any Riemannian metric (not necessarily hyperbolic) on a surface:

The integral over a closed orientable Riemannian surface of the Gaussian curvature K is connected with the Euler characteristic of the surface by the formula:

$$\chi = 2 - 2g = \frac{1}{2\pi} \int K d\sigma.$$

Since in our case we have assumed $K = -1$, we immediately obtain $\int K d\sigma = \int(-1)d\sigma = -$ area, and therefore the area is $4\pi(g-1)$, where $g \geq 1$ is the genus. (The sphere and the torus, which have genus 0 and 1, respectively, cannot be given a hyperbolic metric.)

Corollary 14.2.3. *The area of a closed hyperbolic surface is a topological invariant: two compact hyperbolic surfaces are homeomorphic if and only if their areas coincide.*

14.3.   Three-Dimensional Hyperbolic Geometry

Definition 14.3.1. A smooth n-manifold M is called a *hyperbolic* if it has a complete Riemannian metric with constant negative sectional curvature (by scaling, we can assume that the curvature is -1).

We can consider the *volume* of a hyperbolic manifold (with respect to the given hyperbolic metric). The most interesting case for us is that of compact hyperbolic manifolds, which have *finite* hyperbolic volume. (There are also noncompact hyperbolic manifolds of finite volume).

Examples of three-dimensional hyperbolic manifolds can be obtained from the three-dimensional Lobachevskian metric, which is defined in a way analogous to the two-dimensional case.

Three-dimensional hyperbolic geometry is an area of growing interest, and has been investigated by many well-known mathematicians, such as W. P. Thurston, J. W. Milnor, G. D. Mostov, M. Gromov, T. Jorgensen, G. A. Margulis, E. B. Vinberg, V. S. Makarov, L. S. Krushkal, J. R. Weeks, C. C. Adams, W. D. Neuman, D. Zagier, A. Marden, L. V. Ahlfors, B. N. Apanasov, I. S. Guzul, H. B. Lawson, R. Meyerhoff, J. W. Morgan, V. V. Nikulin, D. Sullivan, and many others.

The group of isometries of three-dimensional hyperbolic space L^3 contains many discrete subgroups. Consequently, as in the two-dimensional case, we obtain the quotient manifolds L^3/Π, which are compact smooth closed hyperbolic manifolds.

It follows from a theorem of Mostow that two complete hyperbolic manifolds of finite hyperbolic volume are homeomorphic if and only if they are homotopically equivalent.

Now consider the set of all complete hyperbolic three-manifolds of finite volume; the volumes of such manifolds form a set of real numbers. It turns out that this set is distributed on the real line in a very interesting way (see Figure 14.10):

Theorem 14.3.2 (Jorgensen, Gromov). *The set of volumes of all complete hyperbolic three-manifolds is totally ordered and has ordinal type ω^ω. There exist only finitely many different hyperbolic three-manifolds with the same volume.*

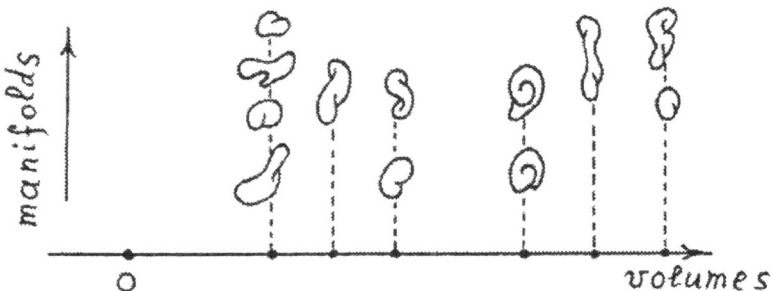

FIGURE 14.10.

Thus, the volume of a hyperbolic three-manifold is an "almost-invariant". Consequently the calculation of hyperbolic volumes is an extremely interesting problem of hyperbolic and computer geometry. It is still an open problem how to determine the list of closed hyperbolic manifolds having a given volume, or even what are the closed hyperbolic manifolds of least volume. Nonetheless, there are algorithms to compute the hyperbolic volume of a given manifold (given by a gluing of polyhedra), and computer programs such as Snappea, by Jeff Weeks (see [Hodgson 1994], for example) have made it possible to catalog and study a tremendous number of interesting hyperbolic manifolds.

14.4. The Complexity of Three-Manifolds

In this section we give a short review of Matveev's theory of complexity. We associate with each compact three-manifold M a nonnegative integer $d(M)$, called its *complexity*. This number has many useful properties and, as we shall see below, conforms closely to the complexity of the three-manifold in an intuitive sense. To describe the complexity $d(M)$ we shall need two notions: *spines* and *almost special polyhedra*. A polyhedron Q *collapses to* a subpolyhedron P if, for any triangulation (K, L) of (Q, P), we can transform K to L by a finite sequence of elementary simplicial collapses, where each elementary simplicial collapse consists in removing a principal open simplex together with a free open face of that simplex (Figure 14.11).

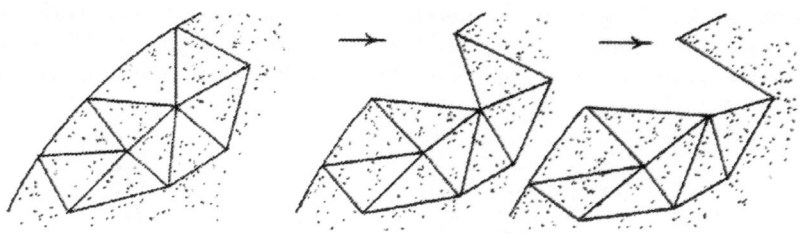

FIGURE 14.11.

Definition 14.4.1. A polyhedron $P \subset \overset{\circ}{M}$ is called a *spine* of the compact manifold M if M (or M with an open ball removed if $M = \varnothing$) collapses onto P.

A spine carries a lot of information about the structure of the manifold, but in general a manifold cannot be uniquely recovered (reconstructed) from its spine. It is useful to bear in mind the following equivalent definition of spine: P is a *spine* of M if M is homeomorphic to the mapping cylinder of a map $p : M \to P$. Let Δ denote the one-skeleton of the standard three-dimensional simplex, that is, a circle with three radii.

Definition 14.4.2. A compact polyhedron P is called *almost special* if each of its points has a neighborhood homeomorphic to the cone on a subpolyhedron of Δ. Points whose links are homeomorphic to Δ are called *vertices* of P.

The singularities that can be seen in the cone $\operatorname{Con}\Delta$ are stable in nature: They do not disappear with small changes in the attaching maps of cells, for any representation of the polyhedron as a cell complex. It is interesting to note that precisely such singularities arise in the theory of minimal surfaces (soap films). It follows easily from the definition of an almost special polyhedron that this property is hereditary, that is, it is preserved on passing to subpolyhedra.

Definition 14.4.3. The *complexity* $d(M)$ of a compact three-manifold M is the smallest number of vertices of an almost special spine of M.

The following theorems were proved by S. V. Matveev.

Theorem 14.4.4 (Finiteness property). *For any integer $k \geq 0$, there exist only finitely many distinct closed irreducible three-dimensional manifolds of complexity k.*

Theorem 14.4.5 (Additive property). *For any compact manifolds M_1 and M_2 we have*
$$d(M_1 \# M_2) = d(M_1) + d(M_2).$$

Experience shows that an almost special spine can easily be constructed for any given manifold (see, for example, [Matveev 1974]). Thus there is usually no difficulty in finding an upper bound for the complexity. Finding a nontrivial lower bound is a difficult problem. There is as yet no known algorithm for computing the complexity.

Definition 14.4.6. A compact polyhedron P is called *special* if the link of each of its points is homeomorphic to a circle or to a circle with two or three radii, and if each connected component of its set of nonsingular points (that is, points whose links are homeomorphic to a circle) is homeomorphic to an open disk.

The set of singular points of a special polyhedron P is a regular graph of degree four. The components of its complement (the open disks of the definition) are called the two-components of P, and the graph itself is called the *singular graph* of P. We shall require the following important fact. In almost all interesting cases, a minimal almost special spine of a closed three-manifold is special. More precisely:

Theorem 14.4.7 (S. V. Matveev). *If a closed orientable irreducible three-manifold M is distinct from S^3 and $\mathbb{RP}^3$, then a minimal almost special spine of M is special.*

A three-manifold can be uniquely recovered from a special spine. There is a close connection between the complexity of an orientable three-manifold and the number of three-simplexes that have to be glued together to construct it (not to be confused with the number of three-simplexes in a triangulation!). We take a collection of three-simplexes, partition the faces into pairs, and identify the faces in each pair via one of the three possible orientation-reversing linear homeomorphisms. If we remove regular neighborhoods of the vertices (the images of the vertices of the simplexes) from the resulting polyhedron. then we obtain a compact orientable manifold M. In other words, M is obtained by gluing together three-simplexes with truncated vertices.

Proposition 14.4.8 (S. V. Matveev). *A compact orientable three-manifold M with boundary can be obtained by gluing n simplexes with truncated vertices if and only if it has a special spine with n vertices.*

The proof is based on the following observation: if we consider the standard three-simplex as a simplicial complex, and take the two-skeleton of its dual cell decomposition, we obtain a polyhedron homeomorphic to Con Δ. Under gluing of simplexes. these polyhedra combine to form a special spine of M. Conversely, if the vertices of a special spine are replaced by simplexes with truncated vertices. and these simplexes are glued together as dictated by the spine. we obtain the manifold M.

14.5. Constant-Energy Sets of Integrable Hamiltonian Systems Cannot be Hyperbolic

We now cite some results from [Fomenko 1992] on the computer geometry of integrable Hamiltonian differential equations.

Define (H) as the class of all compact orientable three-manifolds that are constant-energy sets (sets of constant energy) of integrable Hamiltonian systems of differential equations with two degrees of freedom. For a more detailed definition, see [Fomenko 1989; 1992].

A compact orientable three-manifold lies in (H) if and only if it can be obtained by gluing together copies of $D^2 \times S^1$ and $N^2 \times S^1$ via homeomorphisms between some of their boundary components. Here D^2 is a two-dimensional disk and N^2 is a two-dimensional disk with two holes (Figure 14.12).

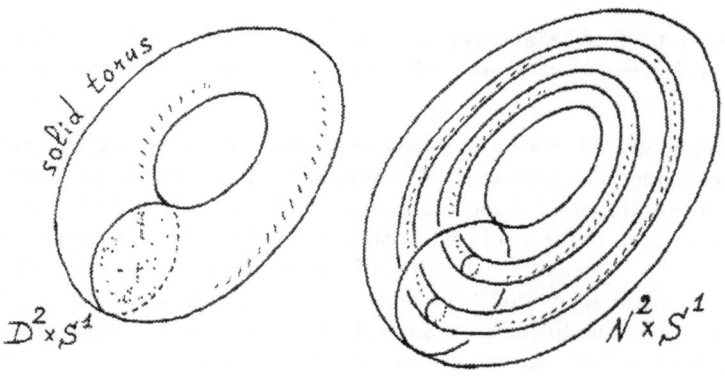

FIGURE 14.12.

Following Thurston [1981], we say that a three-manifold has a *geometric structure* if it admits a complete locally homogeneous Riemannian metric. There exist eight types of such geometric structures: three are isotropic (spherical, Euclidean, and hyperbolic), two are based on products ($S^2 \times \mathbb{R}$ and $L^2 \times \mathbb{R}$), and three are based on twisted products ($SL_2(\mathbb{R})$, Nil, and Sol). A manifold that admits one type of geometric structure cannot have any other type.

Thurston's *geometrization conjecture* asserts that the interior of every compact three-manifold admits a canonical decomposition into pieces having geometric structures. It turns out that the class (H) is closely connected with seven of these types of geometries.

Theorem 14.5.1 (S. V. Matveev, A. T. Fomenko). *A compact orientable three-manifold with (possibly empty) torus boundary belongs to the class* (H) *of constant-energy surfaces of integrable Hamiltonian systems if and only if its* *interior admits a canonical decomposition into pieces having geometries of any of the types of geometry, except hyperbolic. In particular, (H) contains no hyperbolic manifolds.*

One direction of the proof can be found in [Matveev et al. 1989], where the topological structure of manifolds in (H) is described. The other direction follows from [Scott 1983], where geometric manifolds with structures other than hyperbolic are classified.

14.6. The Computer Classification of Manifolds of Small Complexity

As we know, the algorithmic classification of the compact connected surfaces is well-known and very simple: the topological type of the surface is determined by its orientability and its Euler characteristic (or genus). Thus, there is an effective computer procedure that can recognize the topological type of a surface given as a gluing of triangles. In other words, the computer recognition and algorithmical classification of surfaces is a solved mathematical problem.

The next step is the algorithmical classification of three-dimensional manifolds. This problem is much harder and is not yet solved. Many researchers, in both pure and applied mathematics, have worked toward its solution, and many important results have been proved. For example, today there are effective and simple computer algorithms that allow one to recognize the simplest three-manifolds (in different subclasses inside the class of all three-manifolds).

An exposition of the theory of hyperbolic manifolds is given by Thurston [1981; 1982], and in particular a procedure for effectively constructing a large series of hyperbolic three-manifolds. Particular examples of hyperbolic manifolds had already been constructed by Seifert and Weber in 1933, but until recently there have been no very large known classes of *closed* hyperbolic manifolds. The construction of closed hyperbolic three-manifolds is far more complicated than that of noncompact three-manifolds. See details in [Matveev and Fomenko 1988], where we review the main ideas for constructing hyperbolic manifolds and for calculating their volume.

We have approached the problem of understanding the "simplest" three-manifolds from a fundamentally different point of view, using a combination of Fomenko's topological theory of classification of the integrable Hamiltonian systems [1988; 1989; 1992] and Matveev's complexity theory for three-dimensional manifolds [Matveev and Fomenko 1988; Matveev 1975].

The finiteness property of the complexity stated above—or, more precisely, a constructive proof of this property—enables one to organize a search for all closed irreducible three-manifolds of complexity not exceeding a given number d. We present here some of the results from such a search, carried out by computer.

It turns out that the three-manifolds of lowest complexity lie in the class (H) introduced in the preceding section (and therefore, by Theorem 14.5.1, cannot be hyperbolic); and that the "first two" three-manifolds that do not belong to the class (H) are remarkable hyperbolic manifolds. The first of them was discovered by Thurston, and the second by Weeks (and independently by Fomenko and Matveev). More detailed statements will be given later.

Theorem 14.6.1 (S. V. Matveev). *The number $N(d)$ of closed orientable irreducible three-manifolds of complexity d, for $d \leq 6$, is given by the following table (the result of computer calculations):*

d	0	1	2	3	4	5	6
$N(D)$	3	2	4	7	14	31	74

We now list all closed irreducible orientable three-manifolds of complexity $d \leq 4$: (see [Matveev and Fomenko 1988] for the cases $d = 5, 6$. The symbols Q_8, Q_{12}, Q_{16}, Q_{24}, and P_{24} stand for certain discrete subgroups of isometries of S^3.

- Complexity 0: The sphere S^3, the projective space RP^3, the *lens space* $L_{3,1}$.
- Complexity 1: The two lens spaces $L_{4,1}$ and $L_{5,2}$.
- Complexity 2: The quotient space S^3/Q_8 and three other lens spaces.
- Complexity 3: The quotient S^3/Q_{12} and six other lens spaces.
- Complexity 4: The quotients S^3/Q_{16}, S^3/Q_{24}, S^3/P_{24}, $S^3/(Q_8 \times \mathbb{Z}_3)$, and ten other lens spaces.

Unsolved problem of computer geometry. Describe the distribution of the number of three-manifolds for greater values of complexity.

An analysis of the manifolds produced by the computer has enabled us to establish the following facts:

1) All the manifolds of complexity ≤ 5 and most of those of complexity 6 are elliptic.
2) All six Euclidean three-manifolds have complexity 6.
3) The remaining six manifolds of complexity 6 have Nil-geometry. Hence by Proposition 14.2.2 all closed orientable three-manifolds of complexity not greater than 6 belong to the class (H) of integrable constant-energy surfaces.

A further computer search (for $d > 6$) for manifolds of complexity d is difficult, since it requires a fairly large amount of machine time. On the other hand, the class (H) introduced in the preceding section is well understood and completely classified. Thus it makes sense to carry out a search "modulo (H)".

For this we have to teach the computer, in constructing in turn all special spines with a given number of vertices, how to recognize at an early stage of the construction whether or not the corresponding manifold belongs to the class (H). We present a theoretical basis for such a (partial) recognition.

We shall say that a two-component α of a special spine P of a closed orientable three-manifold M *has defect* k if its closure contains all the vertices of P with precisely k exceptions. Let M^α denote the three-manifold with torus boundary whose spine is obtained from P by deleting an open disk interior to α. An important fact is that the complexity of M^α does not exceed k. This is easy to prove, since on puncturing α and collapsing it, all the vertices in its closure also disappear. We note also that if M^α belongs to (H), then so does M.

A computer search of manifolds with torus boundary and a subsequent manual analysis has enabled us to establish the following theorem.

Theorem 14.6.2. *Every compact orientable irreducible atoroidal three-manifold of complexity $d \leq 3$ with torus boundary belongs to the class* (H), *with the exception of precisely two remarkable manifolds* Q_1^d *and* Q_2^d *of complexity 2 and nine manifolds* Q_3^d, ... Q_{11}^d *of complexity 3. These 11 manifolds admit hyperbolic structures, and hence cannot belong to the class* (H).

Each of the manifolds Q_i^d, $1 \leq i \leq 11$, generates a series of closed compact hyperbolic manifolds $(Q_i^d)_{p,q}$, where p and q are coprime integers. Geometric methods were then used to compute the volumes of over 100 manifolds in each series, that is, *around a thousand in all.* The results can be seen in Figure 14.13. We remark that the search revealed many nonhomeomorphic closed hyperbolic three-manifolds with equal volume.

The minimum volume among the manifolds of the type $(Q_i^d)_{p,q}$ is that of the remarkable manifolds $Q_1 = (Q_1^d)_{5,-2}$, and the next is that of $Q_2 = (Q_2^d)_{5,1}$. The details of this computer calculation are described in [Matveev and Fomenko 1988].

The manifold Q_2 was studied by Thurston, who proposed it as a candidate for the compact hyperbolic manifold of minimum volume. This conjecture turned out to be false, since the volume of Q_1 (≈ 0.94) is less than that of Q_2 (≈ 0.98). After we ran the calculations described above, we became aware of the work of J. Weeks, who had investigated this remarkable manifold Q_1 and calculated its volume. A description of Q_1^d and Q_2^d is given in the next section.

We note that the boundary of each manifold Q_1^d, ..., Q_{11}^d, consists of a single torus. If a closed three-manifold M is obtained from one of the Q_i^d by attaching a solid torus to its boundary, then M may or may not belong to the class (H). A special spine for M is easily constructed from a special spine P_i of Q_i^d : essentially, a new two-component (a meridional disk of the attached solid torus) is added to P_i. It turns out that every manifold of complexity $d \leq 8$ obtained from the Q_i^d in this way belongs to (H). This is no longer true for complexity 9: there exist at least two manifolds of complexity 9 that can

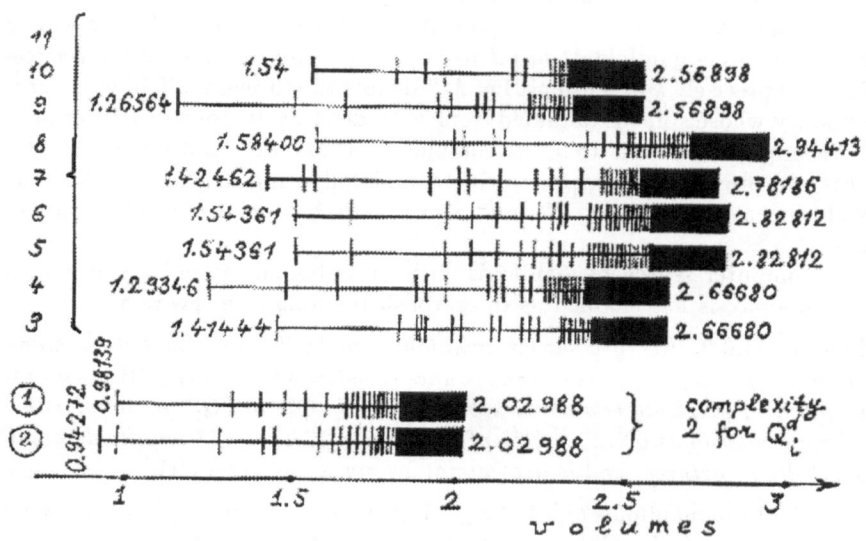

FIGURE 14.13.

be obtained from Q_1^d and Q_2^d by attaching solid tori, and that do not belong to (H).

Collecting together the above facts, we arrive at the following criterion for a closed three-manifold M to belong to the class (H): *if a special spine P of M has no more than 8 vertices and contains a two-component of defect $k \leq 3$, then $M \in$ (H).*

The use of this criterion allowed us to complete the search modulo (H) for all closed orientable three-manifolds of complexity not greater than 8. The following result was obtained:

Theorem 14.6.3 (S. V. Matveev, A. T. Fomenko). *Every closed orientable three-manifold of complexity not greater than 8 belongs to the class (H) of integrable constant-energy surfaces.*

We now state some conjectures.

Conjecture 14.6.4. The manifold Q_1 has the least volume of all closed hyperbolic manifolds.

Conjecture 14.6.5. The distribution of volumes of the manifolds $(Q_i^d)_{p,q}$ given in [Matveev and Fomenko 1988, Theorem 14] forms the "main part" of an initial segment of the sequence of volumes for all closed hyperbolic three-manifolds (Figure 14.13).

Conjecture 14.6.6. Within any sequence of manifolds of the form $M_{p,q}$ (where M is a manifold with torus boundary, see [Matveev and Fomenko 1988]), volumes increase with increasing complexity.

Within each sequence of three-manifolds of the form $M_{p,q}$ that were investigated in [Matveev and Fomenko 1988], where M is a manifold with torus boundary, the volume increases with complexity. We have verified our conjecture for a large number of the manifolds in the sequence $(Q_i^d)_{p,q}$, for which we have been able to find the complexity exactly. It is worth remarking that there is no such correlation between manifolds from distinct sequences, since by Matveev's theorem there are finitely many manifolds of a given complexity, and hence to the left of every accumulation point (of volumes of manifolds in a given sequence) there appear manifolds of arbitrarily large complexity.

14.7. Description of Q_1^d and Q_2^d

We now describe in more detail the two manifolds Q_1^d and Q_2^d, the simplest manifolds (in the sense of Matveev complexity) that do not belong to the class (H).

They are uniquely determined by their special spines; neighborhoods of whose singular graphs are shown in Figure 14.14. As a topological manifold, Q_1^d can be represented as the complement of a certain knot in the lens space $L_{5,1}$, and Q_2^d as the complement to the figure-eight knot in S^3. The interiors of Q_1^d and Q_2^d admit complete hyperbolic structures.

It is extremely useful to produce a clear concrete description of the Q_1 and Q_2. Figure 14.15 shows neighborhoods of the singular graphs of minimal special spines of Q_1 and Q_2 respectively. As can be seen from the Figure 14.15, the number of singular vertices of each of these special spines is equal to 9. Hence these manifolds have complexity not greater than 9. But since in the theorems mentioned above there are no hyperbolic manifolds of complexity 8 or less, the complexity is equal to 9.

Artistic representations by Fomenko of some fundamental notions of this theory are shown in Figures 14.16, 14.17, and 14.18.

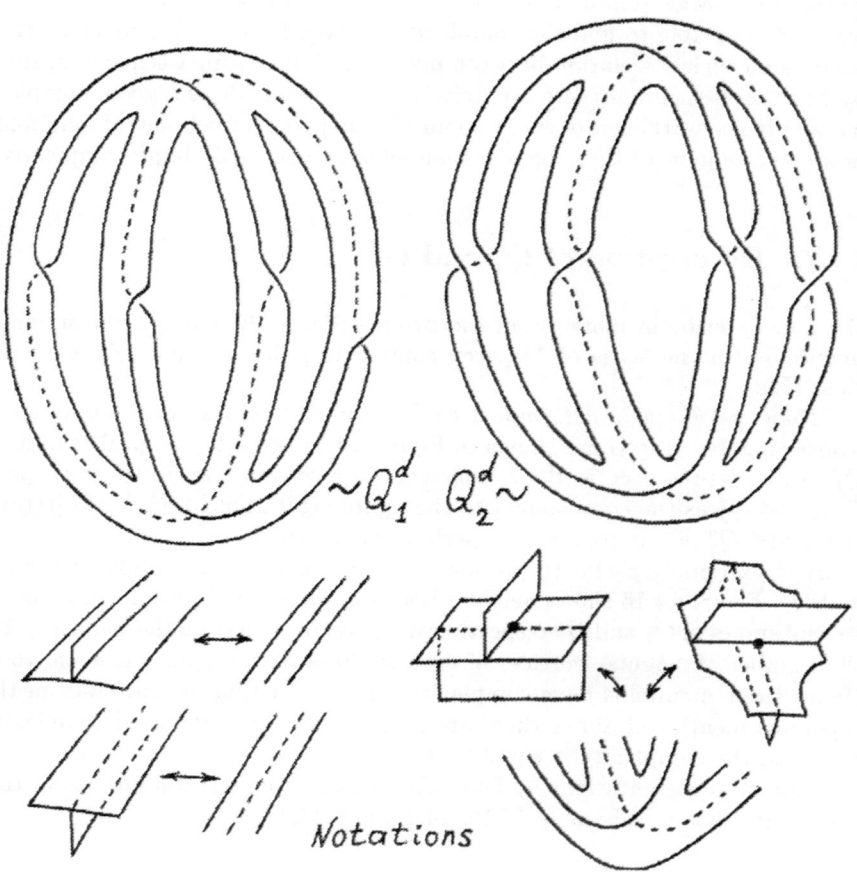

FIGURE 14.14.

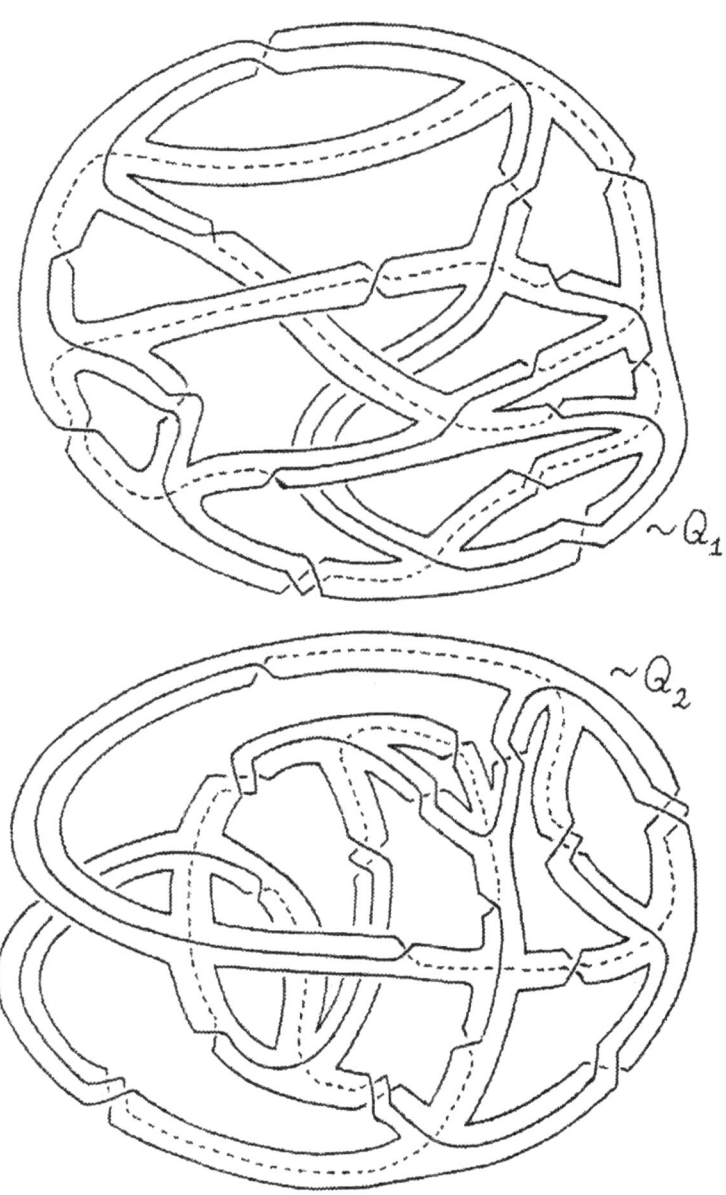

FIGURE 14.15.

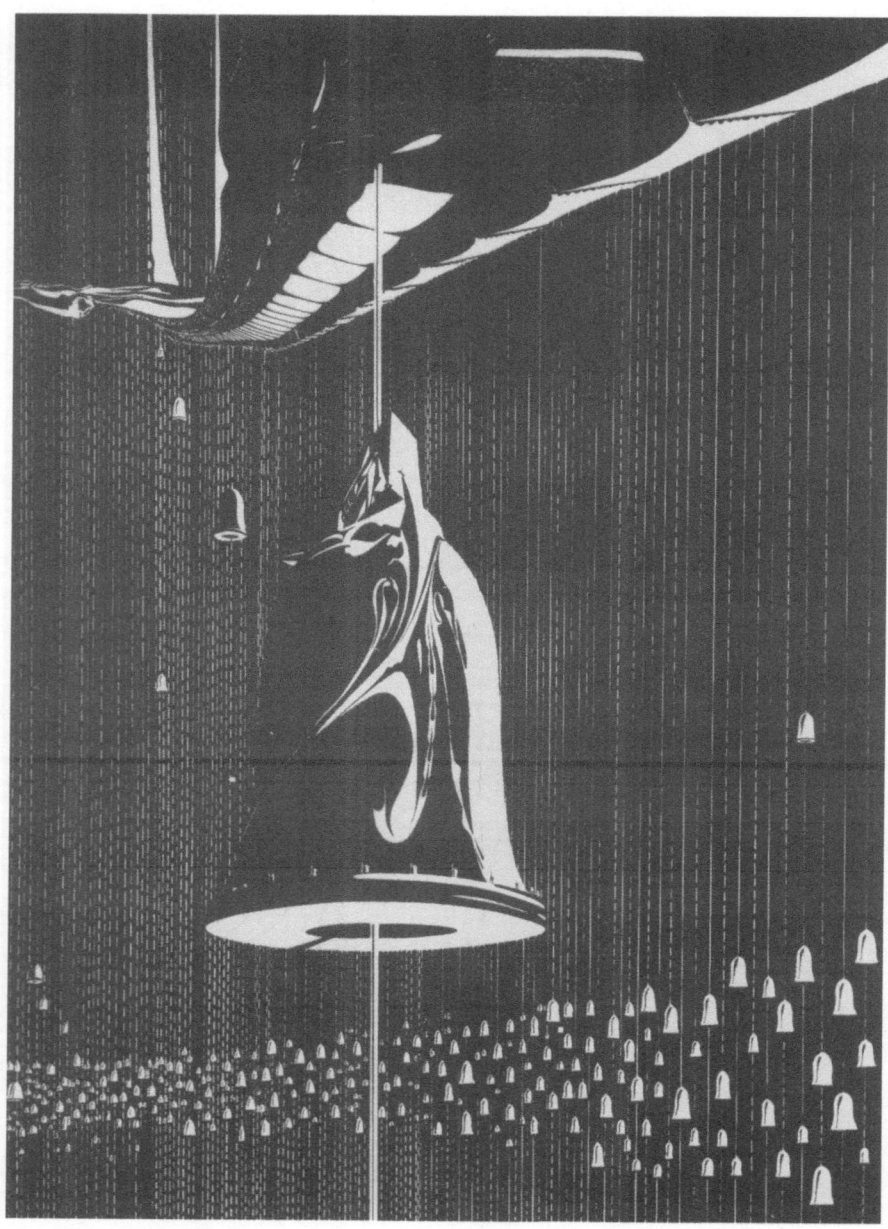

FIGURE 14.16. Spines of the two remarkable hyperbolic manifolds Q_1 and Q_2 (foreground, upper half).

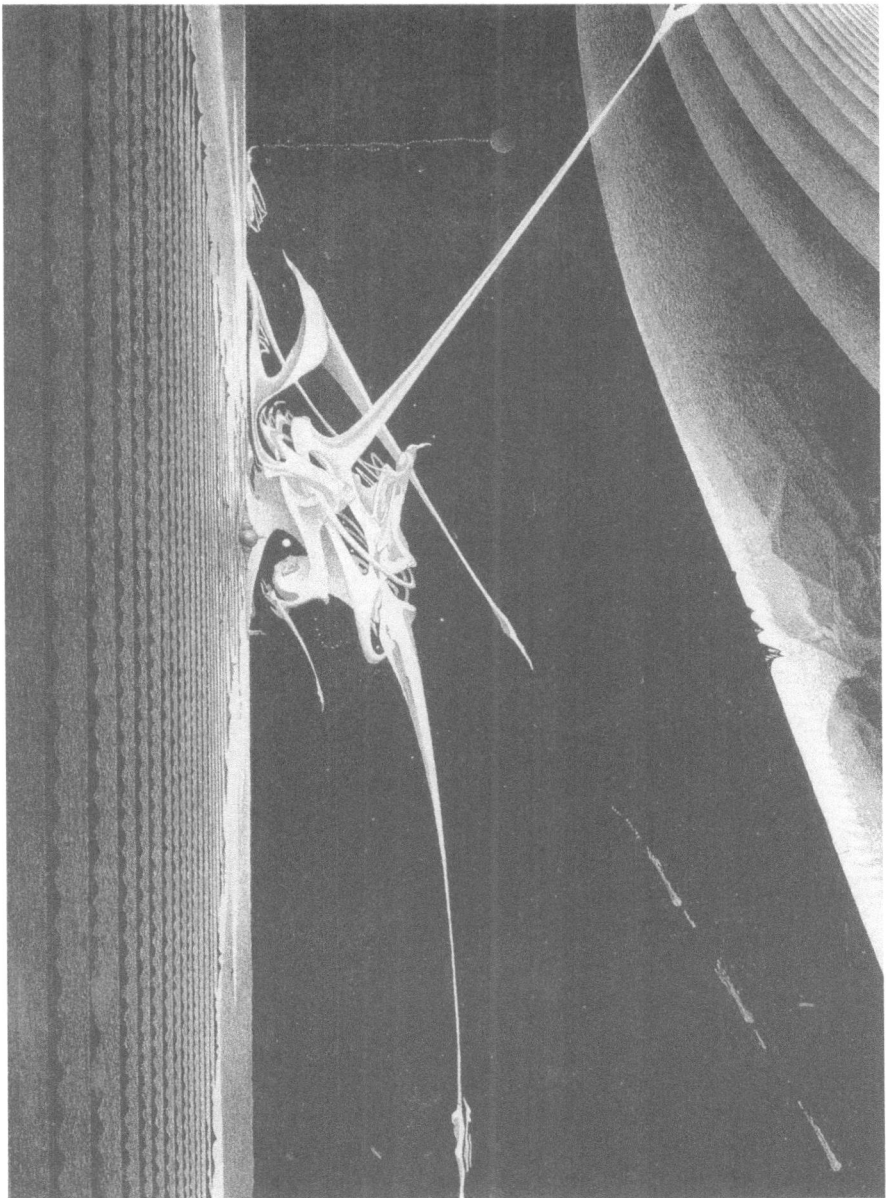

FIGURE 14.17. Visualization of the nondegenerate critical points (saddle points) of a smooth function on three-manifolds. The number of critical points measures the complexity of the Morse function.

FIGURE 14.18. Representation of the two-dimensional spine of a particular three-manifold. and simultaneously of the structure of the singular points of an algebraic surface in Euclidean three-space.

15. Integrable Hamiltonian Systems with Two Degrees of Freedom

A. V. Bolsinov

15.1. Introduction

Our main goal in this chapter is to describe the structure of the Liouville foliation near a singular fiber of an integrable Hamiltonian system (definitions follow). A similar problem was formulated by S. Smale [1970]. Here we discuss some new results in this direction obtained by L. Lerman and Ya. Umanskii [1987; 1993], by A. Bolsinov [1991], and by V. Matveev [1993].

At this point the reader may profit from reviewing the material on Morse theory in Chapter 6, in order to compare the ideas and results of this chapter with those discussed there.

15.2. Basic Notions

We start by introducing some concepts from the theory of integrable Hamiltonian systems: details can be found in [Dubrovin et al. 1985; Fomenko 1989], for example.

Consider a four-dimensional smooth manifold M and a differential two-form ω on it. The two-form $\omega = \sum \omega_{ij}\, dx_i \wedge dx_j$ is called a *symplectic structure* on M if the following conditions are satisfied:

 – ω is nondegenerate, that is, the matrix $\Omega(x) = (\omega_{ij}(x))$ is nondegenerate;
 – ω is closed, that is, $d\omega = 0$.

If ω is a symplectic structure, we call the pair (M, ω) a *symplectic manifold*.

The *skew-symmetric gradient* sgrad f of a smooth function f on a symplectic manifold (M, ω) is the unique vector field that satisfies

$$\omega(v, \text{sgrad}\, f) = v(f)$$

for every smooth vector field v on M, where $v(f)$ denotes the derivative of f along the field v.

Suppose that (M, ω) is a symplectic manifold. There exists a local regular coordinate system $(p_1, \ldots, p_n, q_1, \ldots, q_n)$ in an open neighborhood of $x \in M$ such that ω is represented by $\omega = \sum dp_i \wedge dq_j$. These are called *symplectic coordinates.* In this coordinate system the skew-symmetric gradient sgrad f has the form

$$\text{sgrad } f = \left(-\frac{\partial f}{\partial q_1}, \ldots, -\frac{\partial f}{\partial q_n}, \frac{\partial f}{\partial p_1}, \ldots, \frac{\partial f}{\partial p_n} \right).$$

A smooth vector field v on a symplectic manifold (M, ω) is called a *Hamiltonian field* if it can be written as $v = \text{sgrad } H$, for some smooth function H defined on all of M. The function H is called the *Hamiltonian*, or *energy function*.

The *Poisson bracket* $\{f, g\}$ of two functions f and g is defined as

$$\{f, g\} = \omega(\text{sgrad } f, \text{sgrad } g) = \sum w^{ij} \frac{\partial f}{\partial x_i} \frac{\partial g}{\partial x_j},$$

where $\Omega^{-1}(x) = (w^{ij}(x))$ is the matrix inverse to $\Omega(x) = (\omega_{ij}(x))$. The functions f and g are *in involution* on M if $\{f, g\} = 0$.

We can now state *Liouville's theorem*, which yields several important properties of integrable Hamiltonian systems. Let $v = \text{sgrad } H$ be a Hamiltonian system on a symplectic manifold (M, ω). The system is called *integrable* if there exists a function F that is constant on the integral trajectories of the system, and such that F and H are functionally independent (i.e., the differentials dF and dH are independent almost everywhere) and in involution. The function F is then called *an additional integral*. From now on we assume given a integrable Hamiltonian system (M, ω, H, F).

Consider the level surface $M_{H=H_0, F=F_0}$, where H_0 and F_0 are constants. We assume that this surface is compact, and that the restrictions of H and F to it are independent (this means that $\lambda \text{ sgrad } H + \mu \text{ sgrad } F = 0$ implies $\lambda = \mu = 0$ on the surface). Liouville's theorem says that:

- $M_{H=H_0, F=F_0}$ is diffeomorphic to a set of two-dimensional tori, called *Liouville tori*.
- In an open neighborhood of the level surface $M_{H=H_0, F=F_0}$, there exist regular coordinates $(I_1, I_2, \varphi_1, \varphi_2)$ called *action-angle variables*, satisfying the following conditions:
  - The form ω can be written $\omega = dI_1 \wedge d\varphi_1 + dI_2 \wedge d\varphi_2$.
  - For $i = 1, 2$, we have

  $$\dot{I}_i = -\frac{\partial H}{\partial \varphi_i} = 0, \qquad \dot{\varphi}_i = \frac{\partial H}{\partial I_i} = b_i(I_1, I_2).$$

  - The φ_i are periodic of period 2π: $M_{H=H_0, F=F_0} = \{(\varphi_1, \varphi_2) \bmod 2\pi\}$.

The integral trajectories of the Hamiltonian field sgrad H on a Liouville torus are periodic or almost periodic windings, because the field is uniform in the action-angle coordinates (Figure 15.1):

$$\text{sgrad } H = (\dot{I}_1, \dot{I}_2, \dot{\varphi}_1, \dot{\varphi}_2) = \left(-\frac{\partial H}{\partial \varphi_1}, -\frac{\partial H}{\partial \varphi_2}, \frac{\partial H}{\partial I_1}, \frac{\partial H}{\partial I_2} \right) = (0, 0, b_1, b_2).$$

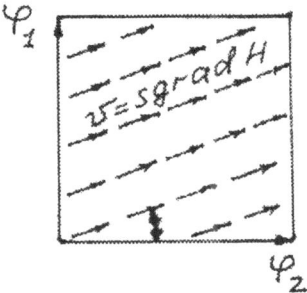

FIGURE 15.1.

Now consider the mapping $\Phi : M \to \mathbb{R}^2$ that associates to $x \in M$ the values of the two independent integrals H and F:

$$\Phi(x) = (H(x), F(x)).$$

We call this the momentum map. The Liouville tori are level sets of this map. Let K be the set of singularities of the momentum map, that is, points where rank $d\Phi(x) < 2$. The *bifurcation diagram* Σ of the momentum map as $\Sigma = \Phi(K)$.

Consider a constant-energy surface $Q_{H_0}^3 = \{x \in M | H(x) = H_0\}$. An additional equation $F(x) = F_0$ restricts $Q_{H_0}^3$ to a Liouville torus $M_{H=H_0, F=F_0}$. The Liouville torus $M_{H=H_0, F=F_0}$ is foliated according to F_0. This is called the *Liouville foliation* of the Hamiltonian system $v = \operatorname{sgrad} H$.

If the Hamiltonian system is nonresonant, the Liouville foliation does not depend on the choice of the additional integral F. Two integrable Hamiltonian systems on M and $\bar{M}$ are called *fiber-equivalent* if there exists a diffeomorphism $\xi : M^4 \to \bar{M}$ such that the fibers of the Liouville foliation on M are transformed into fibers of the Liouville foliation on $\bar{M}$ [Fomenko 1989].

As mentioned earlier, our goal is to describe the structure of the Liouville foliation near a singular fiber. At this point, the reader may profit from reviewing the material on Morse theory in Chapter 6, in order to compare the ideas and results of this chapter with those discussed there.

In contrast with the case of classical Morse theory, we are dealing here not with one smooth function F, but with the momentum map, which is a *pair* of smooth functions (H, F). Because of the way these two functions arise, they cannot be arbitrary; in particular, they must commute with respect to the Poisson bracket. As a result we can introduce a new, and quite natural, definition of nondegeneracy, and a new version of the Morse lemma. Because the symplectic manifold M is assumed to be four-dimensional, we can distinguish exactly four types of critical points. Then, instead of two singular levels L_ε and $L_{-\varepsilon}$ on the two sides of a critical value, we must consider the inverse image Q_γ of a circle γ in the plane $\mathbb{R}^2(H, F)$ with center $y = (H(x), F(x))$, where x is a critical point of the momentum map Φ. This inverse image consists of common level surfaces of H and F. Most of them are two-dimensional Liouville tori that transform one into another (inside Q_γ). As in Morse theory, we can describe these transformations. It turns out that one of the most effective ways to do this is to use a notion "atom" parallel to the one introduced in Chapter 8. In that chapter we used atoms and molecules to describe Morse singularities of smooth functions on surfaces; here we will demonstrate how the same notion can be used in a much more complex situation.

In general, the structure of the singularities of a momentum map may be rather complicated. Therefore we will, first of all, define a special class of singularities, called nondegenerate, that are easier to work with.

Let $L(x)$ denote the fiber of the Liouville foliation passing through $x \in M$ or, which is the same, the connected component of the inverse image $\Phi^{-1}(\Phi(x))$, containing x. Since the flows sgrad H and sgrad F commute, we can define the Poisson action of the abelian group $\mathbb{R}^2$ on M, generated by flowing along the vector fields sgrad H and sgrad F :

$$\mathfrak{X} : \mathbb{R}^2 \to \text{Diff}(M).$$

By $O(x)$ we denote the orbit of this action through $x \in M$. The orbit $O(x)$ must be homeomorphic to one of the following manifolds: $\mathbb{R}^0$, $\mathbb{R}^1$, $\mathbb{R}^2$, S^1, $S^1 \times \mathbb{R}^1$, or $T^2 = S^1 \times S^1$.

Consider a point $x_0 \in K$ such that rank $d\Phi(x_0) = 1$. Let $dH(x_0) \neq 0$ (the case $dF \neq 0$ is analogous). It is clear that in this case the orbit $O(x_0)$ is one-dimensional, and x_0 is critical for the restriction of F to the level surface $\{H = H(x_0)\}$.

Definition 15.2.1. We call the one-dimensional orbit $O(x_0)$ *nondegenerate* if the Hessian of the restriction $F|_{\{H=H(x_0)\}}$ at x_0 has rank 2. This condition means exactly that locally $O(x_0)$ is a nondegenerate critical submanifold of the function F restricted to the level $\{H = H(x_0)\}$.

Comment. This definition corresponds to the notion of the Morse–Bott integral introduced by A. T. Fomenko [1987b]. But sometimes nondegeneracy of orbits is defined in another way (see, for example, [Lerman and Umanskii 1987; 1993]). By our definition, nondegeneracy means that the orbit is elliptic or hyperbolic. According to Lerman and Umanskii, parabolic orbits are also considered nondegenerate.

Proposition 15.2.2 [Lerman and Umanskii 1987; Bolsinov 1991]. *Let $O(x_0)$ be a nondegenerate one-dimensional closed orbit, and suppose $dH(x_0) \neq 0$. Then there exists a tubular neighborhood U of $O(x_0)$ such that*

(1) *there exists a pair $(\lambda, \mu) \in \mathbb{R}^2$ such that $(\lambda dH + \mu dF) = 0$;*
(2) *the set $U \cap K = P$ is a smooth symplectic submanifold in M;*
(3) *the function $H : P \to \mathbb{R}$ has no singularities and stratifies P into one-dimensional closed nondegenerate orbits (in particular, $P \approx S^1 \times (0, 1)$);*

(4) *the image of P under the momentum map is a smooth curve* $\Phi(P) \subset \Sigma$.

Comment. The fiber-classification of integrable Hamiltonian systems near nondegenerate one-dimensional orbits reduces, in essence, to the classification of two-atoms [Bolsinov et al. 1990; Fomenko 1989].

Consider $p \in K_1$ such that rank $d\Phi(p) = 0$. Recall that the Poisson bracket $\{.\}$ allows one to define a Lie algebra structure on the space of symmetric bilinear forms $\text{Sym}(T_p^*M \otimes T_p^*M)$, in the following natural way. If g_1 and g_2 are two smooth functions having singularities at $p \in M$, we put by definition

$$[d^2 g_1(p), d^2 g_2(p)] = d^2 \{g_1, g_2\}(p).$$

It is easily seen that the Lie algebra defined on $\text{Sym}(T_p^*M \otimes T_p^*M)$ by the operation $[.]$ is isomorphic to the symplectic Lie algebra $\text{sp}(2, \mathbb{R})$. Since F and H commute, the Lie subalgebra $L_{F,H}$ generated by $d^2F(p)$ and $d^2H(p)$ is commutative.

Definition 15.2.3 A singular point $p \in K$ is called nondegenerate if $L_{F,H}$ is a Cartan subalgebra.

Remark. Henceforth, when speaking about nondegenerate critical points we mean points $p \in K$ such that rank$(d\Phi(p)) = 0$. If rank$(d\Phi(p)) = 1$, we speak about degeneracy or nondegeneracy of the one-dimensional orbit $O(p)$.

The local structure of singularities of a momentum map $\Phi = (H, F)$: $M \to \mathbb{R}^2$ at nondegenerate singular points has been explored in [Lerman and Umanskii 1987]. We recall briefly some basic results from this paper that will be needed later.

Let $x_0 \in M$ be a nondegenerate critical point of Φ, and let $L_{H,F}$ be the Cartan subalgebra generated by $d^2F(x_0)$ and $d^2H(x_0)$. It is known that there are four different types of Cartan subalgebras in the real Lie algebra $\text{sp}(2, \mathbb{R})$. In accordance with this, we can distinguish four types of nondegenerate critical points: (a) a *center-center*, (b) a *center-saddle*, (c) a *saddle-saddle*, and (d) a *focus-focus*.

Theorem 15.2.4. *Let $x_0 \in M$ be a nondegenerate critical point of the momentum map $\Phi = (H, F)$. Then in some neighborhood of x_0 there exists a canonical coordinate system (p_1, q_1, p_2, q_2) such that the functions H and F have the following form:*

1) *center-center type:*

$$H = h_0 + (p_1^2 + q_1^2)H_1 + (p_2^2 + q_2^2)H_2,$$
$$F = F_0 + (p_1^2 + q_1^2)F_1 + (p_2^2 + q_2^2)F_2;$$

2) *saddle-center type:*

$$H = h_0 + p_1 q_1 H_1 + (p_2^2 + q_2^2)H_2,$$
$$F = F_0 + p_1 q_1 F_1 + (p_2 + q_2)F_2;$$

3) *saddle-saddle type:*

$$H = h_0 + p_1 q_1 H_1 + p_2 q_2 H_2,$$
$$F = F_0 + p_1 q_1 F_1 + p_2 q_2 F_2;$$

4) *focus-focus type:*

$$H = h_0 + p_1 q_1 H_{11} + p_1 q_2 H_{12} + p_2 q_1 H_{21} + p_2 q_2 H_{22},$$
$$F = F_0 + p_1 q_1 F_{11} + p_1 q_2 F_{12} + p_2 q_1 F_{21} + p_2 q_2 F_{22};$$

where the H_i, H_{ij}, F_i, F_{ij} are smooth functions satisfying $H_{11}(x_0) = H_{22}(x_0)$. $F_{11}(x_0) = F_{22}(x_0)$, $H_{12}(x_0) = -H_{21}(x_0)$, $F_{12}(x_0) = -F_{21}(x_0)$,

$$H_1(x_0)F_2(x_0) - H_2(x_0)F_1(x_0) \neq 0,$$
$$H_{11}(x_0)F_{12}(x_0) - H_{12}(x_0)F_{11}(x_0) \neq 0.$$

Corollary 15.2.5 [Lerman and Umanskii 1987]. *Let $x_0 \in M$ be a nondegenerate critical point of the momentum map Φ of saddle-saddle, saddle-center, or center-center type, and let K be the set of critical points of Φ. Then there exists a neighborhood $U(x_0)$ of the point x_0 such that*

1) *$K \cap U(x_0) = P_1 \cup P_2$, where P_1 and P_2 are two-dimensional symplectic submanifolds that intersect transversally at the point x_0; and*
2) *all one-dimensional orbits lying in P_1 and P_2 are nondegenerate.*

Definition 15.2.6 A smooth curve $\gamma(t)$ in $\mathbb{R}^2$ is called *admissible* if the following conditions hold:

1) γ intersects Σ transversally;
2) if A is a point of intersection then critical points in $\Phi^{-1}(A)$ form a collection of nondegenerate orbits;
3) for any intersection point A there exists a neighborhood U_A such that $\Sigma \cap U_A$ can be represented as the graph of some smooth function of H.

Proposition 15.2.7. *Let $\gamma(t)$ be an admissible curve. Then its inverse image $Q_\gamma = \Phi^{-1}(\gamma)$ is a smooth three-dimensional submanifold in M and the function $f_\gamma : Q_\gamma \to \mathbb{R}$ defined by $\gamma(f_\gamma(x)) = \Phi(x)$ is a Morse–Bott function, that is, its critical points form nondegenerate critical submanifolds.*

Following [Bolsinov 1991; Matveev 1993], for each admissible curve $\gamma(t)$ we can define the marked molecule $W^*(Q_\gamma)$, which describes the structure of the Liouville foliation on the constant-energy surface $Q_\gamma = \Phi^{-1}(\gamma)$. On can easily see that a continuous deformation of the curve $\gamma(t)$ within the class of admissible curves does not change the structure of the Liouville foliation on Q_γ, and therefore neither does it change the molecule $W^*(Q_\gamma)$.

Definition 15.2.8. A point $y \in \Sigma$ is called an *isolated singular point* of Σ if any circle γ_ε of sufficiently small radius ε centered at this point is admissible.

Definition 15.2.9. Let y be an isolated singular point of Σ. The marked molecule $W^*(Q_{\gamma_\varepsilon})$ for some small circle γ_ε with center y is called a *circle molecule* and is denoted by $W^*(y)$.

It is clear that $W^*(y)$ is a fiber invariant of the Hamiltonian system $v = \operatorname{sgrad} H$ (that is, an invariant of the Liouville foliation) near the singular fiber $\Phi^{-1}(y)$. It turns out that in many important cases this invariant allows us to classify singularities of the momentum map Φ (up to fiber-equivalence).

Let x be a nondegenerate critical point of the momentum map Φ, and set $y = \Phi(x)$. Different concrete examples of integrable Hamiltonian systems show that, as a rule, y is an isolated singular point of the bifurcation diagram Σ. Moreover, the bifurcation diagram has one of the forms shown in Figure 15.2. We will assume this condition to be valid, although in general a neighborhood of y on the bifurcation diagram can be more complicated (for example, Σ may consist of more than two smooth curves intersecting at y).

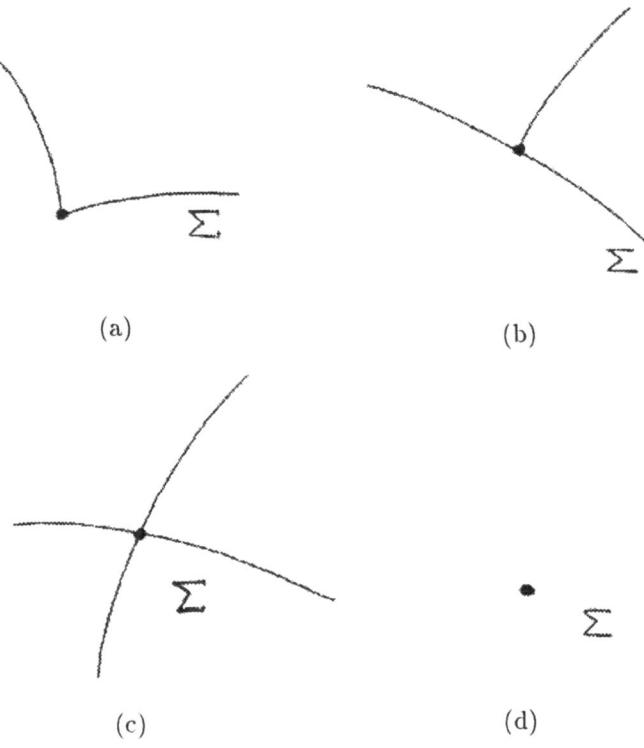

(a)   (b)

(c)   (d)

FIGURE 15.2. Bifurcation diagrams for (a) center-center, (b) center-saddle, (c) saddle-saddle, and (d) focus-focus nondegenerate critical points.

Recall that our goal is to describe and classify the fiber structure of an integrable Hamiltonian system in some invariant neighborhood $U(x)$ of a singular fiber $L(x)$. We will assume that the following natural conditions hold.

(1) The singular fiber $L(x)$ is connected.
(2) The number of critical points $x_i \in L(x)$ is finite and they are all nondegenerate (here $x = x_1$).
(3) $U(x)$ is a regular neighborhood of $L(x)$.
(4) For any point of $U(x)$, the whole fiber of the Liouville foliation passing through this point is contained in $U(x)$. In other words, $U(x)$ can be represented as $\Phi^{-1}(V(y))$, where $V(y)$ is a small open neighborhood of $y = \Phi(x)$.
(5) The form of the bifurcation diagram Σ near the singular point $y = \Phi(x)$ is as shown in Figure 15.2.

A neighborhood $U(x)$ satisfying all these conditions is called a *center-center*, *center-saddle*, *saddle-saddle*, or *focus-focus* neighborhood, depending on the type of the critical points of the critical fiber $L(x)$ (or, which is the same, depending on the form of Σ).

15.3. The Saddle-Saddle Case

We begin with the topologically most interesting case. Let $x \in M^4$ be a saddle-saddle point and $U(x)$ its saddle-saddle neighborhood.

The topology of the singular fiber.   We start by describing the structure of the singular fiber $L(x) = \Phi^{-1}(y)$.

Lemma 15.3.1. *The singular fiber $L(x) = \Phi^{-1}(y)$ is a two-dimensional CW-complex C whose cells are all orbits of the Poisson action $\mathfrak{X}$.*

This lemma is proved, for example, in [Lerman and Umanskii 1987; Bolsinov 1991].

If n is the number of the saddle-saddle points x_i in $L(x)$, then the number of two-cells and the number of one-cells (edges of C) are both $4n$ and the number of zero-cells (vertices of C) is n. Moreover, the vertices of C are exactly the saddle-saddle points $x_1, \ldots, x_n$. We can imagine that the CW-complex C is glued from squares, each of which is a two-cell. The edges of the squares that should be glued together are denoted by the same letters (or numbers). For every letter there are, before gluing, exactly four edges bearing this letter. The local structure of this CW-complex is quite simple; it is a surface with singular points that can be of two types. Near any interior point of an edge, C is the product of a cross by an interval (see Figure 15.3), while a small neighborhood of any of its vertices is the product of two crosses.

FIGURE 15.3.

Note that the vector field sgrad H has no singularities on edges of C. Hence we can naturally orient the edges of C by sgrad H. Thus, if we have a saddle-saddle neighborhood, we can naturally construct the CW-complex C with an oriented one-skeleton.

Definition 15.3.2. The CW-complex C with the orientation on its one-skeleton is called the *saddle complex* of the saddle neighborhood.

The simplest example of a saddle complex can be obtained by gluing the squares shown in Figure 15.4. Topologically, this complex is the product of two figures of eight.

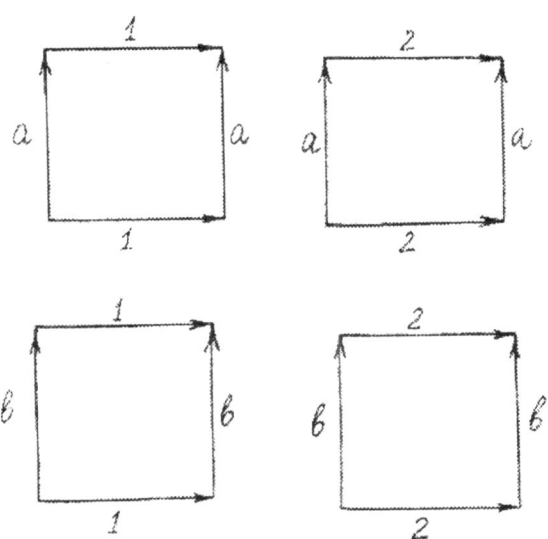

FIGURE 15.4.

L-type and CL-type. Consider the set $K \cap U(x)$. Using Corollary 15.2.5, it is easy to show that $K \cup U(x)$ is a pair of transversally intersecting two-dimensional smooth symplectic surfaces B_w and B_b. Consider the two sets $G_w = B_w \cap L(x)$ and $G_b \cap L(x)$. Since all two-cells of $L(x)$ consist of regular points of the momentum map Φ, the union $G_w \cup G_b$ coincides with the one-skeleton of $L(x)$; in particular, G_w and G_b are both one-dimensional graphs.

It is possible to show that the pairs (B_w, G_w) and (B_b, G_b) have the natural structure of a (possibly disconnected) two-atom.

Definition 15.3.3. The pair of atoms $((B_w, G_w), (B_b, G_b))$ is called the *L-type* of the saddle neighborhood.

For the case shown in Figure 15.4, the graphs G_b and G_w are formed by the edges a, b and $1, 2$ respectively. Both graphs are just figures of eight. Therefore, the atoms (B_w, G_w) and (B_b, G_b) are of the simplest form and denoted, as usual, by the letter B. So in this case the L-type is (B, B).

Important Remark. Not every pair of atoms is admissible. On the other hand, the same L-type can correspond to different saddle-saddle neighborhoods.

It is clear that the L-type and saddle complex $L(x)$ are both topological fiber-invariants of a saddle neighborhood. But there are examples of nonequivalent saddle neighborhoods with the same L-types and different complexes, or, conversely, with the same complexes and different L-types.

Definition 15.3.4. The triple $[C, (B_w, G_w), (B_b, G_b)]$, where C is the saddle complex and the pair $((B_w, G_w), (B_b, G_b))$ is the L-type of the saddle neighborhood, is called the *CL-type* of the saddle neighborhood. Here we assume that the embeddings of graphs G_w and G_b to the one-skeleton of C are given.

Note that $G_w \cap G_b$ consists of saddle-saddle points.

Let $[C, (B_w, G_w), (B_b, G_b)]$ and $[\tilde{C}, (\tilde{B}_w, \tilde{G}_w), (\tilde{B}_b, \tilde{G}_b)]$ be CL-types of saddle neighborhoods U^4 and $\tilde{U}^4$ respectively.

Definition 15.3.5. Two CL-types $[C, (B_w, G_w), (B_b, G_b)]$ and $[\tilde{C}, (\tilde{B}_w, \tilde{G}_w), (\tilde{B}_b, \tilde{G}_b)]$ are called *equivalent* if there exists a homeomorphism $X : C \cup B_w \cup G_b \to \tilde{C} \cup \tilde{B}_w \cup \tilde{B}_b$ such that $X(C) = \tilde{C}$ and X preserves the orientation of edges of C.

Theorem 15.3.6. *The CL-type is a complete topological fiber-invariant of saddle-saddle neighborhoods; that is, two saddle neighborhoods are fiber-equivalent if and only if their CL-types are equivalent.*

Not every triple $[C, (B_w, G_w), (B_b, G_b)]$ is admissible, i.e., can be realized as the CL-types of some saddle-saddle neighborhoods. However, it is possible to formulate some additional formal conditions in order for such a triple to be admissible [Bolsinov 1991].

Counting the types of saddle neighborhoods. Let n be the number of the singular saddle-saddle points in a singular fiber $L(x)$. It is easy to see that, for any fixed n, the number of all saddle neighborhoods $U(x)$ is finite. One of the most interesting problems in the theory of momentum map singularities is finding out how many types there are.

This problem can clearly be reduced to a combinatorial problem, namely, to the classification of all admissible CL-types. For small $n = 1, 2$ this can be done by hand, by considering step by step all possible cases (for $n = 1$, this has been done by L. Lerman and Ya. Umanskii; for $n = 2$, by A. V. Bolsinov). If $n > 2$ the problem becomes more complicated and it is reasonable to use computer methods. A computer algorithm has been proposed by N. Maximova, who solved this problem for $n = 3$.

Theorem 15.3.7. *Let n be the number of saddle-saddle points in a saddle-saddle neighborhood.*

- *If $n = 1$ then there exist exactly 4 saddle-saddle neighborhoods.*
- *If $n = 2$ then there exist exactly 39 saddle-saddle neighborhoods.*
- *If $n = 3$ then there exist exactly 256 saddle-saddle neighborhoods.*

For $n > 3$ the question remains open. It would be very interesting to try to solve it exactly for $n = 4, 5$, and to obtain asymptotic estimates for large n.

15.4. The Focus-Focus Case

Let x be a singular point of focus-focus type, let $y = \Phi(x)$ be the corresponding singular point of the bifurcation diagram. Let $L(x) = \Phi^{-1}(y)$ be the singular fiber and $U(x)$ its focus-focus neighborhood.

Definition 15.4.1. The number of critical points in $L(x)$ is called the *weight* of the focus neighborhood $U(x)$.

Theorem 15.4.2. *The weight is a complete fiber-invariant of focus-focus neighborhood: that is, two focus-focus neighborhoods are fiber-equivalent if and only if they have the same weight. For any integer $n > 0$ there exists a focus-focus neighborhood of weight n.*

Sketch of the proof. First we describe the topological structure of the singular fiber $L(x)$.

Let $S_1, S_2, \ldots, S_n$ be two-dimensional spheres. It is useful to suppose that the index belongs to $\mathbb{Z}_n$, so that $(n-1) + 1 = 0$. Choose two points A_i and B_i on each sphere S_i, and glue A_i and B_{i+1} for any i. As a result we obtain some topological space V that we will called an *n-chain*.

Lemma 15.4.3 [Matveev]. *Let the weight of the focus-focus neighborhood be equal to n. Then the inverse image $\Phi^{-1}(A)$ is the n-chain.*

Thus, if U_1 and U_2 are focus-focus neighborhoods with equal weights, their singular fibers are homeomorphic. Using the fact that, locally, all singular points of focus-focus type have the same structure, we can extend the homeomorphism between the singular fibers to a global homeomorphism between U_1 and U_2. Moreover, this extension can be chosen in such a way that it preserves the Liouville foliation structure.

To prove that any number n can be realized as the weight of some focus-focus neighborhood it is sufficient to note that $n = 1$ is realized in many concrete examples from classical mechanics and mathematical physics (for instance, in the Lagrange case in rigid body dynamics). After that, it is possible to "glue" the desired focus-focus neighborhood from several copies having weight 1. □

15.5. The Center-Center and Center-Saddle Cases

Let $U(x)$ be a center-center neighborhood.

Theorem 15.5.1. *Any two center-center neighborhoods are fiber-equivalent. From the topological point of view, a center-center neighborhood $U(x)$ is an open four-ball, the singular fiber $L(x)$ contains the only point x, and there exist regular local coordinates p_1, q_1, p_2, q_2 such that the Liouville foliation structure is given on $U(x)$ by common level surfaces of two functions $H = p_1^2 + q_1^2$ and $F = p_2^2 + q_2^2$.*

This theorem is a simple corollary of the theorem by Eliasson [1990] describing the action-angle variables of integrable Hamiltonian systems near singular points of elliptic type.

Now let $U(x)$ be a center-saddle neighborhood. Consider the singular fiber $L(x)$ containing one or several nondegenerate critical points $x = x_1, x_2, \ldots, x_n$. From the local structure of a center-saddle point it is easy to see that $L(x)$ is a one-dimensional graph whose vertices are all of degree 4. They are exactly the points $x_1, \ldots, x_n$, and its edges are nondegenerate one-dimensional orbits.

The fiber structure of $U(x)$ can be described in the following way. Consider an arbitrary connected two-atom $V = (P^2, L)$, that is, a two-dimensional compact surface P^2 with boundary and an embedded graph $L \subset P^2$. Recall that, by definition, P is a regular neighborhood of L, and L can be defined as the critical level of some Morse function h on P. Thus, the pair (P, L) defines an one-dimensional Liouville foliation (with one singular fiber) for a Hamiltonian system of one degree of freedom (here h is thought of as its Hamiltonian). Consider another Hamiltonian system with one degree of freedom defined on a 2-disc P', whose Hamiltonian is just $f = p_2^2 + q_2^2$. On P' there is a natural structure of a one-dimensional Liouville foliation, namely the foliation into concentric circles. Thus, we have two different one-dimensional Liouville foliations on P and P', the first having saddle singularities, and the second a singularity of center type. Now from these two foliations we can construct the natural two-dimensional Liouville foliation (into 2-tori, but with singularities) on the direct product $P \times P'$ of P and P'.

It is possible to show that we have really obtained a center-saddle neighborhood, and that any center-saddle neighborhood can be obtained by this procedure. As a result we can formulate the classification theorem for the center-saddle case as follows.

Theorem 15.5.2. *The set of all fiber-nonequivalent center-saddle neighborhoods is in one-to-one correspondence with the set of all two-atoms.*

Note that here we mean the fiber structure of Hamiltonian systems (that is, structure of the foliation into Liouville tori) and say nothing about the foliation into trajectories of the system.

Problem. Classify integrable Hamiltonian systems in their center-center and center-saddle neighborhoods up to orbital equivalence (i.e., up to homeomorphisms preserving the trajectories).

15.6. Circle Molecules

In this section we give the list of circle molecules corresponding to the momentum map singularities described above. It is clear that the circle molecule is a fiber-invariant of an integrable Hamiltonian system. In [Bolsinov 1991] it was shown how information about circle molecules can be used to calculate the classical marked molecules for some concrete examples of integrable Hamiltonian systems. On the other hand, circle molecules are very useful because of the following "experimental fact", which we shall see below: two momentum map singularities are fiber-equivalent if and only if their circle molecules coincide.

Theorem 15.6.1. *The circle molecule of a center-center point is $A\!-\!\!\!\overset{0}{\rule{1.2em}{0.4pt}}\!\!\!-\!A$.*

Theorem 15.6.2. *The circle molecule of a center-saddle neighborhood corresponding to an atom V has the form shown in Figure 15.5.*

These theorems follow directly from the description of the topology of the corresponding Liouville foliations (see the preceding section).

For saddle-saddle neighborhoods, there exists an algorithm for reconstructing the circle molecule from the given CL-type [Bolsinov 1991; Matveev 1993]. The result of the calculation is as follows.

Theorem 15.6.3. *The list of circle molecules for saddle-saddle neighborhoods with 1 and 2 vertices is given in Table 15.1.*

Finally, let $U(x)$ be a focus-focus neighborhood. Its circle molecule is not a molecule in the usual sense because it has no atoms. Indeed, the inverse image of a small circle on the bifurcation diagram is a fiber bundle over the

Table 15.1.

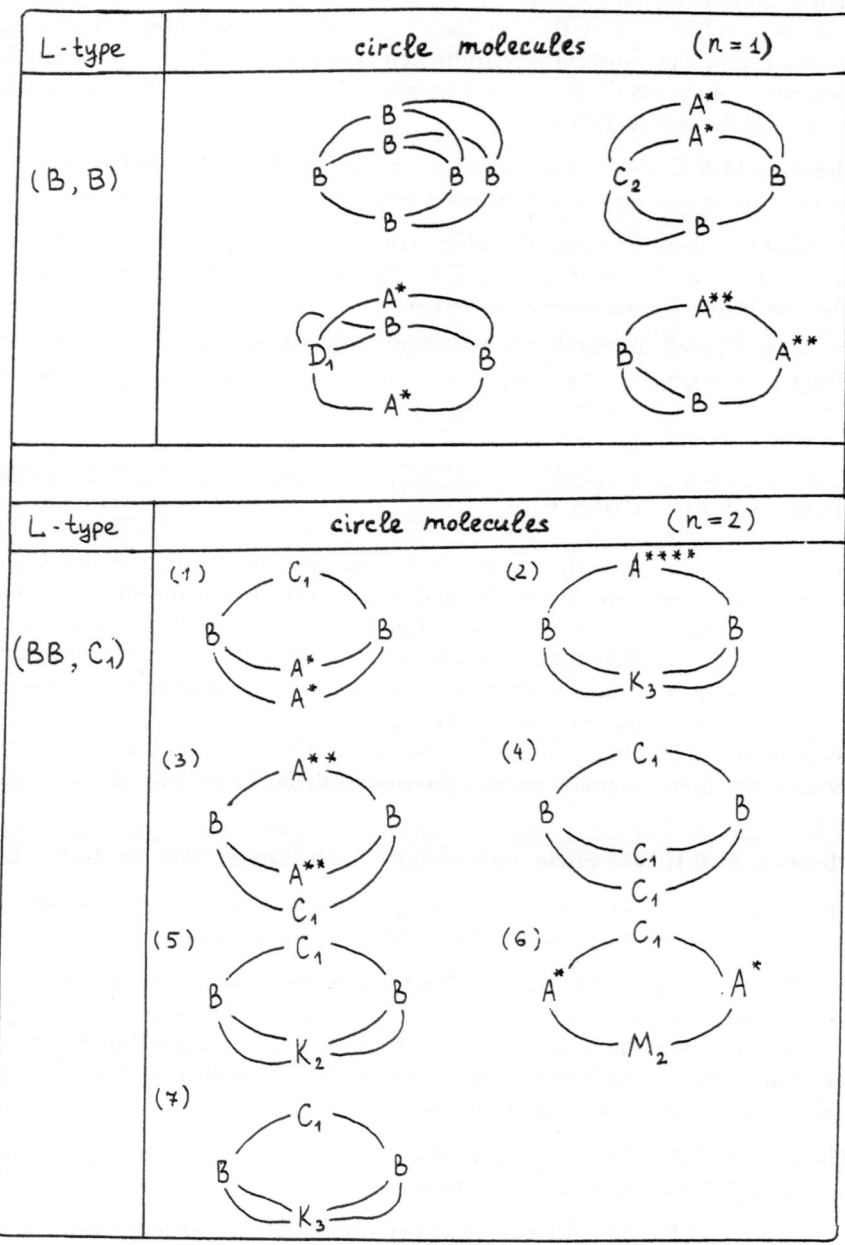

L·type	circle molecules
(BB, C_2)	
(BB', C_1)	
(BB^i, C_2)	

<table>
<tr><th>L-type</th><th>circle molecules</th></tr>
<tr><td>(BB, D₁)</td><td>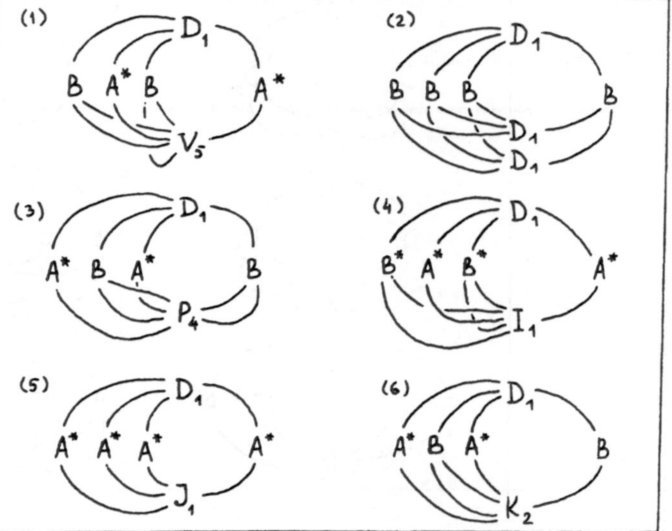
</td></tr>
<tr><td>(BB, D₂)</td><td>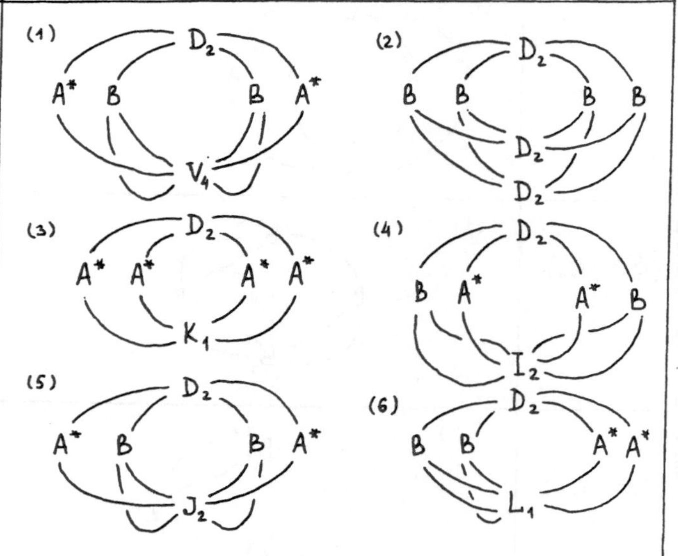
</td></tr>
</table>

Table 15.2.

atom	embedding of the graph	atom	embedding of the graph
I_1		L_1	
I_2		L_2	
J_1		M_1	
J_2		M_2	
K_1		P_4	
K_2		V_5	
K_3		V_4	

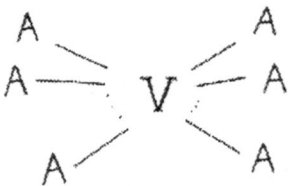

FIGURE 15.5.

circle whose fibers are two-dimensional tori. Every such bundle can be given by a conjugacy class of 2×2 matrices.

Theorem 15.6.4. *For a focus neighborhood with weight n, this fiber bundle is determined by the conjugacy class of the matrix $\left(\begin{smallmatrix} 1 & n \\ 0 & 1 \end{smallmatrix}\right)$.*

This theorem can be proved by standard methods of covering map theory.

Thus, we can see that the circle molecules corresponding to the singularities of different types are all different. This allows us to formulate the following conjecture.

Conjecture 15.6.5. For nondegenerate singularities of the momentum map (satisfying the conditions described above), the circle molecule is a *complete* fiber-invariant of an integrable Hamiltonian system.

Moreover, as can be seen from [Bolsinov 1991], this probably remains true under more general assumptions. In other words, this conjecture can be reformulated as follows. Let $L(x)$ be a singular fiber of the Liouville foliation of an integrable Hamiltonian system. Let $y = \Phi(x)$ be an isolated singular point of the bifurcation diagram Σ (in order for the circle molecule to be well-defined). It is clear from the definition that the circle molecule describes the structure of the Liouville foliation on the three-dimensional boundary $\partial U(x)$ of some four-dimensional regular invariant neighborhood $U(x)$ of $L(x)$. The conjecture states that using this information about the structure on the boundary we can uniquely reconstruct the Liouville foliation structure on the whole of $U(x)$.

16. Topological and Orbital Analysis of Integrable Lagrange and Goryachev–Chaplygin Problems

O. E. Orël

16.1. Introduction

A. T. Fomenko and A. V. Bolsinov [1994] obtained a new topological invariant for systems with two degrees of freedom, which classifies integrable Hamiltonian systems on constant-energy surfaces up to orbital equivalence. (Two smooth dynamical systems on manifolds M_1 and M_2 are called *orbitally equivalent* if there exists a homeomorphism from the first manifold onto the second one that transforms trajectories of the first system to trajectories of the second one with preservation of the orientation. More precisely, this is called *continuous* orbital equivalence; *smooth* orbital equivalence is defined analogously, with "homeomorphism" replaced by "diffeomorphism". Notice that the transformation need not preserve the time parameter along trajectories.)

Let M be a symplectic manifold, and let $v = \operatorname{sgrad} H$ be a Hamiltonian system on it that is integrable with additional integral F (see the preceding chapter, and also the next chapter for the definition of a symplectic structure). According to the new theory of orbital classification, the invariant W^{*t}, or *t-molecule*, is associated with the restriction of the dynamical system to each level surface(or constant-energy surface) $Q^3 = \{H = \text{const}\}$. The central result of the Bolsinov–Fomenko theory is the following: *Two integrable systems are orbitally equivalent if and only if their t-molecules coincide.*

The theory of orbital classification is based on the theory of fiber-classification of integrable Hamiltonian systems with two degrees of freedom discovered by Fomenko, Zieschang, Matveev and Bolsinov [Bolsinov et al. 1990]. Two systems v_1 and v_2 on constant-energy surfaces Q_1 and Q_2 are called *fiber-equivalent* if there exists a diffeomorphism transforming Q_1 onto Q_2 that preserves the Liouville foliations defined by level surfaces of the additional integral. The topological fiber-invariant $W^*(Q)$, called the Fomenko–Zieschang invariant or marked molecule, is a graph that describes the Liouville foliation of the constant-energy surface Q. The edges of the graph correspond to one-parameter families of Liouville tori, and its vertices (atoms) describe bifurcations of the tori at singular levels of the integral F. In addition, the molecule includes a set of numerical marks r, n, and ε that prescribe the gluing of Liouville foliation of Q^3 from elementary bricks.

Besides the marked molecule, a t-molecule includes some other invariants, the most important of which is the rotation vector constructed from the rotation function. Therefore the analysis of the rotation function on constant-energy surfaces Q^3 is an important problem in the theory of orbital classification.

As discussed in the preceding chapter, Liouville's theorem states that a nonsingular connected compact level surface of the integrals H and F is diffeomorphic to the two-torus T^2. Moreover, in its neighborhood one can determine coordinates I_1, I_2, φ_1, and φ_2, called *action-angle variables*, possessing the following properties:

(1) the symplectic structure ω has the form

$$\omega = \sum_i dI_i \wedge d\varphi_i;$$

(2) the Hamiltonian system is written down in the form

$$\dot{I}_i = -\frac{\partial H}{\partial \varphi_i} = 0, \qquad \dot{\varphi}_i = \frac{\partial H}{\partial I_i} = b_i(I_1, I_2);$$

(3) the functions φ_i are periodic of period 2π.

Liouville's theorem implies that I_1 and I_2 are determined up to constants for a given Liouville torus, and therefore they parametrize the family of Liouville tori, while the *angle variables* φ_i yield coordinates on each torus. It was shown in [Arnold 1978] that in the neighborhood of a nonsingular Liouville torus the action variables can be calculated by the formula

16.1.1. $$I_i = \frac{1}{2\pi} \int_{\gamma_i} \kappa_i, \quad i = 1, 2,$$

where γ_1, γ_2 are basis cycles on tori whose homotopy classes in the neighborhood of a torus do not depend on the torus, and where κ is a one-form such that $d\kappa = \omega$ in the neighborhood in question. It was also proved in [Arnold 1978] that the right-hand side of 16.1.1 does not depend on γ_i as a representative of a fixed element of the fundamental group.

As we saw, the vector field $v = \text{sgrad}\, H$ is constant on a Liouville torus when expressed in terms of action variables. Thus we can consider its *rotation number* with respect to these variables:

16.1.2. $$\rho = \frac{\dot{\varphi}_1}{\dot{\varphi}_2} = \frac{b_1}{b_2} = \frac{\partial H/\partial I_1}{\partial H/\partial I_2}.$$

If we consider the collection of Liouville tori we obtain the *rotation function* $\rho = \rho(H, F)$.

Consider a three-manifold $Q^3(H)$ and a family of the rotation functions $\rho(F)$ on the edges of the corresponding molecule. We can associate with each edge the *rotation vector* R, consisting of

(1) the finite critical values of the function ρ (its minima and maxima),
(2) the two limits of the function ρ on the ends of the edge,
(3) the symbols $\pm\infty$, which denote the left and right limits of the function ρ at the poles.

All the components of the rotation vector R should be written down in the order in which they are encountered as one moves along the edge. Figure 16.1 shows an example. To compare two integrable Hamiltonian systems on the edges of the molecules, we need to compare only the critical values of the rotation functions of these systems.

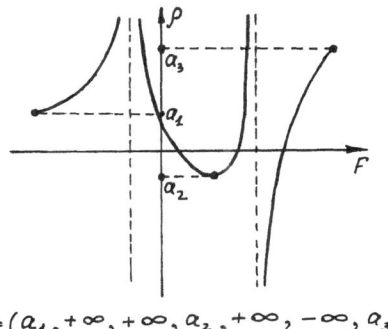

$$R = (a_1, +\infty, +\infty, a_2, +\infty, -\infty, a_3)$$

FIGURE 16.1. $R = (a_1, +\infty, +\infty, a_2, +\infty, -\infty, a_3)$.

What kinds of difficulties arise when we try to calculate action variables and the rotation function for specific problems of mechanics? It would appear that equations (16.1.1) and (16.1.2) can easily furnish the desired result. However, in actual physical systems—for example, in problems of rigid body dynamics—exact formulas for action variables and the rotation function can be obtained only for the simplest integrable systems. The main difficulty is to describe the basis cycles on tori in a formal way. Not only is Liouville's theorem local in nature, but also for the topological analysis of integrable systems as a whole we clearly must find out what basis to work with.

Let's formulate this problem more precisely. Suppose we have basis cycles on a nonsingular torus. Then we can consider the continuous trace of this basis on the family of tori, as the given torus runs along the edge of the molecule. The problem is to find out what happens with these cycles when the torus bifurcates. Another reformulation of the problem is the following. It is well-known that there exists a uniquely defined cycle on every atom: it is the vanishing cycle when the torus becomes the circle, or the critical hyperbolic periodic cycle otherwise. The problem is to find the continuous image of this cycle on the given Liouville torus; in other words, to find out how it can be expressed in terms of the basis cycles, all cycles being considered as the elements of their homology classes. Note that the marks r, n and

ε included in the Fomenko–Zieschang invariant reflect the answer to this question expressed in a formal way.

This problem can be solved for different physical systems in different ways. Here we will show how it is solved for two integrable cases of rigid body dynamics: the Lagrange case and the Goryachev–Chaplygin case. The two cases require quite different approaches.

A direct approach, expressing the torus in terms of the standard Euler–Poisson coordinates, will be used in the Lagrange case.

The approach used in the Goryachev–Chaplygin case can be briefly described as follows. We can consider the symplectic manifold and the Hamiltonian system on it as a finite-sheeted covering over another manifold with a Hamiltonian system. The new system preserves the general information about the initial one, but is in a sense easier to analyze. The main idea is to attempt to "bring" some information from the base space of the covering to the initial system.

Remark. Euler–Poisson dynamical systems in the Lagrange and Goryachev–Chaplygin cases contain certain parameters that characterize physical properties of the rotating body. It is possible to avoid these parameters by means of linear transformations of coordinates. This leads to loss in the initial physical meaning of the coordinates, but it is more convenient to apply them to the topological analysis of the systems.

The author is grateful to A. T. Fomenko and A. V. Bolsinov for bringing the problem to the author's attention.

16.2. The Goryachev–Chaplygin Problem

In this section we consider the Goryachev–Chaplygin case of rigid body dynamics. The corresponding dynamical system describes the rotation of a heavy rigid body about a fixed point, the principal moments of inertia satisfying the condition $A = B = 4C$ and the center of gravity lying in the equatorial plane of the ellipsoid of inertia. In addition the value of the area integral is zero.

The method of topological analysis presented in this section can be applied to some other cases of rigid body dynamics expressed in the form of a system of the Abel equations. The class of these problems is wide, including, for example, the problems of Sretenskii, Kovalevskaya, Clebsch, and Steklov.

The Euler–Poisson equations for the Goryachev–Chaplygin case have the form

$$4\dot{s}_1 = 3s_2 s_3,$$
$$4\dot{s}_2 = -3s_1 s_3 + r_3,$$
$$4\dot{s}_3 = -r_2,$$
$$4\dot{r}_1 = 4s_3 r_2 - s_2 r_3,$$
$$4\dot{r}_2 = s_1 r_3 - 4s_3 r_1,$$
$$4\dot{r}_3 = s_2 r_1 - s_1 r_2.$$

These equations have the four independent integrals

$$H = \tfrac{1}{2}(s_1^{\,2} + s_2^{\,2} + 4s_3^{\,2}) + r_1,$$
$$F = s_3(s_1^{\,2} + s_2^{\,2}) - s_1 r_3,$$
$$f_1 = r_1^{\,2} + r_2^{\,2} + r_3^{\,2} = 1,$$
$$f_2 = s_1 r_1 + s_2 r_2 + s_3 r_3 = 0.$$

Since the values of the last two integrals are fixed, all trajectories of the system belong to the four-dimensional manifold $M = \{f_1 = 1, f_2 = 0\}$. One can define a symplectic structure on M so that the system becomes an integrable Hamiltonian system, H being the Hamiltonian and F being the additional integral.

Recall that it is impossible to pass from the standard coordinates to the action-angle variables immediately in Goryachev–Chaplygin case. However we can pass to so-called Abelian variables, which are intermediate between the Euler–Poisson and the action-angle ones. This transformation was studied in [Kozlov 1980]. This is not a one-to-one correspondence, but a covering map $G : M \to M_0$. given by the formulas

$$z_1 + z_2 = s_3.$$
$$z_1 z_2 = -\tfrac{1}{4}(s_1^{\,2} + s_2^{\,2}),$$
$$H = \tfrac{1}{2}(s^2_1 + s^2_2 + 4s_3^{\,2}) + r_1,$$
$$F = s_3(s_1^{\,2} + s_2^{\,2}) - s_1 r_3.$$

Note that H, F parametrize the Liouville torus and z_1, z_2 are coordinates on the torus. The variables z_1, z_2 are defined up to permutation and we assume, for definiteness, that $z_1 \geq z_2$. As a result the system of the Euler–Poisson equations transforms to the system of Abel equations (the roots are considered as algebraic)

$$\frac{dz_1}{dt} = \frac{1}{2}\frac{\sqrt{P(z_1)}}{z_1 - z_2}, \qquad \frac{dz_2}{dt} = \frac{1}{2}\frac{\sqrt{P(z_2)}}{z_1 - z_2},$$

where

16.2.1. $P(z) = -(z^3 - \tfrac{1}{2}(H+1)z - \tfrac{1}{4}F)(z^3 - \tfrac{1}{2}(H-1)z - \tfrac{1}{4}F).$

With this change of variables, a torus transforms to the square $\{p_1 \leq z_1 \leq q_1\} \times \{p_2 \leq z_2 \leq q_2\}$. where p_i, q_i, for $i = 1, 2$, are different roots of the polynomial $P(z)$ such that $P(z) > 0$ in the interior of each segment $[p_i, q_i]$. To explain the system of Abel equations, let the initial conditions for z_1 and z_2 belong to the intervals (p_1, q_1) and (p_2, q_2), respectively; both radicals of the system are positive at the initial instant of time. The variables z_1 and z_2 increase until they reach the values q_1 and q_2, respectively. It can be shown easily that this happens in finite time. At that moment the sign of the radical in the corresponding equation changes, the variable begins to decrease, and the process is repeated. Note that the coordinates z_1, z_2, H, F are global on M_0, but they are not smooth. The point is that (z_1, z_2) are local coordinates on the image of Liouville torus, just as x is a local coordinate of the circle $x^2 + y^2 = 1$. It would be more rigorous to map the torus to the direct product of two circles $\{w_i^2 = P(z_i), z_i \in [p_i, q_i]\} \subset \mathbb{R}^2(z, w)$ (Figure 16.2) and to use the coordinates w_i near the points $(p_i, 0)$ and $(q_i, 0)$. One can show that with this approach the system of Abel equations has no singularities on the images of Liouville tori. But we will not dwell on this question here; see [Orel] for details.

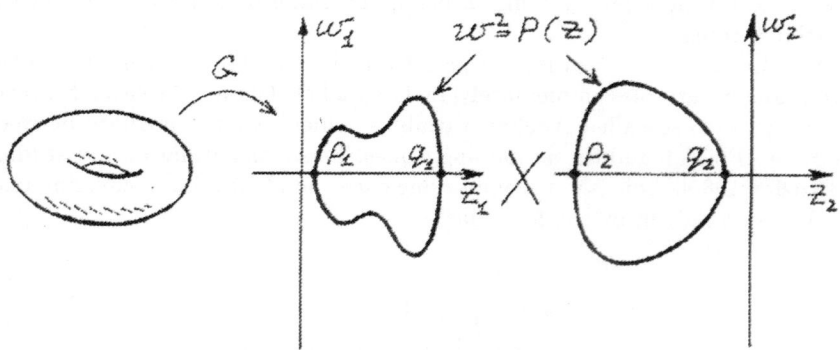

FIGURE 16.2.

The formulas of the inverse transformation from the Abelian to the Euler–Poisson variables are of special interest in this case. In our problem they are useful for the analysis of the covering. But it should be noted that they have one more important application: the exact solution of the Goryachev–Chaplygin problem in terms of two-dimensional theta-functions can be found with the help of these formulas. We now formulate this result.

Theorem 16.2.2. *Formulas of the inverse transformation from the coordinate* z_1, z_2, H, F *to the coordinates* $s_1, s_2, s_3, r_1, r_2, r_3$ *have the form*

$$s_1 = 2\frac{z_1 T(z_2) - z_2 T(z_1)}{\mathrm{Rad}(z_1, z_2)},$$

$$s_2 = -2\frac{z_1 \sqrt{P(z_2)} + z_2 \sqrt{P(z_1)}}{\mathrm{Rad}(z_1, z_2)},$$

$$s_3 = z_1 + z_2,$$

$$r_1 = -2\frac{T(z_1) - T(z_2)}{z_1 - z_2},$$

$$r_2 = -2\frac{\sqrt{P(z_1)} + \sqrt{P(z_2)}}{z_1 - z_2},$$

$$r_3 = 2\frac{\mathrm{Rad}(z_1, z_2)}{z_1 - z_2},$$

where

$$T(z) = z^3 - \tfrac{1}{2}Hz - \tfrac{1}{4}F,$$

$$\mathrm{Rad}(z_1, z_2) = \sqrt{2T(z_1)T(z_2) - 2\sqrt{P(z_1)P(z_2)} - \tfrac{1}{2}z_1 z_2}.$$

The proof of this theorem proceeds by direct analytical calculations.

We turn to a detailed analysis of the Goryachev–Chaplygin case. The momentum map and the bifurcation diagram carry the initial information about the system. Recall from the preceding chapter that the momentum map is a map $\Phi : M \to \mathbb{R}^2$, associating to each point the values of H and F there: $x \mapsto (H(x), F(x))$. Clearly, a Liouville torus maps to a point in $\mathbb{R}^2$. A point of M where H and F are dependent is called a *singularity* of the momentum map. the set of singularities is denoted by $K = \{x \in M : \mathrm{rank}\, d\Phi(x) < 2\}$. The image $\Sigma = \Phi(K)$ of this set is called the *bifurcation diagram*. A connected component of the inverse image $F^{-1}(y)$, for $y \in \Sigma$, is no longer a torus if it contains points of K, but on this surface we see a torus bifurcation.

Such torus bifurcations are classified in [Bolsinov et al. 1990], where the different types are denoted by letters with indices. The simplest bifurcations are marked A, A^*. and B. The letter A means the transformation of a torus to a circle; B means the torus falls apart into two tori; and A^* means the more complicated bifurcation of a torus into another torus.

The bifurcation diagram of the Goryachev–Chaplygin case was described in [Oshemkov 1991]. It was shown that the equations of its critical curves have the form

$$F = 0, \qquad\qquad H > -1,$$

$$H = \tfrac{3}{2}t^2 \pm 1, \quad F = t^3.$$

The momentum map and the bifurcation diagram are shown in Figure 16.3. Torus bifurcations are denoted by letters, and the numbers of

Liouville tori in the inverse image of each point of the plain are also marked. There is no torus in the inverse image of the domain

$$D_0 = \{H < \tfrac{3}{2}F^{2/3} - 1\};$$

there is one torus in the inverse image of every point of the domain

$$D_1 = \{\tfrac{3}{2}F^{2/3} - 1 < H < \tfrac{3}{2}F^{2/3} + 1, \ F \neq 0\};$$

and there are two tori in the inverse image of every point of the domain

$$D_2 = \{H > \tfrac{3}{2}F^{2/3} + 1, \ F \neq 0\}.$$

It can be seen that the range of the integral H is the half-line $[-1, +\infty)$. We distinguish two zones in this domain: $(-1, 1)$ and $(1, +\infty)$. The analysis performed in [Oshemkov 1991] showed that the torus bifurcations are qualitatively distinct depending on which zone H belongs to; for values of H in the same zone, there is no qualitative distinction. For the first zone, the unmarked molecule has the form shown in Figure 16.4(a), and for the second one the form shown in Figure 16.4(b). Tori on the edges of the molecules denoted by I are mapped into D_1, and those on edges II and III are mapped into D_2.

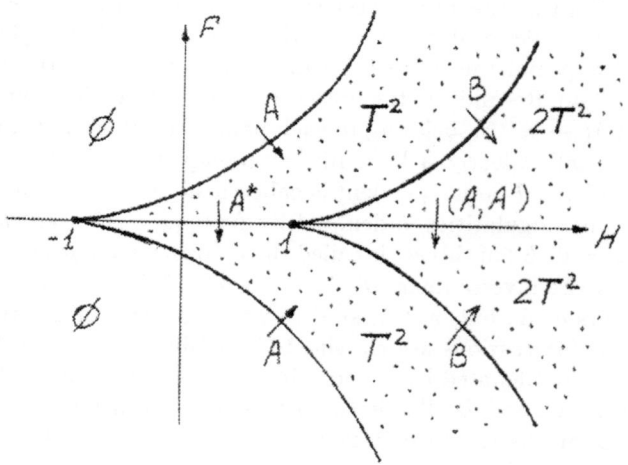

FIGURE 16.3.

A rough analysis shows that Goryachev–Chaplygin systems are not fiber-equivalent (and *a fortiori* not orbitally equivalent) on three-dimensional constant-energy surfaces whose values of energy belong to different zones. We cannot yet answer the question about fiber and orbital equivalence of the

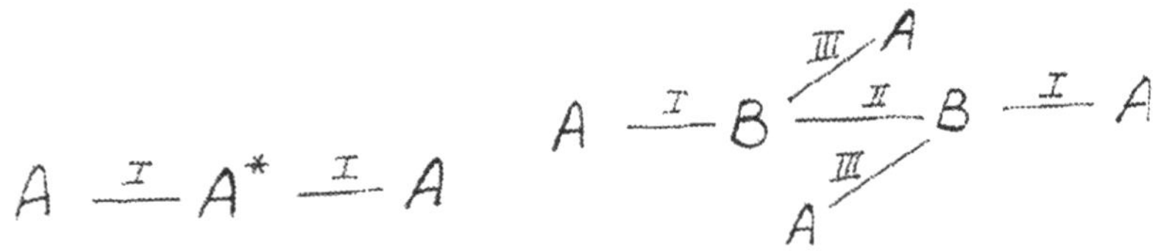

$$A \xrightarrow{\ I\ } A^* \xrightarrow{\ I\ } A \qquad\qquad A \xrightarrow{\ I\ } B \xrightarrow[\ III\ ]{\ II\ } B \xrightarrow{\ I\ } A$$

FIGURE 16.4.

systems with energies in the same zone; this will be a matter of future studies. Note that the analysis of fiber-equivalence can be performed analytically, but in order to solve the problem of orbital equivalence of the systems we have to use computer analysis at the last stage.

We begin with the study of the structure of the covering G. At first we have to find the images of Liouville tori under the covering. An analysis of the number of roots of the polynomial $P(z)$ shows that there are four roots in the domain D_1 and six in D_2. We can describe them in the following way. $P(x)$ is the product of two other polynomials $P_1(z)$ and $P_2(z)$ of third degree; see (16.2.1). The first factor has three real roots u_1, u_2, u_3, with $u_1 < u_2 < u_3$, in both domains, and the second one has three real roots v_1, v_2, v_3, with $v_1 < v_2 < v_3$, in D_2 and only one root v in D_1; all the roots depend on H and F. The polynomial $P(z)$ is greater than zero in the intervals (u_1, u_2) and (v, u_3) if $(H, F) \in D_1$, and in the intervals $(u_1, v_1), (v_2, u_2)$ and (v_3, u_3) if $(H, F) \in D_2$. A detailed analysis shows that the image of a torus on edge I under the covering G is a product of segments $[v, u_3] \times [u_1, u_2]$ if $F > 0$, and a product $[u_2, u_3] \times [u_1, v]$ if $F < 0$; the image of a torus on edge II is $[v_3, u_3] \times [u_1, v_1]$; and the image of a torus on edge III is $[v_3, u_3] \times [v_2, u_2]$ if $F > 0$, and $[u_2, v_2] \times [u_1, v_1]$ if $F < 0$. It can be easily seen that the image of the hyperbolic decomposition of a torus into two tori is the decomposition of the corresponding segment into two segments, and the transformation of a torus into a circle results in the degeneration of a segment to a point. The image of the surface $Q^3(H)$ in terms of local coordinates is shown in Figure 16.5.

In order to avoid numerical cases, we consider the domain $F \geq 0$. Since the dynamical system is symmetric about the line $F = 0$, all obtained results will be valid in the domain $F \leq 0$. Denote the Liouville tori belonging to edges I, II, and III, by T_L^I, T_L^{II}, and T_L^{III}, respectively, and their images by T_J^I, T_J^{II}, and T_J^{III}, respectively. Recall that the coordinates z_1, z_2 are local on a torus T_J. For the following analysis, it will be necessary to work with circles instead of segments. We denote them as follows:

$$a_1 = \{w^2 = P(z), \ z \in [v_3, u_3]\}, \qquad a_2 = \{w^2 = P(z), \ z \in [u_1, u_2]\},$$
$$a_2' = \{w^2 = P(z), \ z \in [u_1, v_1]\}, \qquad a_2'' = \{w^2 = P(z), \ z \in [v_2, u_2]\}.$$

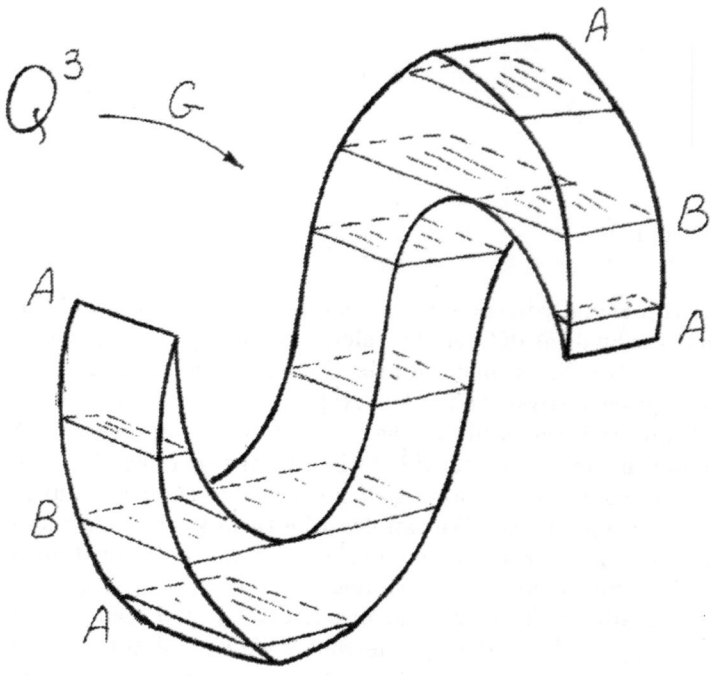

FIGURE 16.5.

Then we have $T_J^{\mathrm{I}} = a_1 \times a_2$, $T_J^{\mathrm{II}} = a_1 \times a_2'$, and $T_J^{\mathrm{III}} = a_1 \times a_2''$. The base space of the covering G is shown in Figure 16.6. Put $\gamma_1 = G^{-1}(a_1)$, $\gamma_2 = G^{-1}(a_2)$, $\gamma_2' = G^{-1}(a_2')$, and $\gamma_2'' = G^{-1}(a_2'')$. One can prove the following theorem by means of the inverse transformation formulas (Theorem 16.2.2).

Theorem 16.2.3. *G is a twofold covering. The curves γ_1, γ_2', and γ_2'' are connected and cover their images with multiplicity two. The curve γ_2 consists of two connected components, covering their images with multiplicity one.*

This theorem, and the structure of the base space of the covering G, make possible the calculation of the marks of the molecules, and lead to classes of topologically equivalent Goryachev–Chaplygin systems. We will not list the exact values for the marks, but we do formulate the theorem on fiber-classification, which is of great interest. Consider two Goryachev–Chaplygin systems on the manifolds $Q_1 = \{H = h_1\}$ and $Q_2 = \{H = h_2\}$, where h_1 and h_2 belong to either of two zones $(-1, 1)$ or $(1, +\infty)$.

Theorem 16.2.4. *Two Goryachev–Chaplygin systems on manifolds Q_1 and Q_2 are fiber-equivalent if and only if the values h_1 and h_2 of the Hamiltonian belong to the same zone.*

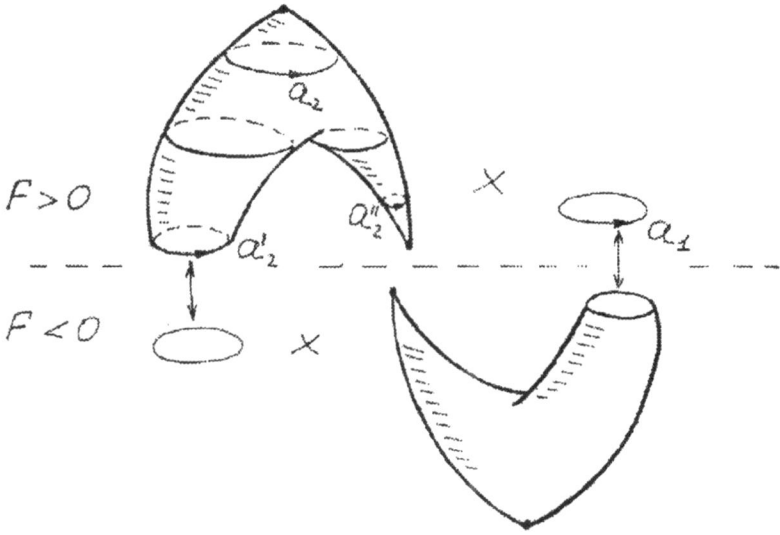

$F > 0$

$F < 0$

FIGURE 16.6.

We turn to the orbital analysis of the Goryachev–Chaplygin systems. Since the molecules contain the atoms A, A^*, and B only, we can apply the theory of orbital classification of simple systems [Bolsinov and Fomenko 1993] to this case. In accordance with the Bolsinov–Fomenko theory, the orbital invariant W^{*t} consists of three components W^*, R, and b, where W^* is a marked molecule, R is a family of rotation vectors on the edges of the molecule, and b is a numerical invariant. In our case the b-invariant carries no additional information because it coincides with the mark n included in W^*. The family of the rotation vectors R is of interest. Recall that the rotation vector is constructed based on the rotation function. The formulas for the functions ρ on the edges were obtained in [Orel]. The main idea of this work is as follows. Using the formulas of the rotation function in the base space of covering, one can extend them to the initial system. An analytical study of the formulas shows that the function ρ is monotone on edges II and III, and in addition we can find the values of the rotation vectors on these edges: $R^{II} = (\infty, 0)$ and $R^{III} = (\infty, 2)$. A computer analysis of the rotation vector on edge I was made by S. Takahashi (see the next chapter). The rotation function is not monotonous, and the values of the rotation vectors are different for different values of the Hamiltonian. Thus we have the following experimental result: *Goryachev–Chaplygin systems restricted to different constant-energy surfaces are orbitally nonequivalent.*

Theorem 16.2.4 and this last statement amount to a complete fiber- and orbital classification of the Goryachev–Chaplygin problem.

16.3. The Lagrange Problem

The Lagrange case of rigid body dynamics describes the rotation of a heavy rigid body about a fixed point. The principal moments of inertia of the body satisfy the condition $A = B > C$, and the center of gravity is located on the positive major semiaxis of inertia (see Figure 16.7). The simplest visual model of this rotating body is the Lagrange top, a toy familiar to every child.

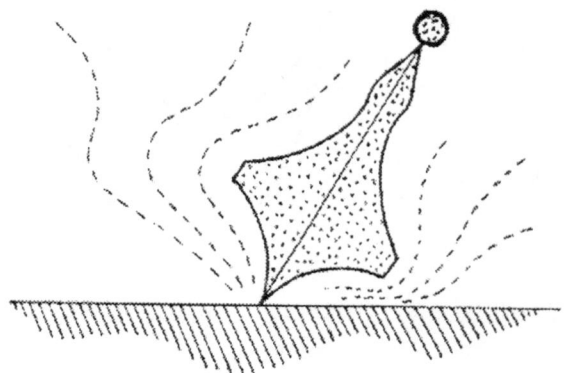

FIGURE 16.7.

The Euler–Poisson equations for the Lagrange case have the form

$$\dot{s}_1 = s_2 s_3 \left(1 - \frac{1}{\beta}\right) - r_2 V'(r_3),$$

$$\dot{s}_2 = s_1 s_3 \left(\frac{1}{\beta} - 1\right) + r_1 V'(r_3),$$

$$\dot{s}_3 = 0,$$

$$\dot{r}_1 = s_2 r_3 - \frac{s_3 r_2}{\beta},$$

$$\dot{r}_2 = -s_1 r_3 + \frac{s_3 r_1}{\beta},$$

$$\dot{r}_3 = s_1 r_2 - s_2 r_1.$$

Here the parameter β depends on the geometry of the body, $V(x)$ is a convex smooth function (potential) that characterizes the force field involved, and $V'(x)$ is its derivative. A linear potential corresponds to the motion of a rigid body in response to a uniform gravitational field.

The integrals of this system have the form

$$H = \frac{1}{2}\left(s_1^2 + s_2^2 + \frac{s_3^2}{\beta}\right) + V(r_3),$$

$$F = s_3,$$

$$f_1 = r_1^2 + r_2^2 + r_3^2 = 1,$$

$$f_2 = s_1 r_1 + s_2 r_2 + s_3 r_3 = g.$$

Here, as in the Goryachev–Chaplygin case, the four-dimensional symplectic manifold M is the level surface $\{f_1 = 1, f_2 = g\}$, where g is an arbitrary constant value. Defining a symplectic structure, we obtain the integrable Hamiltonian system with the Hamiltonian H and the additional integral F. Note that we can consider a family of the Lagrange systems under various values of g and β and various convex functions $V(x)$.

The momentum map for the Lagrange case was investigated in [Oshemkov 1991]. The inverse image of an arbitrary nonsingular point of the range of the momentum map turns out to consist of just one torus. It is sometimes convenient to identify this torus with the corresponding point in the range of the momentum map. Bifurcation diagrams for the four different general types of the Lagrange systems are shown in Figure 16.8. The relation between H and F on the bifurcation diagram is defined implicitly by the system

$$W(x) = 0, \qquad W'(x) = 0, \qquad |x| < 1,$$

where

$$W(x) = H - \frac{F^2}{2\beta} - V(x) - \frac{(g - Fx)^2}{2(1 - x^2)}.$$

The points

$$P_1 = \left(\frac{g^2}{2\beta} + V(1), g\right) \qquad \text{and} \qquad P_2 = \left(\frac{g^2}{2\beta} + V(-1), -g\right)$$

also belong to the bifurcation diagram. As we see in Figure 16.8, these points may or may not lie on the critical curves. We will analyze here the case indicated in Figure 16.8, because, from the viewpoint of topological analysis, it is the general case. The unmarked molecule has the simplest form A————A. The fiber classification of this case is well-known, and can be formulated as follows. Consider three zones for the energy:

$$\left(H_0, \frac{g^2}{2\beta} + v_l\right), \qquad \left(\frac{g^2}{2\beta} + v_l, \frac{g^2}{2\beta} + v_r\right), \qquad \left(\frac{g^2}{2\beta} + v_r, +\infty\right),$$

where H_0 is the least value of H among all values belonging to the range of the momentum map, and where $v_l = \min(V(1), V(-1))$, $v_r = \max(V(1), V(-1))$. (Some of the zones may be empty.) Unlike the Goryachev–Chaplygin case, Lagrange systems depend on three parameters, so we can deal with two

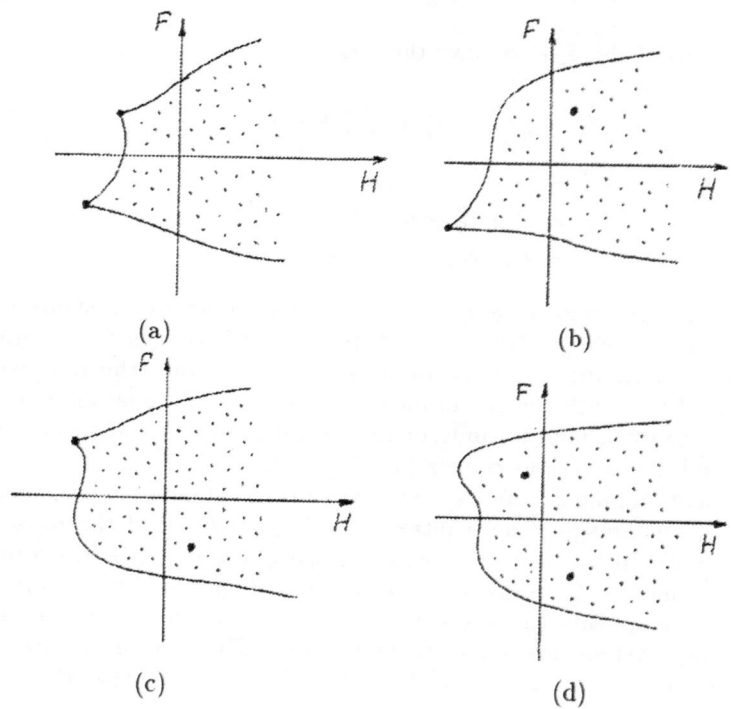

FIGURE 16.8.

different Hamiltonian systems $v_1 = \mathrm{sgrad}\, H_1$ and $v_2 = \mathrm{sgrad}\, H_2$, where $H_i = H_i(\beta_i, V_i)$, for $i = 1, 2$, on the manifolds $M_1(g_1)$ and $M_2(g_2)$, respectively.

Theorem 16.3.1. *Two Lagrange systems on the three-manifolds $Q_1 = \{H_1 = h_1\}$ and $Q_2 = \{H_2 = h_2\}$ are fiber-equivalent if and only if h_1 and h_2 belong to the same zone.*

As in Section 1, for the orbital classification of the Lagrange case it is sufficient to investigate the rotation function, because of the very simple type of the molecules. We will use direct formulas (16.1.1) and (16.1.2) to obtain ρ. Notice that we cannot apply these formulas until we make a more detailed topological analysis of the Lagrange case.

For a topological analysis it is most convenient to operate with the projections of the Liouville tori into $\mathbb{R}^3$. It can be easily shown that with this approach an arbitrary torus transforms to a cylinder $\{a \leq r_3 \leq b\}$ on the sphere $r_1^2 + r_2^2 + r_3^2 = 1$, where a and b are roots of the function $W(r_3)$ whose absolute values are less than 1 and $W(r_3) > 0$ in the interior of the segment $[a, b]$. Define two cycles on the torus $\{H = h, F = f, f_1 = 1, f_2 = g\}$ by the formulas

$$\gamma_1 = \{r_3 = c\}, \quad \gamma_2 = \{s_2 = 0, s_1 > 0\}.$$

Here $c \in [a, b]$ is arbitrary. Their projections into $\mathbb{R}^3$ are shown in Figure 16.9.

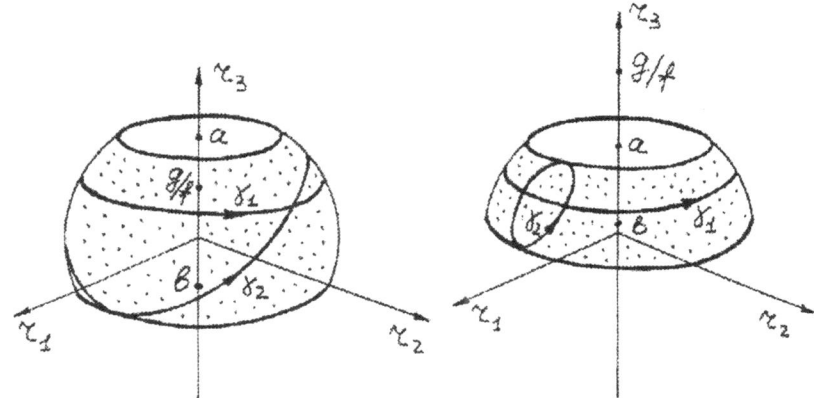

FIGURE 16.9.

The visual representation of these cycles allows us to follow their changes as H and F run through the set of admissible values. It turns out that in the general case, shown in Figure 16.8(d), the cycle γ_1 becomes a critical circle and the cycle γ_2 degenerates into a point as one approaches the bifurcation diagram. Notice that the first cycle is defined smoothly everywhere over the manifold M, but not the second one: it changes abruptly as it goes through the curves

$$H - \frac{F^2}{2\beta} - V\left(\frac{g}{F}\right) = 0, \quad |F| > g,$$

coming out of the points P_1 and P_2 (Figure 16.10). This is not surprising: a detailed topological analysis of the focus-focus singularities of the preceding chapter shows that, as one goes around such a singularity, the cycle γ_2 transforms to the sum $\gamma_1 + \gamma_2$. This clearly means that it is impossible to globally choose the action variable I_2 on the manifold M. Since we are investigating integrable Hamiltonian systems on constant-energy sets, it is more reasonable to make vertical cuts from the points P_1 and P_2 on the bifurcation diagram, instead of the cuts chosen above, and of course to redefine the cycles γ_2 accordingly.

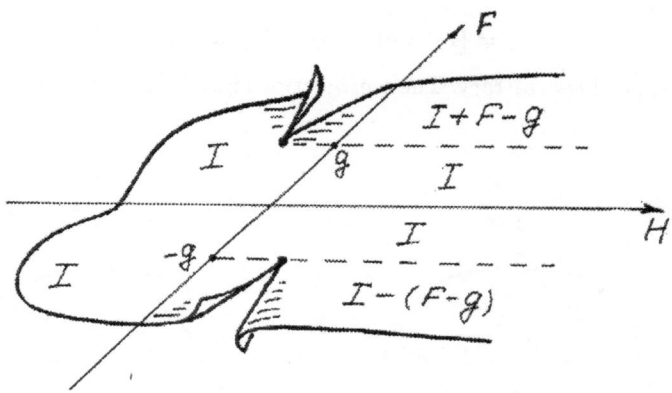

FIGURE 16.10.

It is now possible to apply the formulas (16.1.1) and (16.1.2). Simple calculations lead to the following formulas for the action variables:

$$I_1 = F,$$

16.3.2. $I_2 = \begin{cases} I + F - g & \text{if } H > g^2/(2\beta) + V(1) \text{ and } F > g, \\ I - (F + g) & \text{if } H > g^2/(2\beta) + V(-1) \text{ and } F < -g, \\ I & \text{otherwise,} \end{cases}$

where

16.3.3. $I = \dfrac{1}{\pi} \displaystyle\int_a^b \dfrac{\sqrt{(2H - (F^2/\beta) - 2V(x))(1 - x^2) - (g - Fx)^2}}{1 - x^2}\, dx.$

In the latter formula a and b are the roots of $W(x)$. For visualization, the domains appearing in (16.3.2) are indicated in Figure 16.10. It should be noted that the variable I_2 defined in this way is continuous everywhere over the range of the momentum map, except at the cuts (domain D).

The action variables were obtained earlier by Aksenenkova [1981] in the form $I_1 = F$, $I_2 = I$, where the function I is determined by (16.3.3), but she did not notice that the variable I_2 is not smooth. However, the investigation of the global behavior of these coordinates was not the goal of her work.

Using the formula (16.1.2), we can conclude that the rotation function is

$$\rho = -\frac{\partial I_2}{\partial F}.$$

This function is smooth in the domain D. The behavior of ρ was investigated by S. Takahashi by computer, and the rotation vector was calculated for numerical parameters (see the next chapter). He concluded that *the Lagrange* *systems described by the same parameters, but restricted to different constant-energy surfaces $Q_1 = \{H = h_1\}$ and $Q_2 = \{H = h_2\}$, are not orbitally equivalent.*

The question of orbital equivalence of Lagrange systems described by different parameters is of interest, but remains a matter for further investigation.

17. Numerical Calculation of the Orbital Invariant of Goryachev–Chaplygin and Lagrange Systems

S. Takahashi

17.1. Introduction

This chapter presents numerical calculations of the orbital invariant that classifies integrable Hamiltonian systems on constant-energy surfaces. The Bolsinov–Fomenko theory of orbital classification of integrable Hamiltonian systems is finer than that of fiber-classification, but it is difficult to obtain such an invariant without computers. The orbital invariant W^{*t} consists of some components that can be found and investigated analytically, except for the rotation vector. The rotation vector is an orbital invariant based on the rotation function, which is defined on a family of Liouville tori (the exact definition is given below). Here we consider integrable Hamiltonian systems whose rotation functions have an explicit analytic form: the Goryachev–Chaplygin and Lagrange systems defined in the preceding chapter.

We start by recalling the notions of topological equivalence and orbital equivalence between integrable Hamiltonian systems: We consider two integrable Hamiltonian systems, say v_1 on an constant-energy surface $Q_1 = \{x \in M : H_1 = \text{const}\}$ and v_2 on a constant-energy surface $Q_2 = \{x \in M : H_2 = \text{const}\}$. The two systems are *fiber-equivalent* if there exists a diffeomorphism $\tau : Q_1 \to Q_2$ that transforms the Liouville tori of the system v_1 into the Liouville tori of the system v_2 and preserves the orientation of the constant-energy surfaces [Bolsinov et al. 1990]. The two systems are *orbitally equivalent* if they are fiber-equivalent and in addition there exists a diffeomorphism $\tau : Q_1 \to Q_2$ that transforms the integral trajectories of the system v_1 into the integral trajectories of the system v_2.

While it is possible to check fiber-equivalence analytically, it is difficult to check orbital equivalence without numerical calculation on computers. For an orbital classification of integrable Hamiltonian systems, we defined in the preceding chapter the *rotation vector*, which consists of the following data about the rotation function ρ on an constant-energy surface $Q^3_{H_0} = \{x \in M | H(x) = H_0\}$:

1. minima and maxima,
2. boundary values,
3. $\pm\infty$. representing left and right limits ρ at the poles.

The following sections introduce calculation techniques for these invariants, and also show some results of these calculations for the Goryachev–Chaplygin and Lagrange systems.

17.2. Techniques for Numerical Calculations

This section explains some basic techniques used in the calculations needed for the orbital classification of integrable Hamiltonian systems.

The bisection method.   The bisection method [Press et al. 1992] finds a solution of an equation $\varphi(x) = 0$ when we already know an interval in which there is a solution. Suppose that the equation $\varphi(x) = 0$ has a solution in the interval (a, b). We first split the interval at the midpoint $c = \frac{1}{2}(a + b)$ into the two intervals (a, c) and (c, b). We then look at the signs of $\varphi(a)$, $\varphi(b)$, and $\varphi(c)$, and replace the original interval by one of the two half-length intervals, so that the function still has different signs at the two endpoints of the interval. For example, if $\varphi(a) \cdot \varphi(b) > 0$ and $\varphi(c) \cdot \varphi(b) < 0$, we select the interval (c, b) as a new interval. Note that the function in the above case definitely has a solution in (c, b) according to the intermediate value theorem.

In this way, the interval in which a solution is known to lie becomes twice as short with each step. This bisection process is repeated until the interval is shorter than the desired accuracy for the solution. The method is guaranteed to find a solution to that accuracy.

Calculation of definite integrals.   For the calculation of definite integrals we have used Simpson's rule [Press et al. 1992].

Reduction of improper integrals.   In the computation of the rotation function, we encounter an improper integral that has the form

$$I = \int_a^b \varphi(x)\,dx = \int_a^b \frac{Z(x)}{\sqrt{G(x)}},$$

where a and b are solutions of the polynomial equation $G(x) = 0$. We assume that the degree of the polynomial $G(x)$ is known.

Consider how to reduce this improper integral to a proper one. Since $G(x) = 0$ has two solutions a and b, it can be written as

17.2.1. $$G(x) = (b - x)(x - a)G_1(x)$$

where $G_1(x)$ is a polynomial. With the linear substitution $x = \frac{1}{2}(b - a)y + \frac{1}{2}(b + a)$. the integral becomes

$$I = \int_{-1}^1 \frac{Z(x(y))}{\sqrt{(1 - y^2)G_1(x(y))}}\,dy.$$

Be setting $y = \cos t$, we can obtain the desired proper integral:

$$I = \int_0^\pi \frac{Z(x(t))}{\sqrt{G_1(x(t))}}\, dy.$$

Now our next step is to find $G_1(x)$. If the degree of the polynomial $G(x)$ is m, the degree of $G_1(x)$ is $m-2$ according to (17.2.1). This reduces our problem to finding the coefficients of each term of $G_1(x)$ by comparing the coefficients of both sides of (17.2.1). By substituting $m - 1$ different values for x, we can obtain $m - 1$ independent linear equations from (17.2.1). For example, suppose that $G(x)$ is a polynomial of degree 4. We can then represent the equation (17.2.1) as

$$G(x) = (x_2 - x)(x - x_1)(Ax^2 + Bx + C),$$

where A, B, and C are the coefficients of $G_1(x)$ we have to find. By substituting three different real values for x, we can obtain a system of three independent equations with respect to A, B, and C. We can find these coefficients by solving the system of equations.

17.3. The Goryachev–Chaplygin Case

As described in the preceding chapter, the Goryachev–Chaplygin case has three domains, as follows:

$$D_0 = \{H < \tfrac{3}{2}F^{2/3} - 1\},$$
$$D_1 = \{\tfrac{3}{2}F^{2/3} - 1 < H < \tfrac{3}{2}F^{2/3} + 1,\ F \neq 0\},$$
$$D_2 = \{H > \tfrac{3}{2}F^{2/3} + 1,\ F \neq 0\}.$$

The Liouville foliation of the Goryachev–Chaplygin case is classified into Case I, Case II and Case III, which correspond to each edge of the atom-molecule graph of a constant-energy surface Q^3 (see preceding chapter).

Before examining orbital equivalence, we introduce several functions [Orel]. Set

$$P(x) = -Q(x)\, R(x),$$
$$Q(x) = x^3 - \tfrac{1}{2}(H + 1)x - \tfrac{1}{4}F,$$
$$R(x) = x^3 - \tfrac{1}{2}(H - 1)x - \tfrac{1}{4}F.$$

Let the solutions of $Q(x)$ be x_1, x_2, x_3, with $x_1 < x_2 < x_3$; let the solutions of $R(x)$ in D_2 be y_1, y_2, y_3, with $y_1 < y_2 < y_3$; and let y be the solution of $R(x)$ in D_1. These solutions can be determined by the bisection method. We now find an interval containing each of the solutions. First, we consider the equation $Q(x) = 0$. Since the derivative of $Q(x)$ is $Q'(x) = 3x^2 - \tfrac{1}{2}(H + 1)$, the function $Q(x)$ has slopes as follows:

x	$\ldots$	$-\sqrt{\frac{1}{6}(H+1)}$	$\ldots$	$\sqrt{\frac{1}{6}(H+1)}$	$\ldots$
$Q'(x)$	$+$	0	$-$	0	$+$
$Q(x)$	$\nearrow$	$\frac{1}{3}(H+1)\sqrt{\frac{1}{6}(H+1)}-\frac{1}{4}F$	$\searrow$	$-\frac{1}{3}(H+1)\sqrt{\frac{1}{6}(H+1)}-\frac{1}{4}F$	$\nearrow$

If $H > \frac{3}{2}F^{3/2} - 1$, then $Q(-\sqrt{\frac{1}{6}(H-1)}) > 0$ and $Q(\sqrt{\frac{1}{6}(H-1)}) < 0$. Hence $Q(x)$ has solutions $x_1 < x_2 < x_3$, where $x_1 \in (-\infty, -\sqrt{\frac{1}{6}(H+1)})$, $x_2 \in (-\sqrt{\frac{1}{6}(H+1)}, \sqrt{\frac{1}{6}(H+1)})$, and $x_3 \in (\sqrt{\frac{1}{6}(H+1)}, +\infty)$.

Next we consider the equation $R(x) = 0$. We have $R'(x) = 3x^2 - \frac{1}{2}(H-1)$, so $R(x)$ has slopes as follows:

x	$\ldots$	$-\sqrt{\frac{1}{6}(H-1)}$	$\ldots$	$\sqrt{\frac{1}{6}(H-1)}$	$\ldots$
$Q'(x)$	$+$	0	$-$	0	$+$
$Q(x)$	$\nearrow$	$\frac{1}{3}(H-1)\sqrt{\frac{1}{6}(H-1)}-\frac{1}{4}F$	$\searrow$	$-\frac{1}{3}(H-1)\sqrt{\frac{1}{6}(H-1)}-\frac{1}{4}F$	$\nearrow$

If $H > \frac{3}{2}F^{3/2}+1$, then $Rg(-\sqrt{\frac{1}{6}(H-1)}) > 0$ and $Rg(\sqrt{\frac{1}{6}(H-1)}g) < 0$. Hence $R(x)$ has solutions $y_1 < y_2 < y_3$, where $y_1 \in g(-\infty, -\sqrt{\frac{1}{6}(H-1)}g)$, $y_2 \in g(-\sqrt{\frac{1}{6}(H-1)}, \sqrt{\frac{1}{6}(H-1)}g)$, and $y_3 \in g(\sqrt{\frac{1}{6}(H-1)}, +\infty)$. If $H < \frac{3}{2}F^{3/2} + 1$, then $Rg(-\sqrt{\frac{1}{6}(H-1)}g) < 0$ and $R(\sqrt{\frac{1}{6}(H-1)}g) > 0$. Hence $R(x)$ has only one solution $y \in (-\infty, \infty)$.

In this way, we can confine the interval of each solution, and can find the solution using the bisection method.

Now we are ready to see the calculation of the rotation functions. The following analytic forms of the rotation functions are presented in [Orel].

The Goryachev–Chaplygin case: Case I.   The Liouville tori of Case I appear in the domain D_1. The bifurcation diagram is determined by

$$H = \frac{3}{2}F^{\frac{2}{3}} - 1, \quad H = \frac{3}{2}F^{\frac{2}{3}} + 1, \quad -1 \le H \le 1, \quad \text{and} \quad f = 0.$$

The rotation function is

17.3.1.
$$\rho = \begin{cases} \dfrac{\int_{x_1}^{x_2} \frac{dx}{\sqrt{P(x)}}}{\int_{y}^{x_3} \frac{dx}{\sqrt{P(x)}}} & \text{if } f > 0, \\[2em] \dfrac{\int_{x_2}^{x_3} \frac{dx}{\sqrt{P(x)}}}{\int_{x_1}^{y} \frac{dx}{\sqrt{P(x)}}} & \text{if } f < 0. \end{cases}$$

The first integral in (17.3.1) is improper because x_3 and y are solutions of $P(x)$. Hence, it is necessary to reduce it to a proper integral. Since $P(x)$

has degree 6, we can reduce this integral to a proper integral by finding the coefficients A, B, C, D, and E in the equation

$$P(x) = (x_3 - x)(x - y)(Ax^4 + Bx^3 + Cx^2 + Dx + E).$$

The second integral in (17.3.1) can be reduced to a proper integral by finding the coefficients A, B, C, D, and E in the equation

$$P(x) = (y - x)(x - x_1)(Ax^4 + Bx^3 + Cx^2 + Dx + E).$$

The Goryachev–Chaplygin case: Case II.   The Liouville tori of Case II appear in the domain D_2. The bifurcation diagram is determined by

$$H = \tfrac{3}{2}F^{\frac{2}{3}} + 1.$$

The rotation function is

17.3.2.
$$\rho = \frac{\int_{x_1}^{y_1} \frac{dx}{\sqrt{P(x)}}}{\int_{y_3}^{x_3} \frac{dx}{\sqrt{P(x)}}}$$

The integral (17.3.2) can be reduced to a proper integral by finding the coefficients A, B, C, D, and E in the equation

$$P(x) = (x_3 - x)(x - y_3)(Ax^4 + Bx^3 + Cx^2 + Dx + E).$$

The Goryachev–Chaplygin case: Case III.   The Liouville tori of Case III also appear in the domain D_2. The bifurcation diagram is determined by

$$H = \tfrac{3}{2}F^{\frac{2}{3}} + 1, \quad H \geq 1, \quad \text{and} \quad f = 0.$$

The rotation function is

17.3.3.
$$\rho = \begin{cases} \dfrac{\int_{x_2}^{y_2} \frac{dx}{\sqrt{P(x)}}}{\int_{y_3}^{x_3} \frac{dx}{\sqrt{P(x)}}} & \text{if } f > 0, \\[4ex] \dfrac{\int_{y_2}^{x_2} \frac{dx}{\sqrt{P(x)}}}{\int_{y_1}^{x_1} \frac{dx}{\sqrt{P(x)}}} & \text{if } f < 0. \end{cases}$$

The first integral of (17.3.2) can be reduced to a proper integral by finding the coefficients A, B, C, D, and E in

$$P(x) = (x_3 - x)(x - y_3)(Ax^4 + Bx^3 + Cx^2 + Dx + E).$$

The second integral can be reduced to a proper integral by finding the coefficients A, B, C, D, and E in

$$P(x) = (y_1 - x)(x - x_1)(Ax^4 + Bx^3 + Cx^2 + Dx + E).$$

Results.   The rotation function was calculated for each of the three cases. Figure 17.1(a) shows level curves of the rotation function of Case II in D_2. The shapes of the rotation functions on $H = 2.50$ and $H = 3.50$ are given in Figure 17.1(b) and Figure 17.1(c). The corresponding rotation vectors and the positions of the eigenvalues are shown in Figure 17.1(d) and Figure 17.1(e). Figure 17.2 shows level curves of the rotation function of Case III in D_2. The shapes of the rotation functions for $H = 2.50$ and $H = 3.50$ are given in Figure 17.2(b) and Figure 17.2(c). In this figure, solid lines indicate the case $F < 0$, and dotted lines the case $F > 0$. The corresponding rotation vectors and the positions of the characteristic values are shown in Figure 17.2(d) and Figure 17.2(e). Analytical investigation [Orel] shows that the rotation functions of these two cases are monotonic on $H = $ const. Our results agree with this fact.

Figure 17.3(a) shows level curves of the rotation function of Case I in D_1. This result also shows that the rotation function on $H = $ const is almost monotonic. Let us see the shape of the rotation function on $H = $ const. When $-1.0 \le H \le 1.0$, the rotation function is monotonic on $H = $ const. Figure 17.3(b) and Figure 17.3(c) show the shapes of the rotation functions on $H = 0.50$ and $H = 1.00$. Here, solid lines indicate the rotation function of $F < 0$, and dotted lines indicate the rotation function of $F > 0$. When H is greater than 1.0, the rotation function on $H = $ const has a local minimum. Figure 17.3(d) and Figure 17.3(e) show the shapes of the rotation functions on $H = 1.025$ and $H = 1.225$. Note that these shapes contain local minima. When H becomes greater than some value H_0, the local minima disappear. Figure 17.3(f) and Figure 17.3(g) show the monotonic shapes of the rotation functions on $H = 1.250$ and $H = 1.500$. Parts (h) and (i) of Figure 17.3(h) show the corresponding rotation vectors and the positions of the eigenvalues. It follows from these results that the rotation function on $H = $ const is monotonic except for a small interval of values for H.

Based on these results, we can say that, *in the Goryachev–Chaplygin case, the Hamiltonian systems defined by* sgrad H *on the level sets* $\{H = H_1\}$ *and* $\{H = H_2\}$ *are not orbitally equivalent unless* $H_1 = H_2$.

17.4.   The Lagrange Case

This section presents numerical calculations for the orbital classification in the Lagrange case. The formulas of the Lagrange case are more complicated than those of the Goryachev–Chaplygin case. We describe numerical calculations for the following components: bifurcation diagrams, action variables, and rotation functions.

Calculation of bifurcation diagrams.   As described in the preceding chapter, the critical curves of the bifurcation diagram are given by the equa-

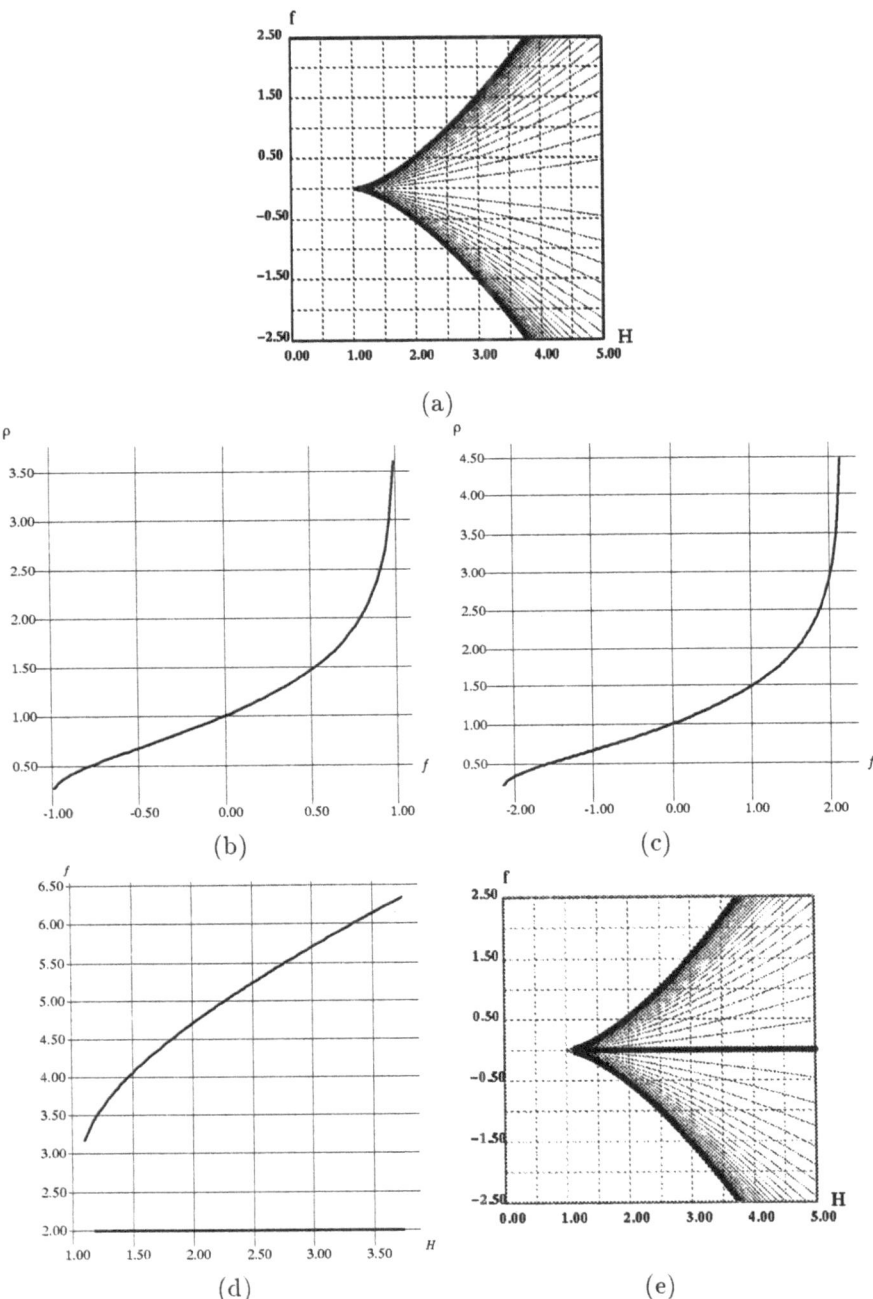

FIGURE 17.1. The rotation function of Case II: (a) level curves of the rotation function, (b) the shapes of the rotation functions on $H = 2.500$ and (c) on $H = 3.500$, (d) the rotation vectors, and (e) the positions of the eigenvalues.

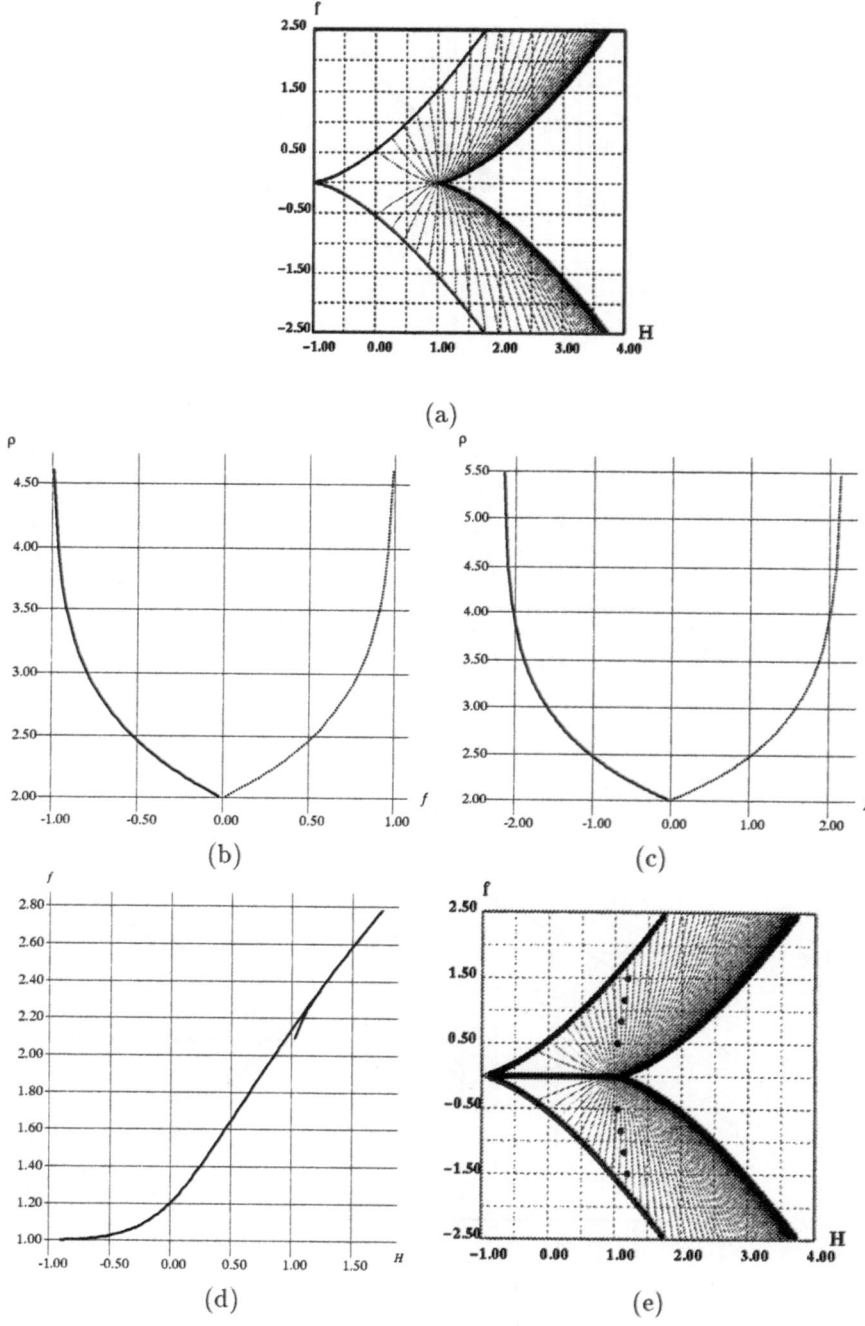

FIGURE 17.2. The rotation function of Case III: (a) level curves of the rotation function, (b) the shapes of the rotation functions on $H = 2.500$, (c) on $H = 3.500$, (d) the rotation vectors, and (e) the positions of the eigenvalues.

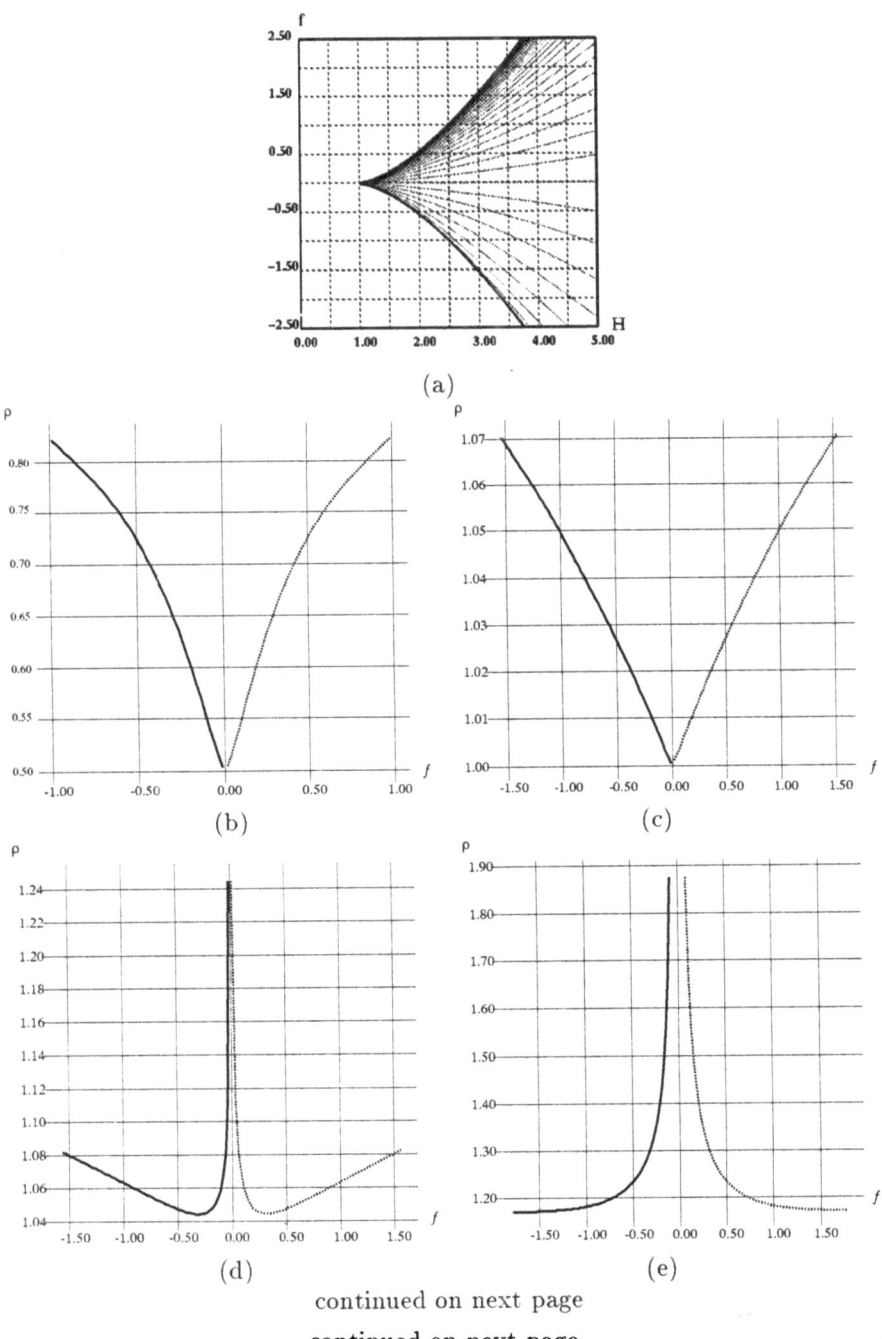

(a)

(b)

(c)

(d)

(e)

continued on next page

continued on next page

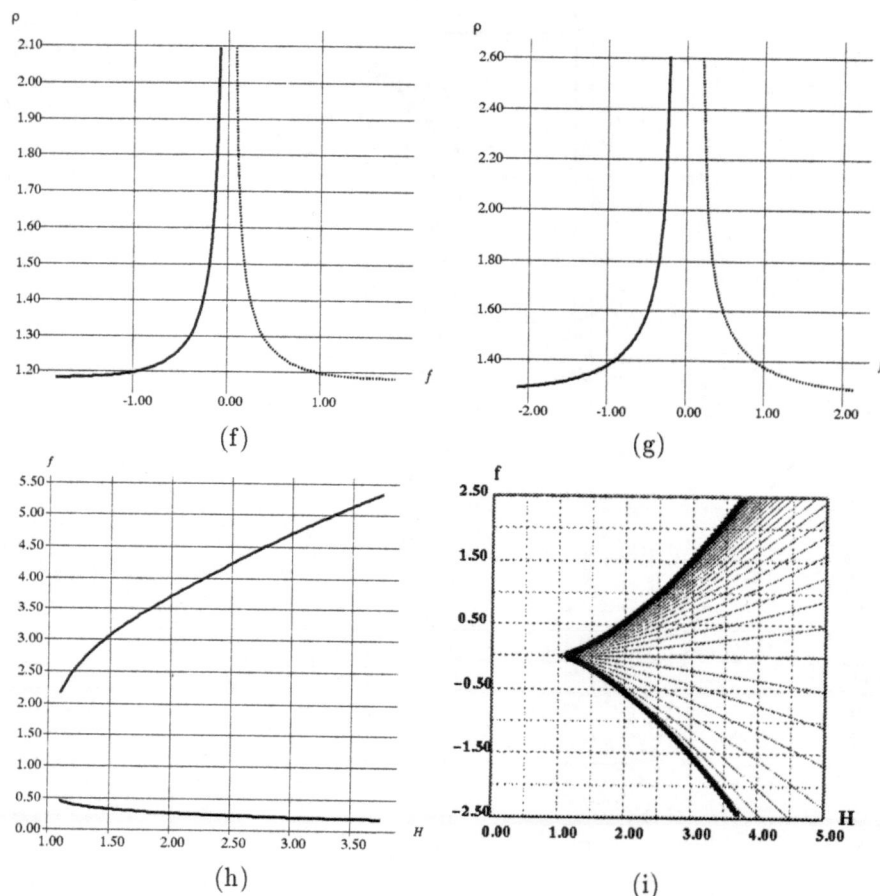

FIGURE 17.3. The rotation of Case I: (a) level curves of the rotation function, (b) the shapes of the rotation functions on $H = 0.500$, (c) on $H = 1.000$, (d) on $H = 1.025$, (e) on $H = 1.225$, (f) on $H = 1.250$, (g) on $H = 1.500$. (h) the rotation vectors, and (i) the positions of the eigenvalues.

tions $W(x) = 0$ and $W'(x) = 0$, where

17.4.1.
$$W(x) = \frac{(g - Fx)^2}{2(1 - x^2)} + \frac{F^2}{2\beta} + V(x) - H$$

and $|x| < 1$. Since

$$\lim_{x \to -1+0} W'(x) = +\infty, \qquad \lim_{x \to +1-0} W'(x) = -\infty, \qquad \text{and} \qquad W''(x) > 0,$$

$W'(x)$ is monotonically increasing in the interval $(-1, 1)$, and there exists only one solution in this interval. The bisection method, with initial interval

$(-1, 1)$, is used to calculate the solution for given values of g, β, and F. Once the solution x_0 has been found, we can then find H from the relation $W(x_0) = 0$.

The critical points of the bifurcation diagram are

$$(H_1, F_1) = \left(\frac{g^2}{2\beta} + V(1), g\right) \quad \text{and} \quad (H_2, F_2) = \left(\frac{g^2}{2\beta} + V(-1), -g\right).$$

Note that the points (H_1, f_1) and (H_2, f_2) are not, in general, contained in the critical curves.

Calculation of the action variables. The action variables are defined by

$$I_1 = F,$$

$$I_2 = \begin{cases} I + F - g & \text{if } H > g^2/(2\beta) + V(1) \text{ and } F > g, \\ I - (F + g) & \text{if } H > g^2/(2\beta) + V(-1) \text{ and } F < -g, \\ I & \text{otherwise}, \end{cases}$$

where

17.4.2. $\quad I = \dfrac{1}{\pi} \displaystyle\int_{\arccos x_2}^{\arccos x_1} \sqrt{2H - \dfrac{F^2}{\beta} - 2V(\cos t) - \dfrac{(g - F\cos t)^2}{1 - \cos^2 t}}\, dt.$

Here, x_1 and x_2 are the solutions of the equation

17.4.3. $\quad \left(2H - \dfrac{F^2}{\beta} - 2V(x)\right)(1 - x^2) - (g - Fx)^2 = 0,$

satisfying $|x_1|, |x_2| < 1$ and $x_1 \leq x_2$.

It is clear that x_1 and x_2 are also solutions of $W(x) = 0$. Let $x_0 \in (-1, 1)$ be the solution of $W'(x) = 0$ with the given parameters g, β, and F. In the region confined by the critical curves, the relation $W(x_0) < 0$ follows from the equality $W(x) = 0$. Since

$$\lim_{x \to -1+0} W(x) \to +\infty, \quad \lim_{x \to +1-0} W(x) \to +\infty, \quad \text{and} \quad W''(x) > 0,$$

it can be shown that $x_1 \in (-1, x_0)$ and $x_2 \in (x_0, 1)$. Here we also use the bisection method to calculate the two solutions x_1 and x_2 of (17.4.3).

The integral (17.4.2) becomes improper when $x_1 = -1$. This occurs if and only if

$$F = -g \quad \text{and} \quad H > \frac{g^2}{2\beta} + V(-1).$$

In this case, (17.4.2) can be reduced to

$$I = \frac{1}{\pi} \int_{\arccos x_2}^{\arccos x_1} \sqrt{2H - \frac{F^2}{\beta} - 2V(\cos t) + \frac{g^2(1 + \cos t)}{1 - \cos t}}\, dt.$$

Similarly, $x_2 = 1$ if and only if

$$F = g \qquad \text{and} \qquad H > \frac{g^2}{2\beta} + V(1).$$

Then (17.4.2) can be reduced to

$$I = \frac{1}{\pi} \int_{\arccos x_2}^{\arccos x_1} \sqrt{2H - \frac{F^2}{\beta} - 2V(\cos t) + \frac{g^2(1 - \cos t)}{1 + \cos t}}\, dt.$$

Calculation of the rotation functions. The rotation function is

$$\rho = -\frac{dI_2}{dF} = \begin{cases} -\dfrac{dI}{dF} - 1, & \text{if } H > g^2/(2\beta) + V(1) \text{ and } F > g, \\[2ex] -\dfrac{dI}{dF} + 1, & \text{if } H > g^2/(2\beta) + V(-1) \text{ and } F < -g, \\[2ex] -\dfrac{dI}{dF}, & \text{otherwise.} \end{cases}$$

Here,

17.4.4.
$$\frac{dI}{dF} = \frac{1}{\pi} \int_{x_1}^{x_2} Z(x)\, dx,$$

where

$$Z(x) = \frac{-F/\beta + x(g - Fx)/(1 - x^2)}{\sqrt{(1 - x^2)(2H - F^2/\beta - 2V(x)) - (g - Fx)^2}},$$

and x_1 and x_2 are the solutions of (17.4.3). As the reader can see, the function $\mathbb{Z}(x)$ has integrable singularities at x_1 and x_2, and the integral (17.4.4) is improper. Thus, we have to reduce it to a proper integral.

Consider the function

$$G(x) = (1 - x^2)(2H - F^2/\beta - 2V(x)) - (g - Fx)^2.$$

$G(x) = 0$ has two solutions x_1 and x_2, and can be written as

$$G(x) = (x_2 - x)(x - x_1)G_1(x).$$

Since $G(x)$ is a polynomial of degree 4, we can see that $G_1(x)$ has the form $G_1(x) = Ax^2 + Bx + C$, where A, B, and C are the coefficients of $G_1(x)$. As described earlier, we can reduce the improper integral by finding the coefficients of $G_1(x)$. Finally, the rotation function ρ can be calculated using the obtained proper integral.

Note that the above integral has singularities at $H > g^2/(2\beta) + V(1)$ and $F = g$, and at $H > g^2/(2\beta) + V(-1)$ and $F = -g$. The rotation functions are calculated from the continuity on these equations.

Results.   Figures 17.4, 17.5, 17.6, and 17.7 show bifurcation diagrams of the Lagrange case. Figure 17.4 is the bifurcation diagram where $g = 2.0$, $\beta = 1.5$, and $V(x) = x$. This diagram has two center-center critical points, i.e., the two critical points are contained in the critical curves. Figure 17.5 is the bifurcation diagram where $g = 1.0$, $\beta = 2.5$, and $V(x) = 2x$. This diagram has one center-center critical point and one focus-focus critical point. Figure 17.6 is the bifurcation diagram where $g = 1.0$, $\beta = 2.5$, and $V(x) = -2x$. This diagram also has one center-center critical point and one focus-focus critical point. Figure 17.7 is the bifurcation diagram where $g = 1.0$, $\beta = 2.0$, and $V(x) = 2x^2 - x$. This diagram has two focus-focus critical points.

Level curves of the action variables are shown in Figures 17.8 (for $g = 1.0$, $\beta = 0.5$, and $V(x) = x + 1$), 17.9 (for $g = 1.0$, $\beta = 0.5$, and $V(x) = x^2$), and 17.10 (for $g = 1.0$, $\beta = 0.5$, and $V(x) = x^2 + \frac{1}{2}x$).

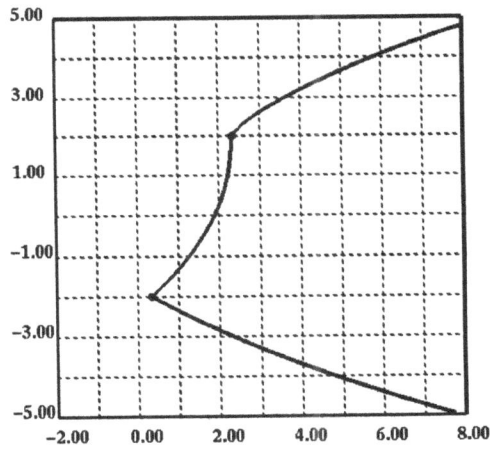

FIGURE 17.4. The bifurcation diagram ($g = 2.0$, $\beta = 1.5$, and $V(x) = x$).

We examine the rotation function of the Lagrange case. Figure 17.11(a) shows level curves of the rotation function, where $g = 1.0$, $\beta = 0.5$, and $V(x) = x + 1$. The shape of the rotation function on $H = $ const changes with H as shown in the following figures. Parts (b) and (c) of Figure 17.11 show the shapes of the rotation functions on $H = 1.80$ and $H = 2.76$. The rotation functions in these figures are monotonic with respect to F. Around the focus-focus critical point, the shape of the rotation function on $H = $ const contains local extrema. Parts (d), (e), and (f) of Figure 17.11 show the shapes of the rotation functions on $H = 2.79$, $H = 3.60$, and $H = 4.11$. For H ranging from some value H_0 to ∞, the rotation function on $H = $ const becomes

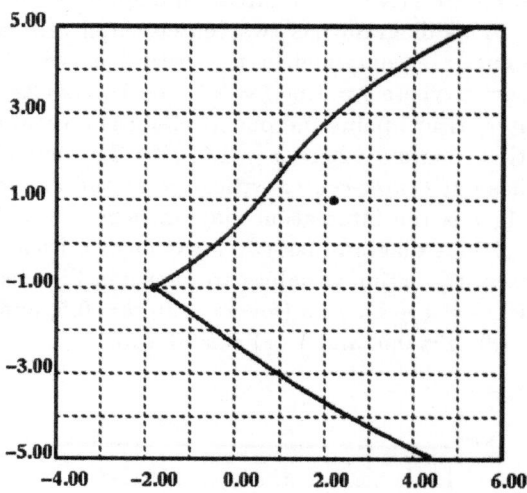

FIGURE 17.5. The bifurcation diagram ($g = 1.0$, $\beta = 2.5$, and $V(x) = 2x$).

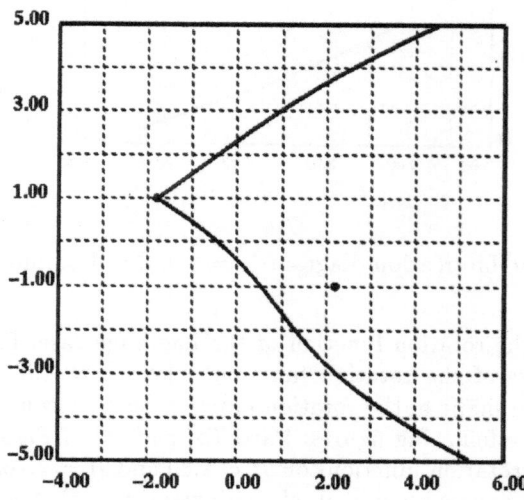

FIGURE 17.6. The bifurcation diagram ($g = 1.0$, $\beta = 2.5$, and $V(x) = -2x$).

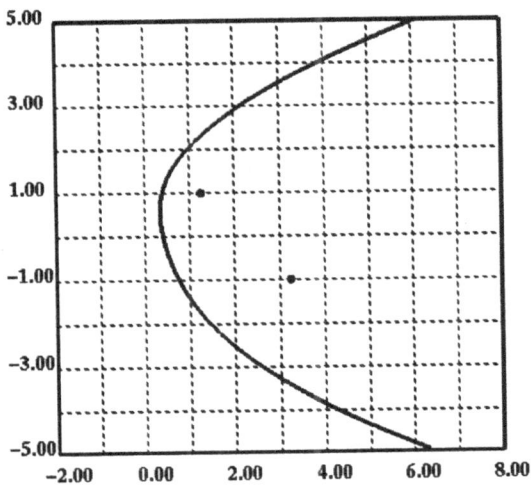

FIGURE 17.7. The bifurcation diagram ($g = 1.0$, $\beta = 2.0$, and $V(x) = 2x^2 - x$).

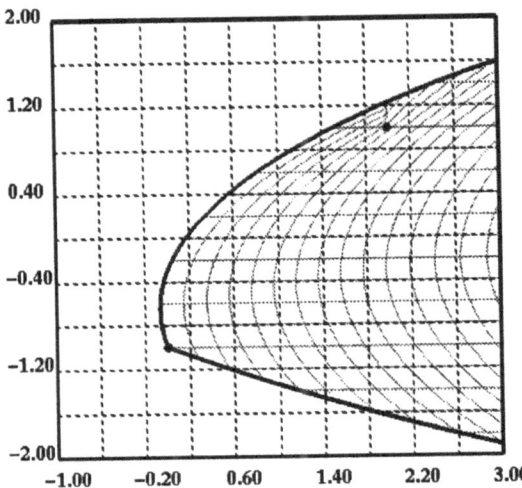

FIGURE 17.8. Level curves of the action variables ($g = 1.0$, $\beta = 0.5$, and $V(x) = x + 1$).

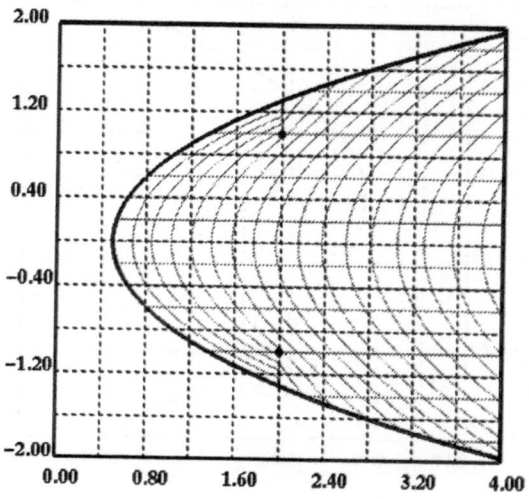

FIGURE 17.9. Level curves of the action variables ($g = 1.0$, $\beta = 0.5$, and $V(x) = x^2$).

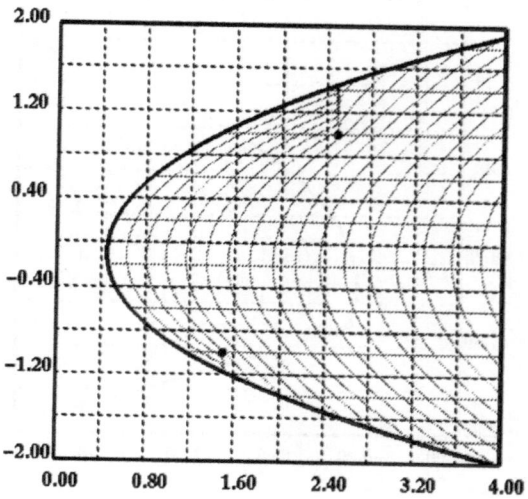

FIGURE 17.10. Level curves of the action variables ($g = 1.0$, $\beta = 0.5$, and $V(x) = x^2 + \frac{1}{2}x$).

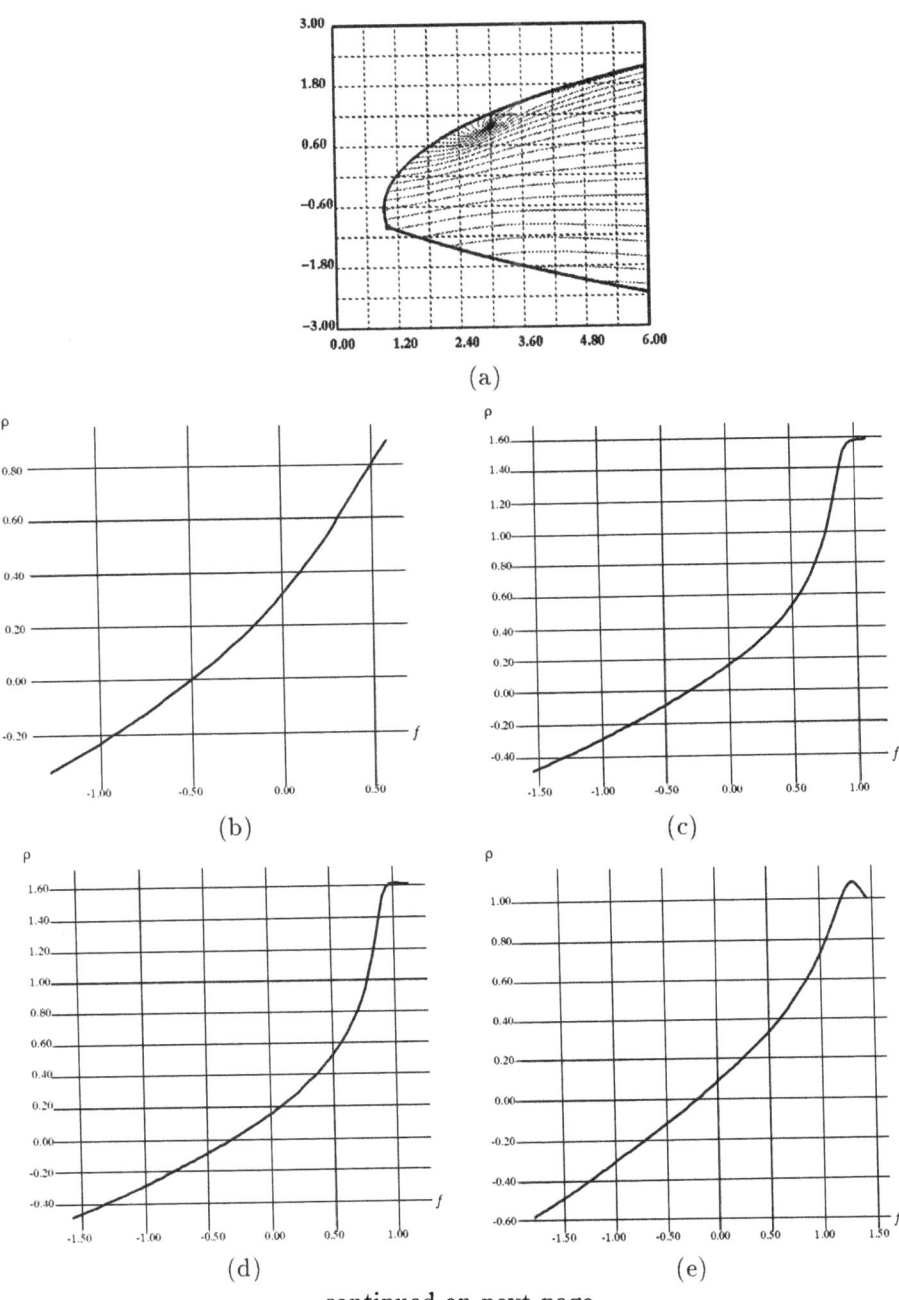

(a)

(b)

(c)

(d)

(e)

continued on next page

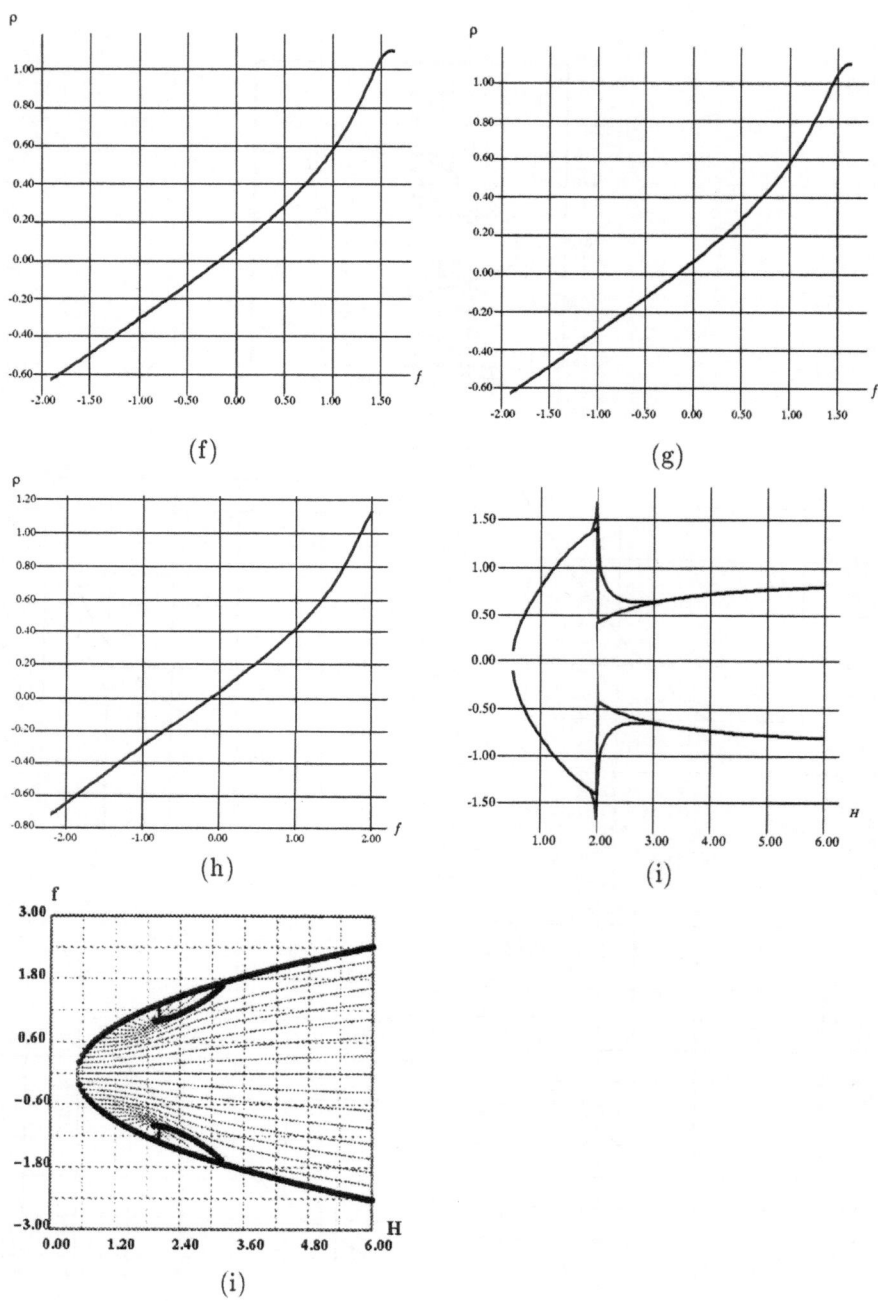

FIGURE 17.11. The rotation function ($g = 1.0$, $\beta = 0.5$, and $V(x) = x +$ 1): (a) level curves of the rotation function, (b) the shapes of the rotation functions on $H = 1.80$, (c) on $H = 2.76$, (d) on $H = 2.79$, (e) on $H = 3.60$, (f) on $H = 4.11$, (g) on $H = 4.14$, (h) on $H = 5.40$, (i) the rotation vectors, and (j) the positions of their eigenvalues.

monotonic again. Parts (g) and (h) show the shapes of the rotation functions on $H = 4.14$ and $H = 5.40$. The corresponding rotation vectors and the positions of the eigenvalues are shown in parts (i) and (j) of the same figure.

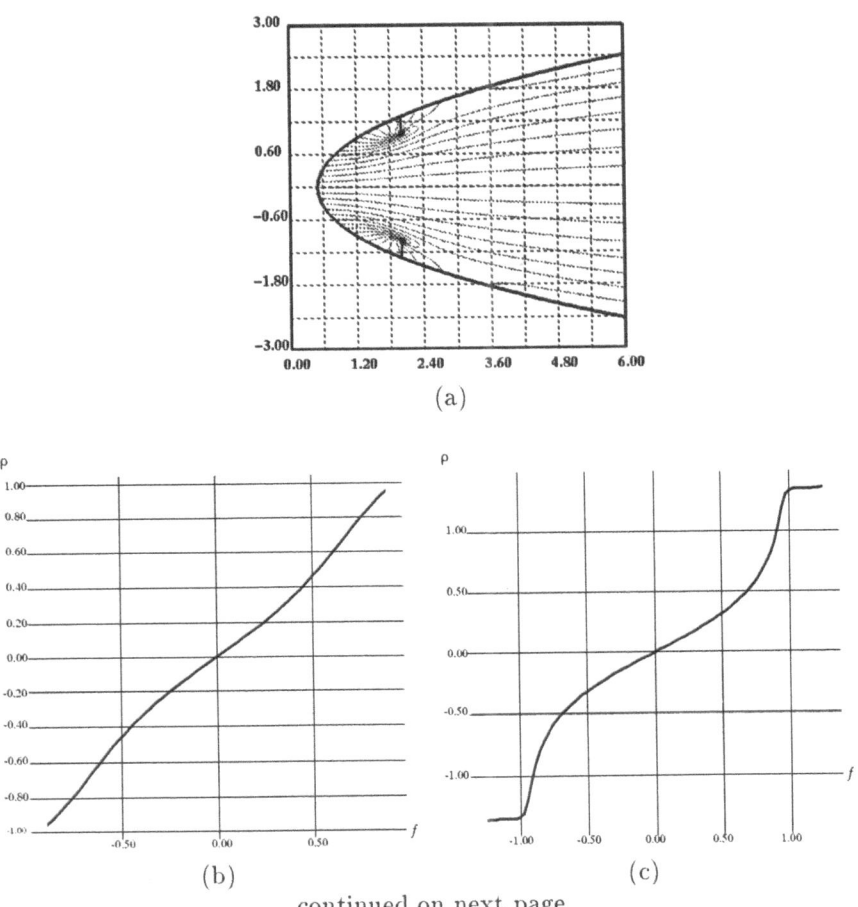

(a)

(b)

(c)

continued on next page

Figure 17.12(a) shows level curves of the rotation function, where $g = 1.0$, $\beta = 0.5$, and $V(x) = x^2$. The shape of the rotation function on $H = $ const is changed according to H as shown in the following figures. Parts (b) and (c) of Figure 17.12 show the shapes of the rotation functions on $H = 1.20$ and $H = 1.86$. The rotation functions in these figures are monotonic with respect to F. Around the two focus-focus critical points, the shape of the rotation function on $H = $ const contains local extrema. Part (d), (e), and (f) show the shapes of the rotation functions on $H = 1.89$, $H = 2.40$, and

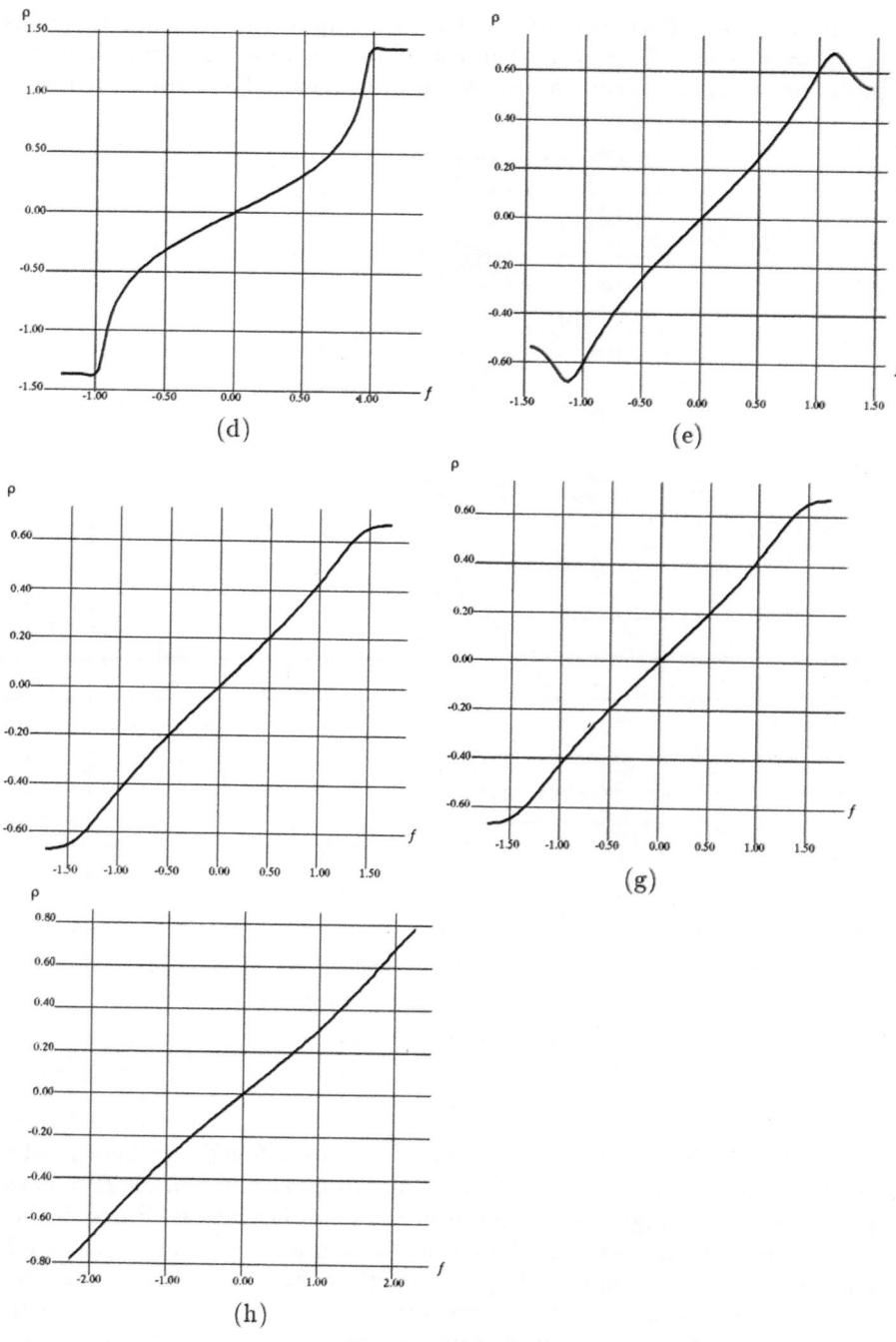

(d)

(e)

(g)

(h)

continued on next page

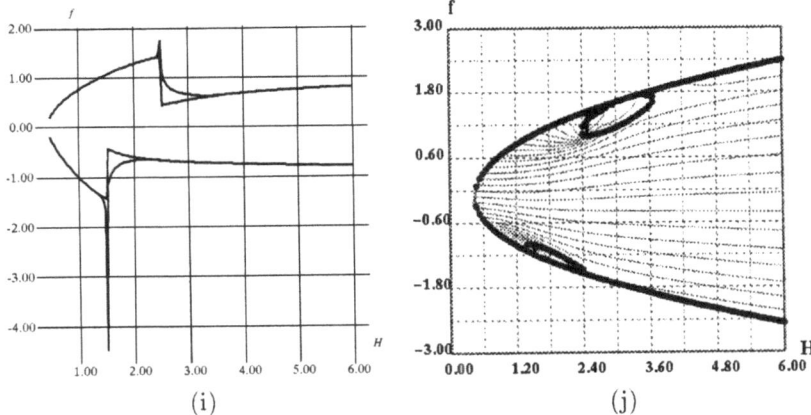

FIGURE 17.12. The rotation function ($g = 1.0$, $\beta = 0.5$, and $V(x) = x^2$): (a) level curves of the rotation function, (b) the shapes of the rotation functions on $H = 1.20$. (c) on $H = 1.86$, (d) on $H = 1.89$, (e) on $H = 2.40$, (f) on $H = 3.21$, (g) on $H = 3.24$, (h) on $H = 5.40$, (i) the rotation vectors, and (j) the positions of the eigenvalues.

$H = 3.21$. From some value H_0 to ∞, the rotation function on $H = \text{const}$ becomes monotonic again. Parts (g) and (h) show the shapes of the rotation functions on $H = 3.24$ and $H = 5.40$. The corresponding rotation vectors and the positions of the eigenvalues are shown in parts (i) and (j) of Figure 17.12.

Figure 17.13(a) shows level curves of the rotation function, where $g = 1.0$. $\beta = 0.5$, and $V(x) = x^2 + \frac{1}{2}x$. Parts (b) through (m) show the shapes of the rotation functions on $H = \text{const}$ with different values of H. The corresponding rotation vectors and the positions of the eigenvalues are shown in parts (n) and (o) of Figure 17.13. It follows from these figures that the rotation function on $H = \text{const}$ is also monotonic when H is larger than some value H_0.

Now we fix the parameters g and β, and also the function $V(x)$. From these experimental graphs, we can conclude that for two different values H_1 and H_2 the rotation vectors do not coincide (see the graphs of rotation vectors as functions of H). Hence, we can say that, *in the Lagrange case. the Hamiltonian systems defined by* sgrad H *on the level sets* $\{H = H_1\}$ *and* $\{H = H_2\}$ *are not orbitally equivalent unless* $H_1 = H_2$.

Finding orbitally equivalent systems with different parameters.
Now assume that $V(x) = x$ and consider two Lagrange-type Hamiltonian systems with different parameters (g_1, β_1) and (g_2, β_2). Another interesting problem is whether there are two values H_1 and H_2 of the Hamiltonian such that these systems, restricted to the corresponding level surfaces, are orbitally

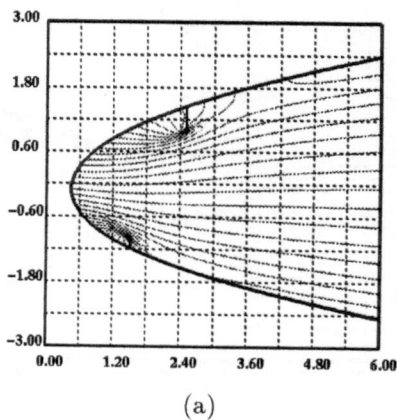

(a)

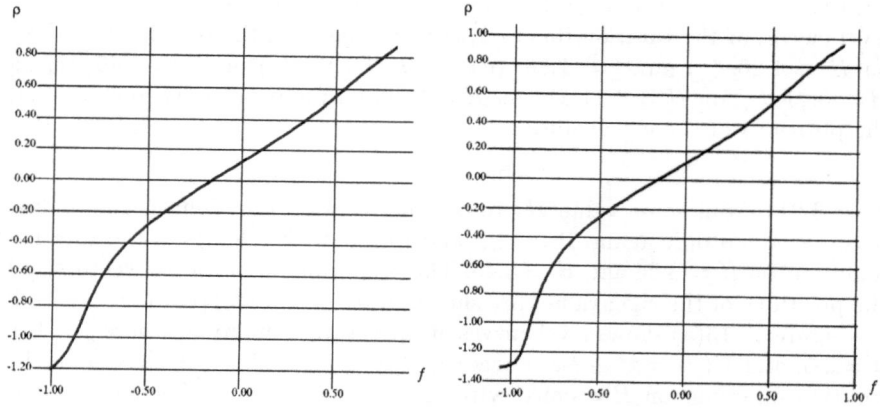

continued on next page

equivalent. This problem also reduces to the problem of comparing rotation vectors $R(H, g_1, \beta_1)$ and $R(H, g_2, \beta_2)$: if there exist two values H_1 and H_2 such that the rotation vectors coincide, the systems are orbitally equivalent on the corresponding level surfaces.

From the results of our calculation of rotation functions, we can see that the rotation function is monotonic if H is larger than some value H_0, when its gradient changes according to the parameter values. In this case, the rotation vector of H consists of two elements: that is, lower and upper boundary values with respect to the additional integral F. Denote the lower and upper boundary values by $R_1(H)$ and $R_2(H)$, where $H > H_0$. Consider the plane spanned by R_1 and R_2 axes. We can draw the curves (R_1, R_2) as H changes, with $H > H_0$. Now the problem is reduced to finding the intersection points

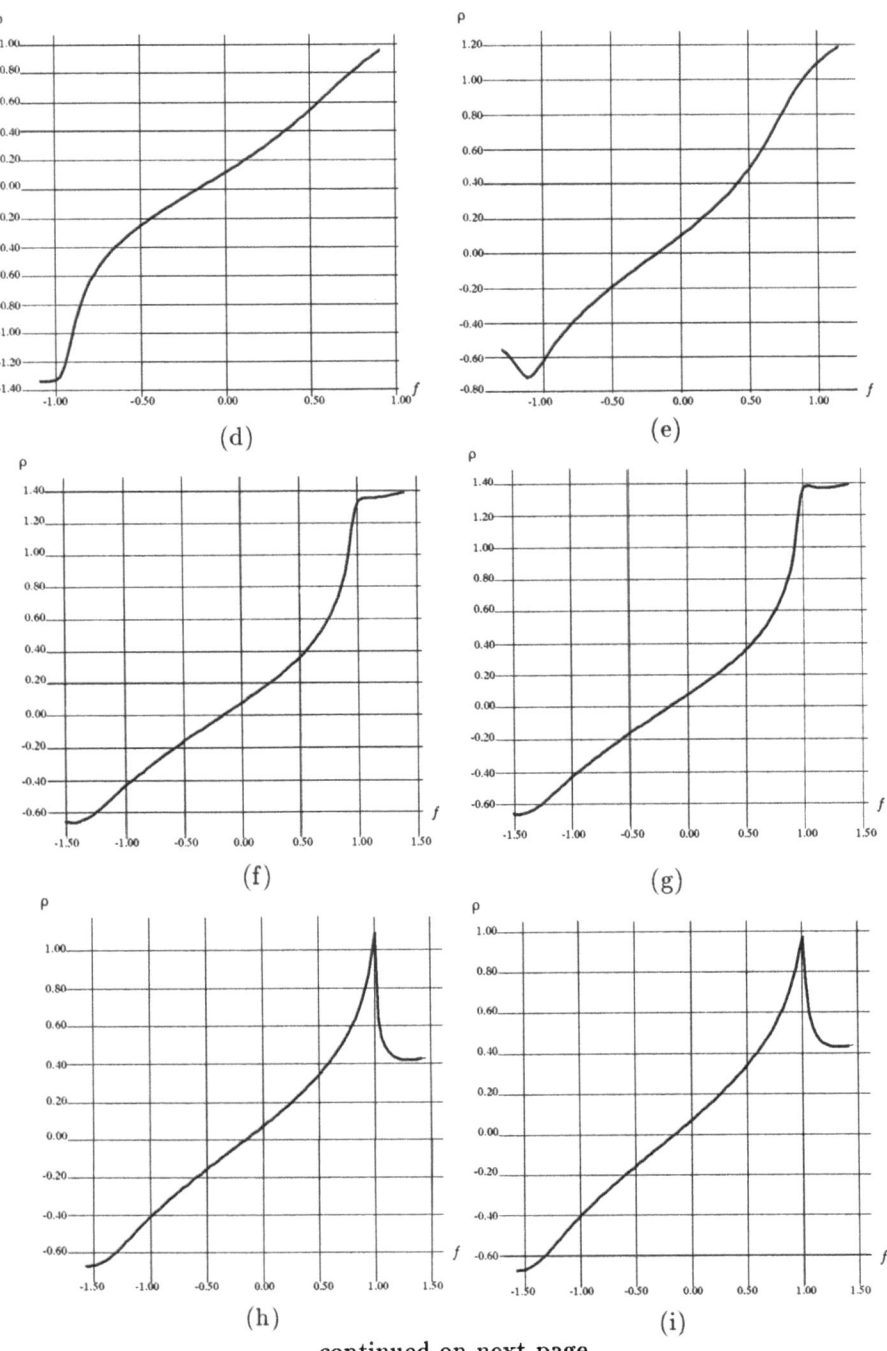

(d)

(e)

(f)

(g)

(h)

(i)

continued on next page

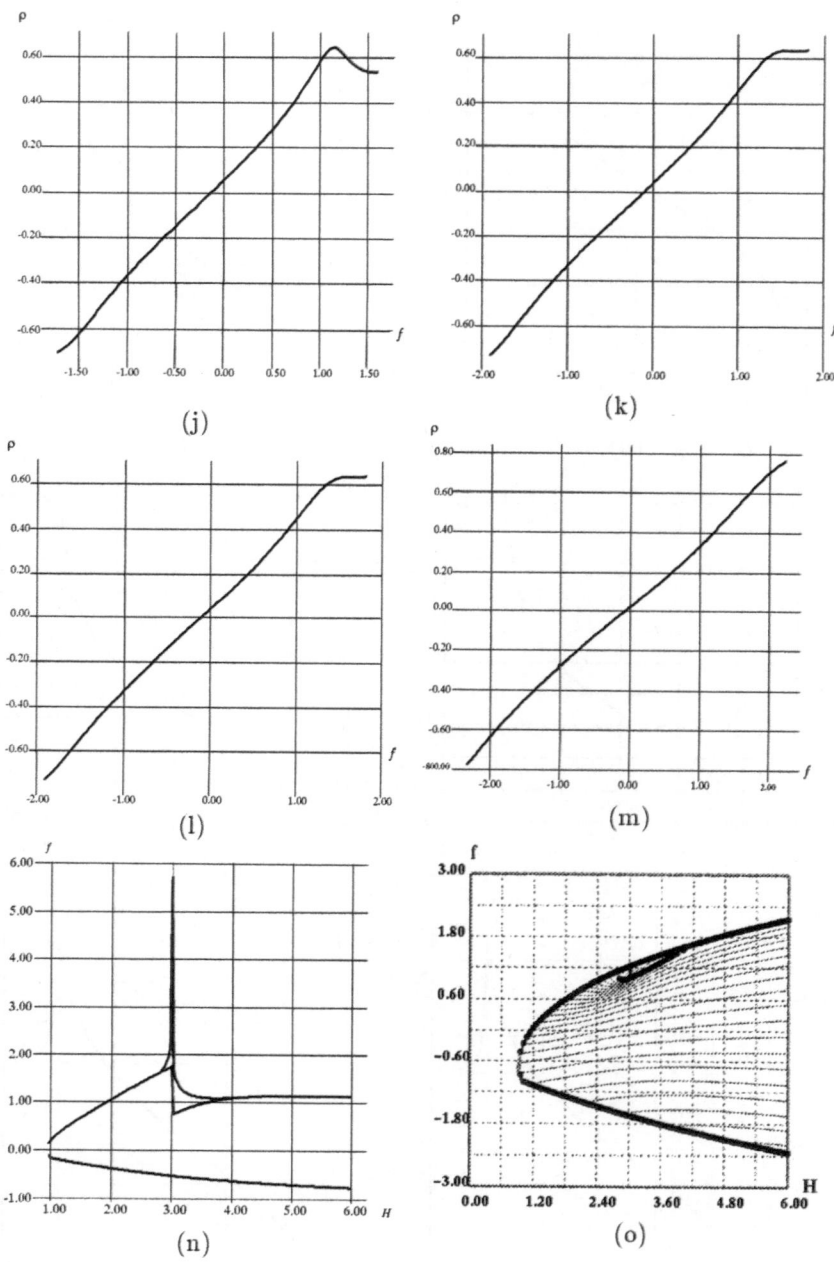

FIGURE 17.13. The rotation function ($g = 1.0$, $\beta = 0.5$, and $V(x) = x^2 + \frac{1}{2}x$): (a) level curves of the rotation function, (b) the shapes of the rotation functions on $H = 1.20$, (c) on $H = 1.32$, (d) on $H = 1.35$, (e) on $H = 1.80$, (f) on $H = 2.37$, (g) on $H = 2.40$, (h) on $H = 2.52$, (i) on $H = 2.55$, (j) on $H = 3.00$, (k) on $H = 3.66$, (l) on $H = 3.69$, (m) on $H = 5.40$, (n) the rotation vectors, and (o) the positions of its corresponding eigenvalues.

of the curves $(R_1(H), R_2(H))$ with different parameters β and g: if they intersect, there exist H_1 and H_2 such that $R(H_1, g_1, \beta_1) = R(H_2, g_2, \beta_2)$.

If the rotation vector has more than two elements, the probability that the rotation vectors coincide is zero by probability theory. Hence, we can confine ourselves to the case when the rotation vector has only two elements.

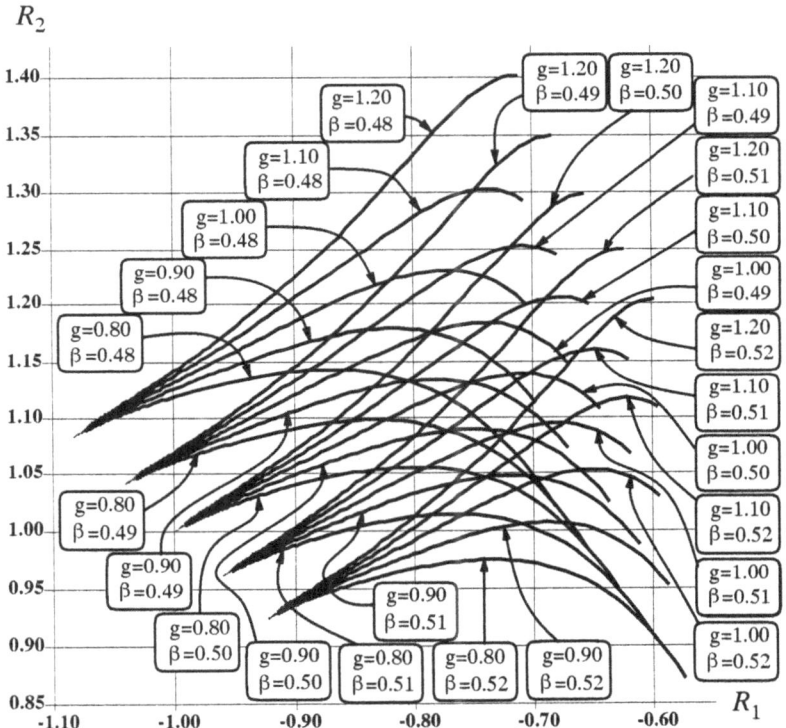

FIGURE 17.14. Curves representing the rotation vector with different parameters g and β ($V(x) = x$). Intersections indicate orbital equivalence of two different systems.

Figure 17.14 shows the curves $(R_1(H), R_2(H))$ with different parameters, where $g = 0.8 \approx 1.2$ and $\beta = 0.48 \approx 0.52$. The values of g and β are also indicated in the figure. An intersection between two curves represents an orbital equivalence between the corresponding two systems. Also note that the figure shows the validity of the statement at the end of the previous section, because each curve has no self-intersections. In this way, we can find orbitally equivalent systems, each of which has different parameters g and

β. More detailed investigation of this orbital equivalence is left for future investigation.

Acknowledgements.   The author expresses his gratitude to Prof. A. T. Fomenko, Prof. A. V. Bolsinov, and Dr. O. E. Orel of Moscow State University. This work was done in collaboration with Dr. O. E. Orel, and it would not have been possible without her help. The author also thanks Prof. T. L. Kunii and Prof. Y. Shinagawa for arranging this collaborative research project.

18. Ridges, Ravines and Singularities

A. G. Belyaev, E. V. Anoshkina, and T. L. Kunii

On a smooth generic surface we define ridges to be the local positive maxima of the maximal principal curvature along its associated curvature line and ravines to be the local negative minima of the minimal principal curvature along its associated curvature line. We investigate relationships between these surface line features, singularities of the caustic generated by the surface normals, and the singularities of the distance function from the surface. We also propose a variational problem to model garment wrinkles and investigate relationships between singularities of a proper solution of the problem and singularities of the distance function.

18.1. Introduction

Recent advances in pattern recognition, computer vision, medical imagery, and free-form shape design inspired a fresh interest in surface features associated with singularities of the caustic generated by the surface normals. Besides the mathematical beauty of such surface features [Porteous 1994], some of them have been studied in connection with applications in accommodation of the eye lens [Gullstrand 1904], structural geology [Ramsay 1967], face recognition [Gordon 1991], quality control of free-form surfaces [Hosaka 1992], analysis of medical images [Bookstein and Cutting 1988; Declerck and Ayache 1995; Guéziec 1992; Monga and Montesinos 1995; Thirion 1993] and satellite data [Monga and Montesinos 1995].

It turns out that some of these features of a smooth surface are in agreement with our intuitive mental perception of ridges and ravines. We introduce ridges and ravines on a given surface via loci of extrema of the principal curvatures along their associated curvature lines and investigate relationships between these surface creases, caustic singularities, and distance function singularities [Anoshkina et al. 1994; Belyaev and Kunii 1995; Belyaev and Kunii 1996; Kunii 1994]. We also propose a variational problem to model garment wrinkles and study relationships between singularities of a proper solution of the problem and singularities of the distance function [Anoshkina et al. 1994].

In our research we deal with generic phenomena. Roughly speaking, a particular property of an object from a particular class of objects is generic if in the space of all the objects of that class the objects exhibiting the property form an open dense set.

18.2.   Ridges, ravines, and caustic singularities

Surface creases can be intuitively defined as sets of sharp variation points of the surface normals. Our mathematical description of such points is based on study of caustic singularities. The caustic of a given surface can be described as the loci where light rays emitted by the surface in the normal directions are concentrated. At the singular points of the caustic the concentration of the light rays is still greater. Thus surface line features associated with caustic singularities may be considered as surface creases.

Let us recall that the *radius of curvature* of a plane oriented curve at a point $\mathbf{p}$ is the radius (with a sign according to a chosen orientation) of the circle best fitted with the curve at $\mathbf{p}$. The center of the circle is called the *center of curvature* and the inverse value of the curvature radius is called the *curvature* at $\mathbf{p}$. Consider a plane passing through a point $\mathbf{p}$ in a given surface M and normal to the surface. We call the intersection curve by the *normal section curve*. The curvature of the normal section curve depends on its tangent vector at $\mathbf{p}$. The maximal and minimal curvatures k_{max} and k_{min} are called the *principal curvatures* of M at $\mathbf{p}$. A point at which one of the principal curvatures vanishes is called *parabolic*. The *principal centers* are defined as the corresponding centers of the tangent circles. The associated tangent directions $\mathbf{v}_{max}$ and $\mathbf{v}_{min}$ are called the *principal directions* of M at $\mathbf{p}$. The integral curves of the principal direction fields are called the *curvature lines*. A point at which the principal curvatures are equal to each other is called *umbilic*.

It is suitable to define the *caustic* as the loci of the principal centers (focal points) of the curvatures of M. Thus the caustic has two sheets corresponding to the principal curvatures.

The caustic is a piecewise smooth surface and contains singularities. The caustic singularities correspond to line and point features on the surface. For a given smooth surface M defined parametrically by the local map $\mathbf{r} : \mathbb{R}^2 \to \mathbb{R}^3$, the caustic is parameterized as follows:

18.2.1. $$\mathbf{c}(u, v) = \mathbf{r}(u, v) + \frac{\mathbf{n}(u, v)}{k(u, v)}, \qquad k = k_{max}, k_{min},$$

where $\mathbf{n}(u, v)$ is the oriented normal (without loss of generality we do not consider parabolic points in (18.2.1)).

Suppose now that at a point $\mathbf{p} \in M$ the coordinates u, v are chosen so that the tangent vectors $\partial \mathbf{r}/\partial u$ and $\partial \mathbf{r}/\partial v$ coincide with the principal directions $\mathbf{v}_{max}$ and $\mathbf{v}_{min}$. If $\mathbf{p}$ corresponds to a singular point on the caustic, then the cross product

$$\frac{\partial \mathbf{c}}{\partial u} \times \frac{\partial \mathbf{c}}{\partial v} = \frac{\partial \mathbf{r}}{\partial u} \cdot \frac{\partial k}{\partial u} \cdot \frac{k - k_{min}}{k^3} + \frac{\partial \mathbf{r}}{\partial v} \cdot \frac{\partial k}{\partial v} \cdot \frac{k - k_{max}}{k^3}, \qquad k = k_{max}, k_{min},$$

vanishes. This proves the following statement (see also [Porteous 1994]).

Proposition 18.2.2. *For a smooth generic surface the caustic singularities correspond to the lines on the surface where the principal curvatures have extrema along their associated principal directions and the points where the principal curvatures are equal (umbilics).*

With the exception of some isolated points the caustic singularities form space curves called the *cuspidal edges*. Near a point of the cuspidal edge the caustic can be locally represented in the parametric form $(c_1 t^3, c_2 t^2, s)$, where $c_1 \neq 0$ and $c_2 \neq 0$, in proper coordinates (s, t).

Now we define ridges and ravines which are supposed to be consistent with our natural mental perception of such shape creases.

Definition 18.2.3. A non-umbilic point $\mathbf{p} \in M$ is called a ridge point if k_{max} attains a local positive maximum at $\mathbf{p}$ along the associated curvature line.

Although in this definition we deal with non-umbilic points the umbilics can be treated by continuity.

Let us note that the definitions of the ridges and ravines are dual: if we change the surface orientation then the ridges turn into the ravines and vice versa. So without loss of generality we can consider only the ridges. For a given non-umbilic point $\mathbf{p} \in M$ let us choose coordinates in the space so that $\mathbf{p}$ is at the origin, the (x, y)-plane is the tangent plane to the surface at $\mathbf{p}$, and the principal directions coincide with x and y axes. The surface is then expressible, in the Monge form, as the graph of a smooth generic function from (x, y)-plane to the z-axis, with the equation

$$z = \frac{1}{2} \left(\lambda x^2 + \mu y^2 \right) + \frac{1}{6} \left(a x^3 + 3b x^2 y + 3c x y^2 + d y^3 \right)$$
$$+ \frac{1}{24} \left(e x^4 + \ldots \right) + O(x, y)^5, \quad \textbf{18.2.4.}$$

where $\lambda = k_{max}(0, 0)$, $\mu = k_{min}(0, 0)$, $\lambda > \mu$.

The Taylor series expansion of k_{max} at $\mathbf{p}$ has the form

$$k_{max}(x, y) = \lambda + a x + b y + O(x, y)^2.$$

Since the vector $(1, 0)$ represents $\mathbf{v}_{max}$ at $\mathbf{p}$ then k_{max} has an extremum along $\mathbf{v}_{max}$ at $\mathbf{p}$ iff $a = 0$.

Let the maximal principal curvature be strictly positive at $\mathbf{p}$

$$k_{max}(0, 0) > 0.$$

The curvature line associated with k_{max} is locally described by the problem

$$\frac{dy}{dx} = \frac{bx + cy}{\lambda - \mu} + O(x, y)^2, \qquad y(0) = 0.$$

Therefore

$$y'(0) = 0, \qquad y''(0) = \frac{b}{\lambda - \mu}$$

and in a neighborhood of the origin the curvature line is approximated by the parabola

$$y = \frac{bx^2}{2(\lambda - \mu)}.$$

It allows to compute the Taylor series expansion of k_{max} at the origin along the associated curvature line:

18.2.5.
$$\lambda + ax + \left(-3\lambda^3 + e + \frac{3b^2}{\lambda - \mu}\right)\frac{x^2}{2} + O(x^3)$$

Analyzing the asymptotic expansion (18.2.5) we obtain that **p** is a generic ridge point iff

18.2.6.
$$a = 0 \quad and \quad -3\lambda^3 + e + \frac{3b^2}{\lambda - \mu} < 0.$$

Computations show that intersection between the caustic sheet associated with k_{max} and the plane $\{y = 0\}$ gives a curve described locally in parametric form as follows

$$x = \frac{1}{3\lambda}\left(3\lambda^3 - e - \frac{3b^2}{\lambda - \mu}\right)t^3 + O(t^4),$$

$$z = \frac{1}{\lambda} + \frac{1}{2\lambda^2}\left(3\lambda^3 - e - \frac{3b^2}{\lambda - \mu}\right)t^2 + O(t^3).$$

Locally at a neighborhood of the point $(0, 0, 1/\lambda)$ this curve is a semicubical parabola. This parametric representation of the intersection curve together with (18.2.6) provide us the following visual description of the ridges via caustic singularities:

Theorem 18.2.7. *The cuspidal edges of the caustic sheet associated with the maximal principal curvature and pointing towards to the surface correspond to the ridges.*

Suppose that a ridge goes through an umbilic. Passing to the limit $\lambda - \mu \to 0$ in (18.2.6) implies $a = 0 = b$ at the umbilic and the umbilic is not generic.

Theorem 18.2.8. *The ridges do not pass through the generic umbilics.*

This result was recently discovered by I. A. Bogaevski from topological considerations. Close results can be found in [Porteous 1994].

In particular, from Theorem 18.2.8 it follows that the ridges on a generic surface do not contain branch points. For applications this pure mathematical statement should be reformulated in the following way: the ridges on a surface with complex geometry are mostly fragmentary, the branch points occur relatively seldom.

We also might consider the curves generated by the closure of the set of non-umbilic points at which k_{max} attains a local positive maximum along the normal section curve associated with $\mathbf{v}_{max}$ [Belyaev and Kunii 1996]. If M is given in the Monge form (18.2.4) at $\mathbf{p}$ then the Taylor series expansion of k_{max} at $\mathbf{p}$ along the normal section curve has the form

$$\lambda + ax + \left(-3\lambda^3 + e + \frac{2b^2}{\lambda - \mu}\right)\frac{x^2}{2} + O(x^3)$$

and the points where k_{max} takes a local positive maximum along the associated normal section curves are characterized by the conditions

$$a = 0 \quad and \quad -3\lambda^3 + e + \frac{2b^2}{\lambda - \mu} < 0.$$

Thus the curves formed by these points contain all the ridge points but also do not pass through the generic umbilics.

18.3. Ridges, ravines, and distance function singularities

The medial axis transformation or skeletonization proposed by Blum [1967] in the mid-1960's is one of the earliest and probably the most widely studied systematic technique for global shape description. Skeletonization reduces input shapes to CW-complex representation (axial in 2D and polyhedral in 3D).

Let us consider a bounded figure in the space. A ball is said to be maximal within the figure if it is contained in the figure but is not a proper subset of any other ball contained in the figure. The *skeleton* of the figure is the locus of centers of balls which are maximal within the figure, together with the limit points of this locus.

A natural generalization of the skeleton is the so-called cut locus. Let M be a piecewise oriented smooth surface in the space. Consider the distance function from M:

$$dist(\mathbf{r}, M) = \inf_{\mathbf{p} \in M} \|\mathbf{r} - \mathbf{p}\|.$$

The distance function is continuous, but generally not smooth even if M is smooth. The *cut locus* of M is the closure of the set of all singular points of the distance function. When M is a closed surface, then the skeleton of the figure bounded by M is the intersection of the figure with the cut locus of M.

A point $\mathbf{p}$ of the cut locus is called *regular* if the cut locus is a manifold at $\mathbf{p}$: there exists a neighborhood of p on the cut locus to be diffeomorphic to a disk. We call a point $\mathbf{p}$ of the cut locus *singular* if the cut locus is not a manifold at $\mathbf{p}$; a singular point is *internal* if there is a set in the cut locus that contains $\mathbf{p}$, and is homeomorphic to a disk; otherwise, the singular point is called *boundary*.

From results of [Bryzgalova 1977; Bryzgalova 1978; Wassermann 1975] it follows that the singularities of the cut locus of a generic surface are locally organized similar to generic singularities of the locus of the nonsmoothness of the maximum function of a three-parameter family. The complete list of these singularities can be found, for example, in [Arnold 1992]. For us it is important that for any boundary point, its neighborhood in the cut locus is homeomorphic to a half-disk.

Consider a point $\mathbf{q}$ in the cut locus. The function $l_q(\mathbf{p}) = \|\mathbf{p} - \mathbf{q}\|^2$ defined on M has two equal absolute minima. Let us move along the cut locus from $\mathbf{q}$ to some boundary point $\mathbf{r}$. At $\mathbf{r}$ two equal absolute minima merge. Thus $l_r(\mathbf{p})$ has a degenerate absolute minimum and $\mathbf{r}$ is a singular point of the caustic. A neighborhood of $\mathbf{r}$ in the cut locus lies inside the cuspidal edge of the caustic at a vicinity of $\mathbf{r}$ in $\mathbb{R}^3$. Thus at $\mathbf{r}$ the cuspidal edge points to the surface and according to Theorem 18.2.8 $l_r(\mathbf{p})$ attains the absolute minimum at a ridge point. The images of Figure 18.1 demonstrate these arguments. We come to

Theorem 18.3.1. *Let M be a generic smooth oriented surface. For any boundary point of the cut locus of M its nearest point on M is either a ridge point or a ravine point.*

An analytical proof of the theorem was recently given in [Belyaev and Kunii 1996].

18.4. Distance function singularities and garment wrinkles modeling

Wrinkles are formed because of excess material covering a body. Let's try to glue a hole in a plane using a patch that's bigger than the hole. We select as the solution the patch that has the minimal Dirichlet integral (an analog of elastic energy in linear elasticity theory). It turns out that the graph of the solution of this variational problem (with area constraints) has singularities that can roughly be considered as wrinkles.

To make this more precise, let the hole Ω be a bounded domain in $\mathbb{R}^2$, and let the graph of a function $f(x) : \Omega \to \mathbb{R}$, $x = (x_1, x_2) \in \mathbb{R}^2$ be a patch

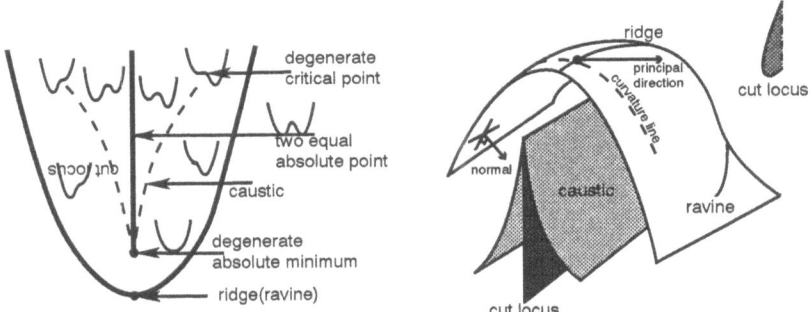

FIGURE 18.1. Left: the boundary points of the cut locus are situated on caustic singularities; zoo of distance functions. Right: geometric idea of ridges, ravines, and related structures.

with the area equal to a given number $S \geq |\Omega|$, where $|\Omega|$ is the area of Ω. Consider the set of functions

18.4.1. $\qquad \left\{ f(x) : f(x) = 0 \text{ on } \partial\Omega, \int_\Omega \sqrt{1 + |\nabla f(x)|^2} \, dx = S \right\}$,

where ∇ is the gradient, and investigate the minimization problem

18.4.2. $\qquad \hat{J} = \inf_{f \in (18.4.1)} J(f), \qquad J(f) = \int_\Omega |\nabla f(x)|^2 \, dx.$

Applying the method of Lagrange multipliers, we get
18.4.3.
$$\hat{J} = \inf_{f \in (18.4.1)} \sup_\lambda \left\{ \int_\Omega |\nabla f(x)|^2 \, dx + \lambda \left(S - \int_\Omega \sqrt{1 + |\nabla f(x)|^2} \, dx \right) \right\}.$$

Changing the order of calculation of the lower and upper bounds, we come to
18.4.4.
$$\hat{J} \geq \sup_\lambda \inf_{f \in (18.4.1)} \left\{ \int_\Omega |\nabla f(x)|^2 \, dx + \lambda \left(S - \int_\Omega \sqrt{1 + |\nabla f(x)|^2} \, dx \right) \right\}.$$

Put

18.4.5. $\qquad \sigma(x) = \sqrt{1 + |\nabla f(x)|^2}, \quad \text{where } f(x) = 0 \text{ on } \partial\Omega;$

we now can rewrite (18.4.4) in the form

$$\hat{J} \geq \sup_\lambda \inf_{\sigma \in (18.4.5)} \left\{ \int_\Omega (\sigma^2 - 1) \, dx + \lambda \left(S - \int_\Omega \sigma(x) \, dx \right) \right\}.$$

Replacing the constraint σ in (18.4.5) by $\sigma \geq 0$, we get

18.4.6. $\qquad \hat{J} \geq \sup_\lambda \inf_{\sigma \geq 0} \left\{ \int_\Omega (\sigma^2 - 1) \, dx + \lambda \left(S - \int_\Omega \sigma(x) \, dx \right) \right\}.$

The calculation of the infimum in (18.4.6) reduces to finding the minimum of the function $\sigma^2 - \lambda\sigma$. Therefore, $\sigma_{min} = \lambda/2$ and

18.4.7.
$$\hat{J} \geq \sup_{\lambda} \left\{ -\tfrac{1}{4}\lambda^2|\Omega| - |\Omega| + \lambda S \right\}.$$

Now the calculation of the supremum in (18.4.7) reduces to finding the maximum of the function $-\lambda^2|\Omega|/4 - |\Omega| + \lambda S$. A simple calculation gives $\lambda_{max} = 2S/|\Omega|$. Thus, we obtain the lower bound for our initial variational problem (18.4.2):

$$\hat{J} \geq \frac{S^2}{|\Omega|} - |\Omega|.$$

We now construct an upper bound for the variational problem (18.4.2). Consider the following nonlinear boundary value problem (the Hamilton–Jacobi equation)

18.4.8.
$$|\nabla \hat{f}(x)|^2 = \frac{S^2}{|\Omega|^2} - 1, \qquad \hat{f}(x) = 0 \text{ on } \partial\Omega.$$

It is easy to see that $\hat{f}(x) \in (18.4.1)$ and $J(\hat{f}) = S^2/|\Omega| - |\Omega|$. Thus $J(\hat{f}) = \hat{J}$, $\hat{f}(x)$ is a minimizer of the initial variational problem (18.4.2), and it remains to construct the solution of the Hamilton–Jacobi equation (18.4.8) and investigate properties of $\hat{f}(x)$.

Our Hamilton–Jacobi equation will have a unique convex solution if we supplement (18.4.8) with the condition

18.4.9.
$$\frac{\partial \hat{f}(x)}{\partial \nu} = \sqrt{\frac{S^2}{|\Omega|^2} - 1} \qquad \text{on } \partial\Omega,$$

where ν is the unit inner normal vector of $\partial\Omega$. Since

$$|\nabla \hat{f}(x)|^2 = \left| \frac{\partial \hat{f}(x)}{\partial \nu} \right|^2 + \left| \frac{\partial \hat{f}(x)}{\partial s} \right|^2 \quad \text{and} \quad \hat{f}(x) = 0 \quad \text{on } \partial\Omega,$$

where s is the arclength parameter on $\partial\Omega$, we get

$$\frac{\partial \hat{f}(x)}{\partial s} = 0 \text{ on } \partial\Omega \quad \text{and} \quad \frac{\partial \hat{f}(x)}{\partial \nu} = \sqrt{\frac{S^2}{|\Omega|^2} - 1} \quad \text{in a neighborhood of } \partial\Omega.$$

Therefore, the solution of the problem (18.4.8), (18.4.9) is arranged as follows: the level lines $\hat{f}(x) = $ const are parallel to the contour $\partial\Omega$ and the derivative $\partial \hat{f}(x)/\partial \nu$ along the normal ν to a level line for $\hat{f}(x)$ has the value $\sqrt{S^2/|\Omega|^2 - 1}$.

The graph of the solution of the problem (18.4.8), (18.4.9) is a ruled surface, and can be described as follows: consider the distance function from

$\partial\Omega$ on our plane $\mathbb{R}^2$; the graph of $\hat{f}(x)$ is obtained from the graph of the distance function by multiplying by $\sqrt{S^2/|\Omega|^2 - 1}$. This means that the singular points of the graph of the solution of the problem (18.4.8), (18.4.9) and the singular points of the graph of the distance function are the same. The projection of the singular points onto the hole plane gives the skeleton of the hole (see Figure 18.2). The fractures of the graph of $\hat{f}(x)$ can be considered

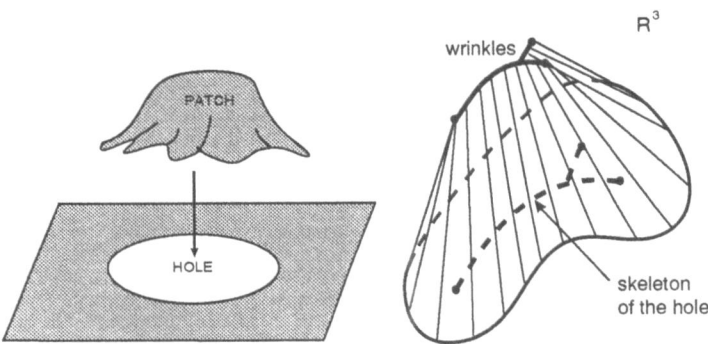

FIGURE 18.2. Left: a hole and a patch. Right: modeling of wrinkles via the fractures of the solution.

as wrinkles. Wrinkle branching and vanishing processes can be also described in the frame of the above variational model [Anoshkina et al. 1994].

Of course, the proposed variational model just a very simple example. Instead of a patch covering a hole, we might consider a surface similar to a sleeve (with arm constraints), and instead of the Dirichlet integral, which mimics the square-law elastic energy, we might use a more complicated functional taking into account the anisotropic, plastic, and rigid properties of materials (it is easy to see that materials having different physical properties form different wrinkle patterns). Such variational problems are more complicated than the one just considered, and can be studied by numerical simulation, or investigated by asymptotic analysis.

Acknowledgments
It is a pleasure to acknowledge fruitful discussions with I. A. Bogaevski.

Bibliography

[Adams 1983] Adams, C. C. (1983). Hyperbolic structures on link complements. Preprint, Univ. of Wisconsin, Madison, WI.

[Aksenenkova 1981] Aksenenkova, I. M. (1981). Canonical variable angle-action in the problem of Lagrange top. *Vestnik Moscov. Univ. Ser. Mat. Mekh* (1), 86–90. (in Russian).

[Almgren Jr. 1976] Almgren Jr., F. J., J. E. Taylor. (1976, Jul.). The geometry of soap films and soap bubbles. *Scientific American*, 82–93.

[Anoshkina et al. 1994] Anoshkina, E. V., A. G. Belyaev, and T. L. Kunii (1994). Detection of ridges and ravines based on caustic singularities. *International Journal of Shape Modeling*.

[Anoshkina et al. 1994] Anoshkina, E. V., A. G. Belyaev, O. G. Okunev, and T. L. Kunii (1994). Ridges and ravines: a singularity approach. In *International Journal of Shape Modeling*.

[Arnold 1978] Arnold, I. V. (1978). *Mathematical methods in classical mechanics.* Springer-Verlag.

[Arnold 1992] Arnold, V. I. (1992). *Catastrophe Theory*. Springer-Verlag.

[Baumgart 1975] Baumgart, B. G. (1975). A polyhedron representation for computer vision. In *AFIPS Proceedings*, pp. 589–596.

[Belyaev and Kunii 1995] Belyaev, A. G., E. V. Anoshkina, and T. L. Kunii (1995, Jul.). Ridges, ravines, and related point features on a surface. In *Vision Geometry IV, Prceedings of SPIE 2573*, San-Diego, CA, pp. 84–95.

[Belyaev and Kunii 1996] Belyaev, A. G., I. A. Bogaevski, and T. L. Kunii (1996, Jun.). Principal direction ridges. Technical Report 96-4-001, The University of Aizu.

[Blum 1967] Blum, H. (1967). A transformation for extracting new descriptors of shape. In *Symposium on Models for the Perception of Speech and Visual Form*. M.I.T. Press.

[Bolsinov 1991] Bolsinov, A. V. (1991). Methods of calculation of the fomenko-zieschang invariant. In A. T. Fomenko (Ed.), *Topological Classification of Integrable Systems* Advances in Soviet Mathematics, Vol. 6, pp. 147–183.

[Bolsinov and Fomenko 1993] Bolsinov, A. V. and A. T. Fomenko (1993). Orbital classification of simple integrable Hamiltonian systems on three-dimensional level. surfaces of constant energy. *Dokl. Akad Nauk 332*(5), 553–555.

[Bolsinov and Fomenko 1994] Bolsinov, A. V. and A. T. Fomenko (1994). Orbital classification of integrable Hamiltonian systems with two degrees of freedom. classification theorem. *Mat.Sb. 185*, 27–80 (Part I, No. 4), 27–78 (PartII, No. 5).

[Bolsinov et al. 1990] Bolsinov, A. V., S. V. Matveev, and A. T. Fomenko (1990). Topological classification of integrable Hamiltonian systems with two degrees or freedom. list of systems of small complexity. *Uspekhi Mat. Nauk 45*(2), 49–77. English transl. in Russian Math. Surveys 45(1990), no.2, pp.59-94.

[Bookstein and Cutting 1988] Bookstein, F. L. and C. B. Cutting (1988). A proposal for the apprehension of curving craniofacial form in three dimensions. *Craniofacial Morphogenesis and Dysmorphogenesis*, 127–140.

[Boyse 1979] Boyse, J. W. (1979, Jan.). Interference detection among solids and surfaces. *Communications of the ACM 22*(1), 3–9.

[Bryzgalova 1977] Bryzgalova, L. N. (1977). Singularities of the maximum of a parametrically dependent function. *Functional Analysis and its Applications 11*, 49–50.

[Bryzgalova 1978] Bryzgalova, L. N. (1978). Maximum functions of a family of functions depending on parameters. *Functional Analysis and its Applications 12*, 50–51.

[Cameron 1989] Cameron, S. (1989, Feb.). Efficient intersection tests for objects defined constructively. *The International Journal of Robotics Research 8*(1), 3–25.

[Chern 1955] Chern, S. (1955). An elementary proof of the existence of isothermal parameters on a surface. *Pro. Amer. Math. Soc. 6*, 771–782.

[Christiansen and Sederberg 1978] Christiansen, H. N. and T. W. Sederberg (1978). Conversion of complex contour line definitions into polygonal element mosaics. *Computer Graphics (Proc. Siggraph) 12*(3), 187–192.

[Clark 1981] Clark, J. H. (1981, Nov.). Parametric curves, surfaces and volumes in computer graphics and computer-aided geometric design. Technical Report 221, Computer Systems Laboratory, Departments of Electrical Engineering and Computer Science, Stanford University, Stanford, California 94305.

[Declerck and Ayache 1995] Declerck, J., G. Subsol, J.-P. Thirion, and N. Ayache (1995). Automatic retrieval of anatomical structures in 3d medical images. Technical report, INRIA, Le Chesnay, France.

[Dubrovin et al. 1985] Dubrovin, B. A., A. T. Fomenko, and S. P. Novikov (1985). *Modern Geometry - Methods and Applications: Part2. The Geometry and Topology of Manifolds*, Volume 104 of *GTM*. Springer-Verlag.

[Dubrovin et al. 1990] Dubrovin, B. A., A. T. Fomenko, and S. P. Novikov (1990). *Modern Geometry - Methods and Applications: Part3. Introduction to Homology Theory*, Volume 124 of *GTM*. Springer-Verlag.

[Dubrovin et al. 1992] Dubrovin, B. A., A. T. Fomenko, and S. P. Novikov (1992). *Modern Geometry - Methods and Applications: Part1. The Geometry of Surfaces, Transformation Groups and Fields* (2nd ed.), Volume 93 of *GTM*. Springer-Verlag.

[Ekoule et al. 1991] Ekoule, A. B., F. C. Peyrin, and C. L. Odet (1991, Apr.). A triangulation algorithm from arbitrary shaped multiple planar contours. *tog 10*(2), 182–199.

[Eliasson 1990] Eliasson, L. H. (1990). Normal form for Hamiltonian systems with poisson commuting integrals - elliptic case. *Comm. Math. Helv. 65*, 4–35.

[Fomenko 1987a] Fomenko, A. T. (1987a). *Differential Geometry and Topology*. Plenum Publ. Corp.

[Fomenko 1987b] Fomenko, A. T. (1987b). The topology of surfaces of constant energy in integrable Hamiltonian systems and obstructions to integrability. *Math. USSR Izvestiya 29*(3), 629–658.

[Fomenko 1988] Fomenko, A. T. (1988). *Integrability and Nonintegrability in Geometry and Mechanics*. Kluwer Academic Publishers.

[Fomenko 1989] Fomenko, A. T. (1989). Topological classification of all integrable Hamiltonian differential equations of general type with two degrees of freedom. In T. Ratiu (Ed.), *The Geometry of Hamiltonian Systems*, pp. 131–339. Springer-Verlag.

[Fomenko 1992] Fomenko, A. T. (1992). Computer geometry and topological classification of integrable Hamiltonian differential equations: Visualization of concrete physical examples. In T. L. Kunii and Y. Shinagawa (Eds.), *Modern Geometric Computing for Visualization*, pp. 3–10. Springer-Verlag.

[Fuchs et al. 1977] Fuchs, H., Z. M. Kedem, and S. P. Uselton (1977, Oct.). Optimal surface reconstruction from planar contours. *Communications of the ACM 20*(10), 693–702.

[Ganapathy and Dennehy 1982] Ganapathy, S. and T. G. Dennehy (1982). A new general triangulation method for planar contours. *Computer Graphics (Proc. ACM SIGGRAPH '82) 16*, 69–75.

[Gordon 1991] Gordon, G. G. (1991, Jul.). Face recognition from depth maps and surface curvature. In *Proceedings of SPIE Conference on Geometric Methods in Computer Vision*, San-Diego, CA, pp. 234–247.

[Gourret et al. 1989] Gourret, J., N. M. Thalmann, and D. Thalmann (1989). Simulation of object and human skin deformations. *Computer Graphics 23*(3), 21–30.

[Guéziec 1992] Guéziec, A. (1992). Large deformable splines, crest lines and matching. Technical report, INRIA, Le Chesnay, France.

[Guillemin 1974] Guillemin, V., A. Polack (1974). *Differential Topology*. Englewook Cliffs, NJ: Prentice-Hall.

[Gullstrand 1904] Gullstrand, A. (1904). Zur kenntnis der kreispunkte. *Acta Mathematica 29*, 59–100.

[Haeckel 1887] Haeckel, E. (1887). Report on the radiolaria collected by h.m.s. challenger during the years 1873–76. *Zoology*.

[Hodgson 1994] Hodgson, C., J. W. (1994). Symmetries, isometires and length spectra of closed hyperbolic three-manifolds. *Experimental Math.*, 261–274.

[Hopf 1983] Hopf, H. (1983). Differential geometry in the large. *Lecture Notes in Mathematics* (1000).

[Hosaka 1992] Hosaka, M. (1992). *Modeling of Curves and Surfaces in CAD/CAM*. Springer-Verlag.

[Huston 1990] Huston, R. L. (1990). *Mutibody Dynamics*. Butterworth-Heinemann.

[Inamoto 1993] Inamoto, N. (1993, Jun.). Hexadecimal-tree:a time-continuous 4d interference check method. In B. Falcidieno and T. L. Kunii (Eds.), *Modeling in Computer Graphics*, pp. 144–156. IFIP: Springer-Verlag.

[Jackins and Tanimoto 1980] Jackins, C. L. and S. L. Tanimoto (1980). Oct-trees and their use in representing three-dimensional objects. *Computer Graphics and Image Processing 14*, 249–270.

[Kaneda et al. 1987a] Kaneda, K., K. Harada, E. Nakamae, M. Yasuda, and A. G. Sato (1987a). Reconstruction and semi-transparent display method for observing inner structure of an object consisting of multiple surfaces. *The Visual Computer 3*(3), 137–144.

[Kaneda et al. 1987b] Kaneda, K., K. Harada, E. Nakamae, M. Yasuda, and A. G. Sato (1987b). Reconstruction and semi-transparent display method for observing inner structure of an object consisting of multiple surfaces. In T. L. Kunii (Ed.). *Computer Graphics 1987*, pp. 367–380. Springer-Verlag.

[Knörrer 1980] Knörrer, H. (1980). Geodisics on the ellipsoid. *Inventions Math 59*, 119–143.

[Kozlov 1980] Kozlov, V. V. (1980). *Methods of quantitative analysis in rigid body dynamics*. Izdat. Moscov. Univ., Moscow. (in Russian).

[Kunii 1994] Kunii, T. L., A. G. Belyaev, and E. V. Anoshkina (1994). Hierarchic shape description via singularity and multiscaling. *Eighteenth Annual International Computer Software & Applications Conference(COMPSAC 94)*, IEEE, 242–251.

[Kunii et al. 1993] Kunii, T. L., H. Hioki, and Y. Shinagawa (1993). Visualizing highly abstract mathematical concepts: A case study in animation of homology groups. In T.-S. Chua and T. L. Kunii (Eds.), *Proc. of the First International Conference on Multi-Media Modeling MULTIMEDIA MODELING*, pp. 3–30.

[Kunii and Sun 1990] Kunii, T. L. and L. Sun (1990). Dynamic analysis-based human animation. In T. S. Chua and T. L. Kunii (Eds.), *Computer Graphics Around the World (Proc. CG International '90)*, pp. 3–15. Springer-Verlag.

[Kunii and Takahashi 1993] Kunii, T. L. and S. Takahashi (1993). Area guide map modeling by cw-complexes and manifolds. In B. Falcidieno and T. L.Kunii (Eds.), *Modeling in Computer Graphics (IFIP Series on Computer Graphics)*, pp. 5–20. Springer-Verlag.

[Kunii et al. 1993] Kunii, T. L., Y. Tsuchida, Y. Arai, H. Matsuda, M. Shirahama, and S. Miura (1993). A model of hands and arms based on manifold mappings. In N. M. Thalmann and D. Thalmann (Eds.), *Communicating with Virtual Worlds (Proc. CG International '93)*, pp. 381–398. Springer-Verlag.

[Landsmeer 1963] Landsmeer, J. M. F. (1963). The coordination of finger-joint motions. *The Journal of Bone and Joint Surgery 45-A*(8), 1654–1662.

[Lee et al. 1990] Lee, P., S. Wee, J. Zhao, and N. I. Badler (1990). Strength guided motion. *Computer Graphics 24*(4), 253–262.

[Lerman and Umanskii 1987] Lerman, L. M. and Y. L. Umanskii (1987). Structure of the poisson action of $\searrow^2$ on a four-dimensional symplectic manifold. *Selecta Mathematica Sovietica 6*(4), 365–396. note = II, ibid. 7 (1988) No. 1 pp.39-48.

[Lerman and Umanskii 1993] Lerman, L. M. and Y. L. Umanskii (1993). The classification of four-dimensional Hamiltonian systems and poisson actions of $\searrow^2$ in extended neighborhoods of simple singular points. *Matem. Sbornik 184*(4), 103–138. in Russian.

[Level and Sharir 1987] Level, D. and M. Sharir (1987). An efficient and simple motion planning algorithm for a ladder amidst polygonal barriers. *Journal of Algorithms 8*, 192–215.

[Leven and Sharir 1987] Leven, D. and M. Sharir (1987). Planning a purely translational motion for a convex objects in two-dimensional space using generalized voronoi diagrams. In *Discrete & Computational Geometry, Vol. 2*, pp. 9–31. Springer-Verlag.

[Lozano-Pérez 1983] Lozano-Pérez, T. (1983, Feb.). Spatial planning: A configuration space approach. *IEEE Transactions on Computers C-32*(2), 108–120.

[Lozano-Pérez and Wesley 1979] Lozano-Pérez, T. and M. A. Wesley (1979, Oct.). An algorithm for planning collision-free paths among polyhedral obstacles. *Communications of the ACM 22*(10), 560–570.

[Mantyla and Sulonen 1982] Mantyla, M. and R. Sulonen (1982, Sep.). GWB: A solid modeler with Euler operators. *IEEE Computer Graphics and Applications 2*(7), 17–31.

[Matveev 1975] Matveev, S. V. (1975). A certain method for the construction of 3-manifolds. In *Vestnik of Moscow Univ. Ser.1. Math. Mech. 30:3*, pp. 11–20.

[Matveev and Fomenko 1988] Matveev, S. V. and A. T. Fomenko (1988). Constant energy surfaces of Hamiltonian systems, enumeration of three-dimensional manifolds in increasing order of complexity, and computation of volumes of closed hyperbolic manifolds. *Russian Math. Surveys 43*(1), 3–24.

[Matveev et al. 1989] Matveev, S. V., A. T. Fomenko, and V. V. Sharko (1989). Round morse functions and isoenergy surfaces of integrable Hamiltonian systems. *Math. USSR Sbornik 63*(2), 319–336.

[Matveev] Matveev, V. S. Integrable Hamiltonian systems with two degrees of freedom. the topological structure of saturated neighborhoods of points of saddle-saddle and focus-focus type. in preparation.

[Matveev 1993] Matveev, V. S. (1993). The calculation of the fomenko invariant for a "saddle-saddle" type point of an integrable Hamiltonian system. In *Trudy Semin. po Vekt. i Tenz. analizu* Vol. 25, pp. 75–104. Moscow State Univ. Publ.

[Matveev 1974] Matveev, S. V., V. V. Savvateev (1974). Three-dimensional manifolds having simple special spines. *Colloq. Math. 32*, 83–97.

[Meager 1982] Meager, D. (1982). Geometric modeling using octree encoding. *Computer Graphics and Image Processing 19*, 129–147.

[Mishchenko and Fomenko 1988] Mishchenko, A. S. and A. T. Fomenko (1988). *A Course of Differential Geometry and Topology*. Moscow: MIR Publishers.

[Monga and Montesinos 1995] Monga, O., N. A. and P. Montesinos (1995). Thin nets and crest lines: application to satellite data and medical images. Technical report, INRIA, Le Chesnay, France.

[Monheit and Badler 1991] Monheit, G. and N. I. Badler (1991). A kinematic model of the human spine and torso. *IEEE Computer Graphics & Applications 11*(2), 29–38.

[Moore and Wilhelms 1988] Moore, M. and J. Wilhelms (1988, Aug.). Collision detection and response for computer animation. *Computer Graphics (proc. of SIGGRAPH '88) 22*(4), 289–298.

[Munkres 1984] Munkres, J. (1984). *Elements of Algebraic Topology*. Addison-Wesley.

[Noborio et al. 1987] Noborio, H., S. Fukuda, and S. Arimoto (1987). A new interference check algorithm using octree representation. *Japan Robot Society Journal 5*, 21–30. (In Japanese).

[Nomura et al. 1989] Nomura, Y., T. Okuno, M. Hara, Y. Shinagawa, and T. L. Kunii (1989. May-Jun.). Walking through a human ear. *Acta Otolaryngolica 107*(5-6), 366–370.

[Okubo 1992] Okubo, Y. (1992). The development of three-dimentional analyzing system of occlusal tooth contacts. *J Jpn Prosthodont Soc. 36*(1), 53–63. (In Japanese).

[Orel] Orel, O. E. Rotation function for integrable problems reducing to the system of Abelian equations. Orbital classification of Goryachev-Chaplygin case. Mat.Sb. (in print).

[Oshemkov 1991] Oshemkov, A. A. (1991). Fomenko invariants for the main integrable cases of the rigid body motion equation. *Adv. in Soviet Math. 6*, 67–145.

[Plateau 1873] Plateau, J. A. F. (1873). *Statique expérimentale et théorétique des liquides soumis aux seules forces moléculaires*. Paris: Gauthiers-Villars.

[Porteous 1994] Porteous, I. R. (1994). *Geometric Differentiation for the intellegence of curves and surfaces*. Cambridge University Press.

[Preparata and Shamos 1985] Preparata, F. P. and M. I. Shamos (1985). *Computational Geometry: An Introduction*. Springer-Verlag.

[Press et al. 1992] Press, W. H., S. A. Teukolsky, W. T. Vetterling, and B. P. Flannery (1992). *Numerical Recipes in C*. Cambridge University Press.

[Ramsay 1967] Ramsay, J. G. (1967). *Folding and fracturing of rocks*. McGraw Hill.

[Reeb 1946] Reeb, G. (1946). Sur les points singuliers d'une forme de pfaff completement integrable ou d'une fonction numerique. *Comptes Rendus Acad. Sciences Paris 222*, 847–849.

[Rijpkema and Girard 1990] Rijpkema, H. and M. Girard (1990). Computer animation of knowledge-based human grasping. *Computer Graphics 25*(4), 339–348.

[Schwartz and Sharir 1983] Schwartz, J. T. and M. Sharir (1983). On the piano movers' problem: I. the case of a two-dimensional rigid polygonal body moving amidst polygonal barriers. *Communications on Pure and Applied Mathematics 36*, 345–398.

[Schwarz 1994] Schwarz, A. S. (1994). Topology for physicists. *Grundlehren der Mathematischen Wissenschaften*.

[Scott 1983] Scott, P. (1983). The geometries of 3-manifolds. *Bull. Amer. Math. Soc. 15*, 401–487.

[Seifert and Weber 1933] Seifert, H. and C. Weber (1933). Die beiden dodekaederräume. *Math. Z(37)*, 237–253.

[Shinagawa et al. 1991] Shinagawa, Y., Y. L. Kergosien, and T. L. Kunii (1991. Sep.). Surface coding based on morth theory. *IEEE Computer Graphics & Applications 11*(5), 66–78.

[Shinagawa and Kunii 1991] Shinagawa, Y. and T. L. Kunii (1991). The homotopy model: A generalized model for smooth surface generation from cross sectional data. *The Visual Computer 7*(2–3), 72–86.

[Shinagawa and Kunii 1992] Shinagawa, Y. and T. L. Kunii (1992). Using surface coding to detect errors in surface reconstruction. In T. L. Kunii and Y. Shinagawa (Eds.), *Modern Geometric Computing for Visualization*, pp. 227–240. Springer-Verlag.

[Shinagawa et al. 1990] Shinagawa, Y., T. L. Kunii, Y. Nomura, T. Okuno, and Y.-H. Young (1990). Automating view function generation for walk-through animation. In N. Magnenat-Thalmann and D. Thalmann (Eds.), *Proc. of Computer Animation '90*, pp. 227–237. Springer-Verlag.

[Smale 1970] Smale, S. (1970). Topology and mechanics. *Invent. Math. 10*(6), 305–331. II Vol. 11 No. 1 pp.45–64.

[Struik 1950] Struik, D. J. (1950). *Lectures on Classical Differential Geometry (Reading, Mass)*. Addison-Wesley.

[Takahashi and Kunii 1994] Takahashi, S. and T. L. Kunii (1994. Aug.). Manifold-based Multiple viewpoint cad: A case study of mountain guide map generation-. *Computer-Aided Design 26*(8). 622–631.

[Tam and Davis 1988] Tam, Y. and W. A. Davis (1988). Display of 3d medical images. In *Graphics Interface '88*, pp. 78–86.

[Thirion 1993] Thirion, J.-P., A. Gourdon (1993). The matching lines algorithm: new results and proofs. Technical report, INRIA, Le Chesnay, France.

[Thompson 1917] Thompson, D. W. (1917). *On Growth and Form*. Cambridge: Cambridge U. Press.

[Thurston 1981] Thurston, W. P. (1981). The geometry and topology of 3-manifolds. Preprint.

[Thurston 1982] Thurston, W. P. (1982). Three-dimensional manifolds, kleinian groups and hyperbolic geometry. *Bull. Amer. Math. Soc. 6*, 357–381.

[Uchiki et al. 1983] Uchiki. T., T. Ohashi, and M. Tokoro (1983). Collision detection in motion simulation. *Computers & Graphics 7*(3-4), 285–293.

[Wassermann 1975] Wassermann, G. (1975). Stability of caustics. *Math. Ann. 216*(1), 43–50.

[Wente 1987] Wente, H. C. (1987). Immersed tori of constant mean curvature in r sp 3. pp. 565–573.

[Wu et al. 1977] Wu, S., J. F. Abel, and D. P. Greenberg (1977, Oct.). An interactive computer graphics approach to surface representation. *Communications of the ACM 20*(10). 703–712.

Index